U0304198

"十二五"普通高等教育本科国家级规划教材

Fundamentals of Nanomaterials

纳米材料基础（双语版）

第二版

张耀君　编著

化学工业出版社
·北京·

本书为“十二五”普通高等教育本科国家级规划教材。

为使学生更好地掌握纳米材料的基本概念及基础知识，增强学生跨文化进行专业交流的能力，亦为学生以后阅读专业文献、撰写科研论文做铺垫，本书以中英双语的形式呈现；每章均以英文部分开篇，随后附有专业词汇的音标及释义，然后是对英文部分的中文翻译，以方便学生理解。全书共分为八章，第1章主要介绍了纳米材料的基本概念及分类，纳米科技的研究进展及最新成果。第2章为纳米效应的相关概念。第3章是关于纳米材料的力学、热学、磁学及光学性能。第4章重点介绍了“自上而下”和“自下而上”纳米材料的制备方法以及纳米材料的自组装。第5章对纳米材料的表征及纳米制造的常用仪器——扫描隧道显微镜和原子力显微镜的基本原理及操作模式进行了简述。第6章是碳纳米材料的合成。第7章为纳米制造的光刻技术。第8章主要论述纳米技术用于太阳能制氢的新能源研究。

本书可作为高等院校材料类、化学类、化工类、环境类、能源类、电子类等专业的本科生和研究生教材，亦可供相关专业工程技术人员、科研人员参考。

图书在版编目 (CIP) 数据

纳米材料基础 (双语版)/张耀君编著. —2版. —北京：化学工业出版社，2015. 9 (2024.6 重印)
“十二五”普通高等教育本科国家级规划教材
ISBN 978-7-122-24343-0

Ⅰ. ①纳…　Ⅱ. ①张…　Ⅲ. ①纳米材料-高等学校-教材-汉、英　Ⅳ. ①TB383

中国版本图书馆 CIP 数据核字（2015）第 129946 号

责任编辑：宋林青　　　　装帧设计：王晓宇
责任校对：宋　玮

出版发行：化学工业出版社（北京市东城区青年湖南街 13 号　邮政编码 100011）
印　　装：北京虎彩文化传播有限公司
787mm×1092mm　1/16　印张 14½　字数 372 千字　　2024 年 6 月北京第 2 版第 6 次印刷

购书咨询：010-64518888　　　　售后服务：010-64518899
网　　址：http://www.cip.com.cn
凡购买本书，如有缺损质量问题，本社销售中心负责调换。

定　　价：39.80 元

前　言

本书第一版于 2011 年正式出版发行后，多所院校选其作为教材，这是我们和出版社都始料未及的。为不辜负读者的厚爱，每次重印时，我们均尽自己最大可能进一步提升本书的质量，但限于版面问题，每次都是“小修小补”。值本书入选“十二五”普通高等教育本科国家级规划教材之际，再加上纳米科技的发展也亟需更新、补充部分内容，所以推出了第二版。

本次再版延续了第一版由浅入深、循序渐进、注重基础、关注前沿的特色；全书内容新颖、简明扼要、知识系统、重点突出，在强化基础知识、基本理论的同时，侧重纳米科技的研究进展及最新成果，体现了基本理论与生产实际相结合的教育理念。

本次再版第 1 章增加了零维、一维、二维及三维系统中电子的运动行为，从理论上阐明了自由电子模型、能量与态密度之间的关系；细化了纳米结构的分类以及纳米结构中各符号的含义；增加了纳米材料在医疗，尤其是再生疗法中的应用。第 4 章增加了超临界水热合成、喷雾冷冻干燥技术、各向异性纳米粒子的合成和组装、一维纳米结构聚合物的合成。第 6 章增加了石墨烯的性能与制备。第 8 章增加了氢能经济的内容。

本书承蒙李聚源教授审阅，提出了许多宝贵的修改意见；王亚超、刘礼才、康乐、柴倩、杨梦阳、张力、张科等在翻译及生词列表等工作中给予了诸多帮助；化学工业出版社的编辑对本书的修订给予了大力支持；在此一并表示衷心的感谢。

限于作者水平，书中疏漏和不妥之处，敬请同行及读者批评指正。

编著者

2015 年 3 月 20 日于西安

前言

第一版前言

作为纳米科技基石的尺度在1～100nm范围内的纳米材料，因其独特的纳米效应，近年来已成为全球高新科技炙手可热的研究领域之一。纳米材料是一门涉及知识面广的新的交叉学科，新概念、新理论、新技术及新方法层出不穷。纳米科技充满着原始创新的机遇与挑战，尤其是纳米科技正在将微制造推向纳制造与纳加工的前沿，各种产品正从微尺度向纳尺度悄然转变，新材料、新产品呼之欲出，这将对信息产业、能源、环境检测、生命科学、军事、材料的生产与加工带来一场革命性的变革。因此，了解纳米科技的发展动态，加强对纳米材料的基本概念和基础知识的学习，掌握纳米材料的特性、制备原理及研究方法就显得十分重要。

本书是在作者多年来为本科生及研究生开设的“纳米材料基础”双语教学讲义的基础上，进行不断的修改、补充及完善后撰写而成的。在编写过程中，作者查阅了大量的国内外相关的文献资料，阅读了诸多的教材及专著，结合本研究小组的科研成果，以纳米材料的基本概念、纳米效应、纳米材料制备、表征、纳米制造以及纳米技术在新能源中的应用为主线，力图条理清楚、结构严谨地将基本概念及基础知识奉献给读者。本书具有以下特色。

① 为了将纳米材料的基础知识学习与阅读外文资料及提升科研能力相融合，双语编著是本书的特色之一。

② 为适应初学者学习，本书由浅入深，循序渐进，着力强化教材的基础性和系统性。

③ 本书内容新颖，简明扼要，知识系统，重点突出，在强化基础知识、基本理论的同时，注重纳米科技的研究进展及最新成果介绍，体现基本理论与研究实践相结合的特色。

④ 为了使读者能对自己感兴趣的内容进一步自学，书中对重要的概念、图表、实例等引注了出处，便于查阅导读；另外，为了便于阅读及掌握章节中的重点内容，每章后附有词汇、复习题及相关章节的译文。

本书共八章，第1章主要介绍了纳米材料的基本概念及分类，纳米科技的研究进展及最新成果。第2章涉及纳米效应的相关概念。第3章是关于纳米材料的力学、热学、磁学、电学及光学性能。第4章重点介绍了“自上而下”和“自下而上”的纳米材料的制备方法以及纳米材料的自组装。第5章对纳米材料的表征及纳米制造的常用仪器——扫描隧道显微镜和原子力显微镜的基本原理及操作模式进行了简述。第6章是碳纳米材料的制备及纳米车的雏形。第7章涉及纳米制造的光刻技术。第8章主要论述纳米技术用于太阳能制氢的新能源研究。

在编写过程中，作者阅读了大量的相关文献资料，从中获得了许多前瞻性的珍贵信息，向本书中引用的文献作者表示深深的谢意。化学工业出版社对本书的出版提供了大力的支持，在此一并表示衷心的感谢。

鉴于作者水平有限，编写时间仓促，本书中、英文疏漏和不足之处在所难免，敬请同行和读者批评指正。

编著者

2010年10月于西安

CONTENTS

1. Introduction to Nanoscale Materials

1.1 Introduction to the nanoworld

The nanoscale material with at least one dimension in the nanometer range is a bridge between isolated atoms or small molecules and bulk materials. Therefore, it is referred to as mesoscopic scale materials. Nanoscale materials as foundation of nanoscience and nanotechnology have become one of the most popular research topics in recent years. The intense interests in nanotechnology and nanoscale materials have paid to several areas by the tremendous economical, technological, and scientific impact: ① with exponential growth of the capacity and speed of semiconducting chips, the key components which virtually enable all modern technology is rapidly approaching their limit of arts, this needs the coming out of new technology and new materials; ② novel nanoscale materials and devices hold great promise in energy, environmental, biomedical, and health sciences for more efficient use of energy sources, effective treatment of environmental hazards, rapid and accurate detection and diagnosis of human diseases; and ③ when a material is reduced to the dimension of nanometer, its properties can be drastically different from those of the bulk material that we can either see or touch even though the composition is essentially the same. Therefore, nanoscale materials prove to be a very fertile ground for great scientific discoveries and explorations.

It has been said that a nanometer is "a magical point on the length scale", for this is the point where the smallest man-made devices meet the atoms and molecules of the natural world [1].

Indeed, nanoscience and technology have been an explosive growth in the last few years. "Nanotechnology mania" is sweeping through essentially all fields of science and engineering, and the public is becoming aware of the quote of the chemist and Nobel laureate, Richard Smally: "Just wait, the next century is going to be incredible. We are able to build things that work on the smallest possible length scales, atom by atom. These little nanothings will revolutionize our industries and our lives [1]."

1.2 Definition of nanoscale materials

1.2.1 Nanometer

The prefix "nano" is from the Greek word "nanos" and it means dwarf. Nanometer is a length unit. A nanometer (nm) equals a billionth of a meter ($1nm = 1\times10^{-9}m$).

Fig.1.1 shows the length scales of some materials synthesized and biology. Beginning at small scales, feature of Au atomic diameter is on the order of 0.1nm in size. The diameter of a carbon nanotube is about 1～2nm, and a double helix of DNA is about 3nm. A HIV virus is about 100nm

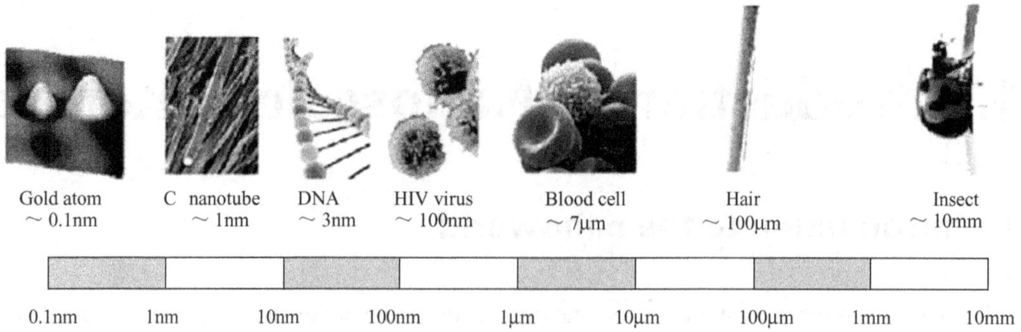

Fig.1.1 Length scale and some examples related [2]

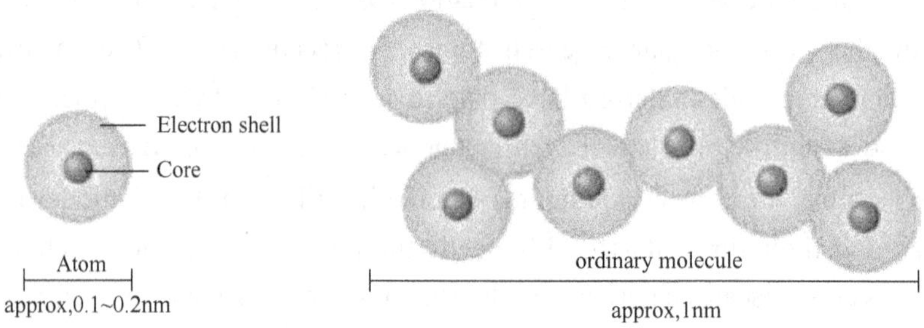

Fig.1.2 Schematic representation for atom and molecule

and so on [2]. The diameter of one atom is about 0.1～0.2nm, and the length of 8～10 atoms is about one nanometer as shown in Fig.1.2.

1.2.2 Definition of nanoscale materials

Nanoscale material is defined as a material having one or more external dimensions in the nanoscale (1～100nm).

Fig.1.3 shows a picture of single-walled carbon nanotubes in comparison to a human hair which is about 80000nm in diameter. The single-walled carbon nanotube is about 1000 times smaller than that of human hair in diameter.

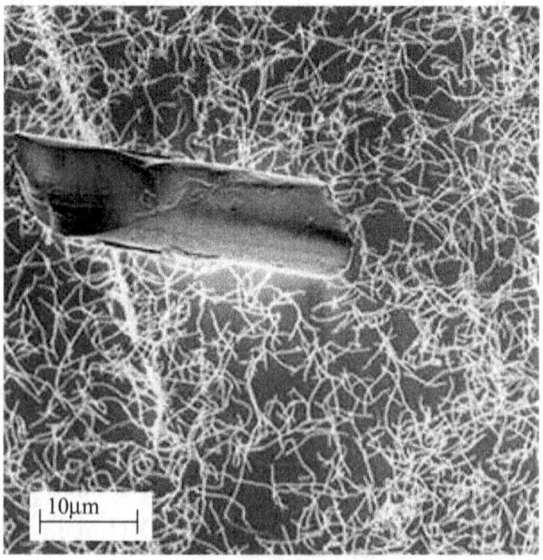

Fig.1.3 Human hair fragment and a network of single-walled carbon nanotubes

1.3 Classification of nanoscale materials

Nanoscale materials are primitively divided into discrete nanomaterials and nanostructured materials, but also there are other classification methods.

The discrete nanomaterial means that the material has an appearance characteristic at least one dimension on the nanoscale, such as nanoparticles, nanofibers, nanotubes and membrane.

The nanostructured material is the material has an appearance characteristic of bulk material, but it may be built up of discrete nanomaterials, such as bulk materials by consolidation nanopowders, or it may be composed of continuously nanostructural units, such as porous materials including microporous (<2nm), mesoporous (2～50nm) and macroporous (>50nm), nanophase and polycrystalline materials.

The technique of consolidation nanopowders is a fabrication method of bulk nanostructured materials. However, because of the very small size of the powder particles, special precautions must be taken to reduce the interparticles frication and minimize the danger of explosion or fire. The powders themselves may have a microscale average particle size, or they may be true nanopowders, depending on their synthesis routes. They would be compacted at low or moderate temperature to produce a so-called green body with a density in excess of 90% of the theoretical maximum. Any residual porosity would be evenly distributed throughout the material and the pores would be fine in scale and have a narrow size distribution. Polycrystalline materials with grain sizes between 100nm and 1μm are made up of many nanocrystals and are conventionally called ultrafine grains.

1.3.1 According to the spatial dimension of materials

A reduction in the spatial dimension or confinement of nanoparticle in a particular crystallographic direction within a structure generally leads to changes in physical properties of the system in that direction. Hence one classification of nanostructured materials and systems essentially depends on the number of dimensions which lie within the nanometer range. The examples of reduced dimensionality systems are shown in Table1.1 and Fig.1.4[1,3].

Table 1.1 Examples of reduced-dimensionality systems

3D confinement
Fullerenes
Colloidal particles
Nanoporous silicon
Activated carbon
Nitride and carbide precipitates in high-strength low-alloy steels
Semiconductor particles in a glass matrix for non-linear optical components
Semiconductor quantum dots(self-assembled and colloidal)
Quasi-crystals
2D confinement
Carbon nanotubes and nanofilaments
Metal and magnetic nanowires
Oxide and carbide nanorods
Semiconductor quantum wires
1D confinement
Nanolaminated or compositionally modulated materials
Grain boundary films
Clay platelets
Semiconductor quantum wells and superlattices
Magnetic multilayers and spin valve structures
Langmuir-Blodgett films

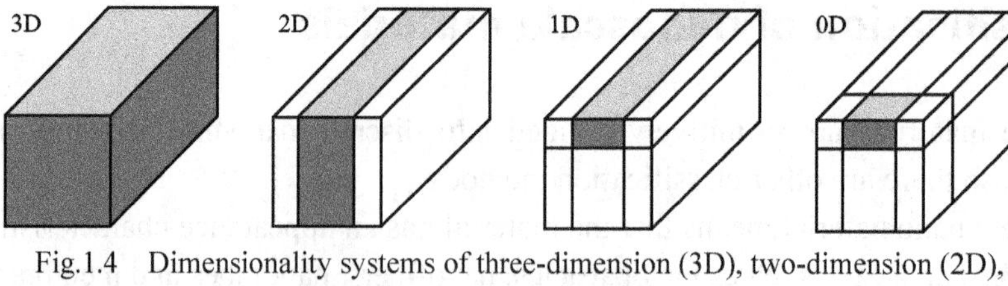

Fig.1.4 Dimensionality systems of three-dimension (3D), two-dimension (2D), one-dimension (1D) and zero-dimension (0D)

(1) Zero dimension (0D) materials

There are three dimensions for material on the nanoscale. This means that the size of material is confined in three dimensions (the material is dimensionless in three directions). This system includes the nanoparticles, nanocrystals and etc.

(2) One dimension (1D) materials

There are two dimensions for material on the nanoscale. This means the size of material is confined in two dimensions (the material is dimensionless in the other two directions). The system includes nanowires, nanorods, nanofilaments, nanotubes and etc. The ratio of the length to the diameter of these structures is called aspect ratio. The aspect ratio for nanorods generally lies in the range of 10～100. If aspect ratio of nanorods becomes more than 100, they are termed as nanowires. Nanowires are hence long nanorods. Nanotubes are on the other hand, nanorods with hollow interiors.

(3) Two dimension (2D) materials

There is one dimension on the nanoscale in material, that is, the size of material is confined in one dimension. The system includes ultrathin films, multilayered films, thin films, surface coatings, superlattices and etc.

1.3.2 According to the quantum properties of materials

(1) Bulk material

The electronic structure of material is strongly related to the nature of material. We now consider the case of a three-dimensional solid in d_x, d_y and d_z directions containing a number of "free" electrons. The "free" means those electrons are delocalized and not bound to individual atom.

In the free-electron model, each electron in the solid moves with a velocity $\vec{v}=(v_x, v_y, v_z)$. The energy of an individual electron is then just its kinetic energy:

$$E=\frac{1}{2}m\vec{v}^2=\frac{1}{2}m(v_x^2+v_y^2+v_z^2) \tag{1.1}$$

According to Pauli's exclusion principle, each electron must be in a unique quantum state. Since electrons can have two spin orientations ($m_s=+1/2$) and ($m_s=-1/2$), only two electrons with opposite spins can have the same velocity $\vec{v}$. This case is analogous to the Bohr model of atoms, in which each orbital can be occupied by two electrons at maximum. In solid-state physics, the wavevector $\vec{k}=(k_x, k_y, k_z)$ of a particle is more frequently used instead of its velocity to describe the particle's state. Its absolute value $k=|\vec{k}|$ is the wavenumber. The wavevector $\vec{k}$ is directly proportional to the linear momentum ($\vec{p}$) and thus also to the velocity ($\vec{v}$) of the electron.

The calculation of the energy states for a bulk crystal is based on the assumption of periodic boundary conditions. Periodic boundary conditions are a mathematical trick to "simulate" an infinite $(d=\infty)$ solid. This assumption implies that the conditions at opposite borders of the solid are identical. In this way, an electron that is close to the border does not really "feel" the border. In other words, the electrons at the borders "behave" exactly as if they were in the bulk. This condition can be realized mathematically by imposing the following condition to the electron wavefunctions: $\psi(x,y,z)=\psi(x+d_x,y,z)$, $\psi(x,y,z)=\psi(x,y+d_y,z)$, and $\psi(x,y,z)=\psi(x,y,z+d_z)$, In other words, the wavefunctions must be periodic with a period equal to the whole extension of the solid [4,5]. Each function describes a free electron moving along one Cartesian coordinate. In the argument of the functions, $k_{x,y,z}$ is equal to $\pm n\Delta k=\pm n2\pi/d_{x,y,z}$ and n is an integer [4~6]. These solutions are waves that propagate along the negative and the positive direction, for $k_{x,y,z}>0$ and $k_{x,y,z}<0$, respectively. An important consequence of the periodic boundary conditions is that all the possible electronic states in the $\vec{k}$ space are equally distributed. There is an easy way of visualizing this distribution in the ideal case of a one-dimensional free-electron model: there are two electrons $(m_s=\pm 1/2)$ in the state $k_x=0$ $(v_x=0)$, two electrons in the state $k_x=+\Delta k$ $(v_x=+\Delta v)$, two electrons in the state $k_x=-\Delta k$ $(v_x=\Delta v)$, two electrons in the state $k_x=+2\Delta k$ $(v_x=+2\Delta v)$ and so on.

For a three-dimensional bulk material we can follow an analogous scheme. Two electrons $(m_s=\pm 1/2)$ can occupy each of the states $(k_x,k_y,k_z)=(\pm n_x\Delta k,\pm n_y\Delta k,\pm n_z\Delta k)$, again with $n_{x,y,z}$ being an integer. A sketch of this distribution is shown in Fig.1.5. We can easily visualize the occupied states in $\vec{k}$-space because all these states are included into a sphere whose radius is the wavenumber associated with the highest energy electrons. At the ground state, at 0 K, the radius of the sphere is the Fermi wavenumber k_F (Fermi velocity v_F). The Fermi energy $E_F\propto k_F^2$ is the energy of the last occupied electronic state. All electronic states with an energy $E\leqslant E_F$ are occupied, whereas all electronic states with higher energy $E>E_F$ are empty. In a solid, the allowed wave numbers are separated by $\Delta k=\pm n2\pi/d_{x,y,z}$. In a bulk material $d_{x,y,z}$ is large, and so Δk is very small. Then the sphere of states is filled quasi-continuously [4].

We need now to introduce the useful concept of the density of states $D_{3d}(k)$, which is the number of states per unit interval of wavenumbers. From this definition, $D_{3d}(k)\Delta k$ is the number of electrons in the solid with a wavenumber between k and $k+\Delta k$. If we know the density of states in a solid we can calculate, for instance, the total number of electrons having wavenumbers less than a given k_{max}, which we will call $N(k_{max})$. Obviously, $N(k_{max})$ is equal to $\int_0^{k_{max}} D_{3d}(k)dk$. In the ground state of the solid, all electrons have wavenumbers $k\leqslant k_F$, where k_F is the Fermi wavenumber. Since in a bulk solid the states are homogeneously distributed in

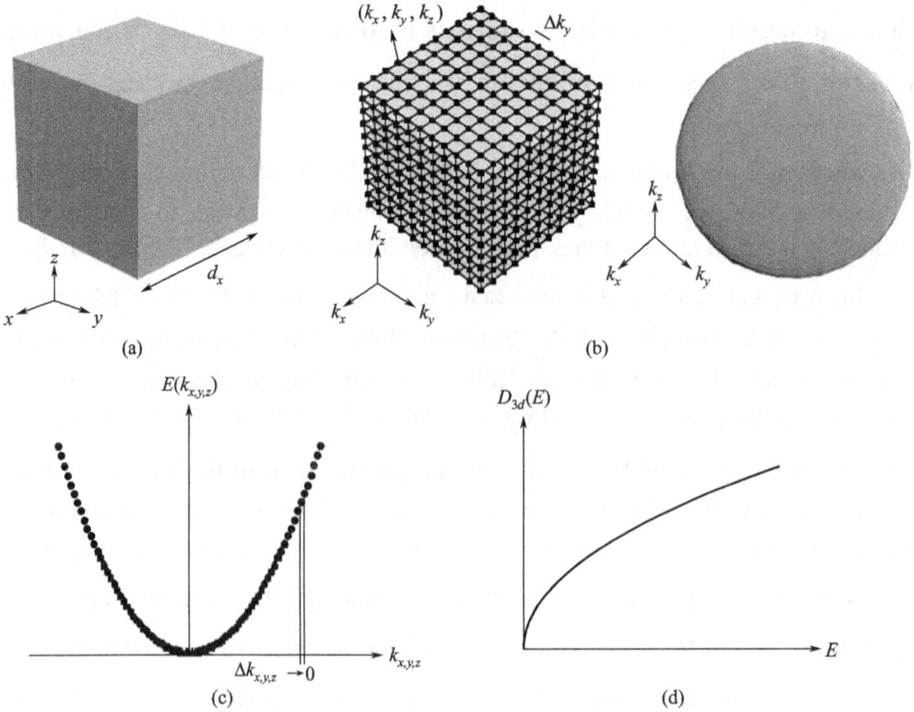

Fig.1.5 Electrons in a three-dimensional bulk solid [5]

$\vec{k}$ -space, we know that the number of states between k and $k+\Delta k$ is proportional to $k^2\Delta k$ (Fig.1.5). This can be visualized in the following way. The volume in three-dimensional $\vec{k}$ -space varies with k^3. If we only want to count the number of states with a wavenumber between k and $k+\Delta k$, we need to determine the volume of a spherical shell with radius k and thickness Δk. This volume is proportional to product of the surface of the sphere (which varies as k^2) with the thickness of the shell (which is Δk). $D_{3d}(k)\Delta k$ is thus proportional to $k^2\Delta k$, and in the limit when Δk approaches zero, we can write:

$$D_{3d}(k) = \frac{\mathrm{d}N(k)}{\mathrm{d}k} \propto k^2 \tag{1.2}$$

Instead of knowing the density of states in a given interval of wavenumbers it is more useful to know the number of electrons that have energies between E and $E + \Delta E$. We know that $E(k)$ is proportional to k^2, and thus $k \propto \sqrt{E}$. Consequently, $\mathrm{d}k / \mathrm{d}E \propto 1/\sqrt{E}$. By using Eq. (1.2), we obtain for the density of states for a three-dimensional electron model [5]:

$$D_{3d}(E) = \frac{\mathrm{d}N(E)}{\mathrm{d}E} = \frac{\mathrm{d}N(k)}{\mathrm{d}k}\frac{\mathrm{d}k}{\mathrm{d}E} \propto E\frac{1}{\sqrt{E}} \propto \sqrt{E} \tag{1.3}$$

This can be seen schematically in Fig.1.5. With Eq. (1.3) we conclude our simple description of a bulk solid. The possible states in which an electron can be found are quasi-continuous. The density of states varies with the square root of the energy.

Fig.1.5 shows electrons in a three-dimensional bulk solid. (a) Such a solid can be modeled as an infinite crystal along all three dimensions *x,y,z*. (b) The assumption of periodic boundary conditions yields standing waves as solutions for the Schrődinger equation for free electrons. The associated wavenumbers (k_x,k_y,k_z) are periodically distributed in the *k*-space [5]. Each of the dots shown in the figure represents a possible electronic state (k_x,k_y,k_z). Each state in k-space can be only occupied by two electrons. In a large solid the spacing $\Delta k_{x,y,z}$ between individual electron states is very small, and therefore the k-space is quasi-continuously filled with states. A sphere with radius k_F includes all states with $k = (k_x^2 + k_y^2 + k_z^2)^{1/2} < k_F$. In the ground state, at 0 K, all states with $k < k_F$ are occupied by two electrons, and the other states are empty. Since the k-space is homogeneously filled with states, the number of states within a certain volume varies with k^3. (c) It is the dispersion relation for free electrons in a three-dimensional solid. The energy of free electrons varies with the square of the wavenumber, and its dependence on k is described by a parabola. For a bulk solid the allowed states are quasi-continuously distributed and the distance between two adjacent states (here shown as points) in k-space is very small. (d) It is the density of states D_{3d} for free electrons in a three-dimensional system. The allowed energies are quasi-continuous and their density of states varies with the square root of the energy $E^{1/2}$ shown in Fig.1.5 [1].

(2) Quantum wells (2D)

When a solid is fully extended along the x- and y-directions, but the thickness along the z-direction (d_z) is only a few nm, electrons can still move freely in the x- and y-directions. However, movement of electrons in the z-direction is restricted and becomes quantized. Such a system is called two-dimensional (2D) system and is also named as quantum well.

When one or more dimensions of a solid become smaller than the De Broglie wavelength associated with the free charge carriers, an additional contribution of energy is required to confine the component of the motion of the carriers along this dimension. In addition, the movement of electrons along such a direction becomes quantized. This situation is shown in Fig.1.6. No electron can leave the solid, and electrons that move in the z-direction are trapped in a "box". Mathematically this is described by infinitely high potential wells at the border $z = \pm\frac{1}{2}d_z$.

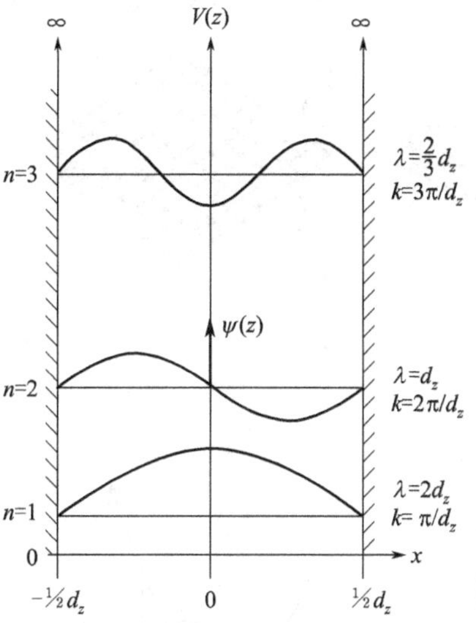

Fig.1.6 Particle-in-a-box model for a free electron moving along in the z-axis. The movement of electrons in the z-direction is limited to a "box" with thickness d: since electrons cannot "leave" the solid (the box), their potential energy $V(z)$ is zero within the solid, but is infinite at its borders.

The solutions for the particle-in-a-box situation can be obtained by solving the one-dimensional Schrödinger equation for an electron in a potential $V(z)$, which is zero within the box but infinite at the borders. As can be seen in Fig.1.6, the solutions are stationary waves with energies

$$E_{nz} = \hbar^2 k_z^2 / (2m) = h^2 k_z^2 / (8\pi^2 m) = h^2 n_z^2 / (8md_z^2), n_z = 1,2,3$$

This is similar to states $k_z = n_z \Delta k_z$ with $\Delta k_z = \pi / d_z$. Again, each of these states can be occupied at maximum by two electrons.

For a two-dimensional solid that the states in the k-space is extended in the x-y-plane only discrete values are allowed for k_z. The thinner the solid in the z-direction, the larger is the spacing Δk_z between these allowed states. On the other hand, the distribution of states in the k_x-k_y plane remains quasi-continuous. Therefore one can describe the possible states in the k-space as planes parallel to the k_x- and k_y-axes, with a separation Δk_z between the planes in the k_z-direction. We can number the individual planes with n_z. Since within one plane the number of states is quasi- continuous, the number of states is proportional to the area of the plane. This means that the number of states is proportional to $k^2 = k_x^2 + k_y^2$. The number of states in a ring with radius k and thickness Δk is therefore proportional to $k\ \Delta k$. Integration over all rings yields the total area of the plane in k-space. Here, in contrast to the case of a three-dimensional solid, the density of states varies linearly with k:

$$D_{2d}(k) = \frac{\mathrm{d}N(k)}{\mathrm{d}k} \propto k \tag{1.4}$$

In the ground state, all states with $k \leqslant k_F$ are occupied by two electrons. We now want to know how many states exist for electrons that have energies between E and $E + \Delta E$. We know the relation between k and E: $E(k) \propto k^2$ and thus $k \propto \sqrt{E}$ and $\mathrm{d}k / \mathrm{d}E \propto 1/\sqrt{E}$. By using Eq. (1.4) we obtain the density of states for a 2-dimensional electron gas shown in Fig.1.7.

$$D_{2d}(E) = \frac{\mathrm{d}N(E)}{\mathrm{d}E} = \frac{\mathrm{d}N(k)}{\mathrm{d}k}\frac{\mathrm{d}k}{\mathrm{d}E} \propto \sqrt{E}\frac{1}{\sqrt{E}} \propto 1 \tag{1.5}$$

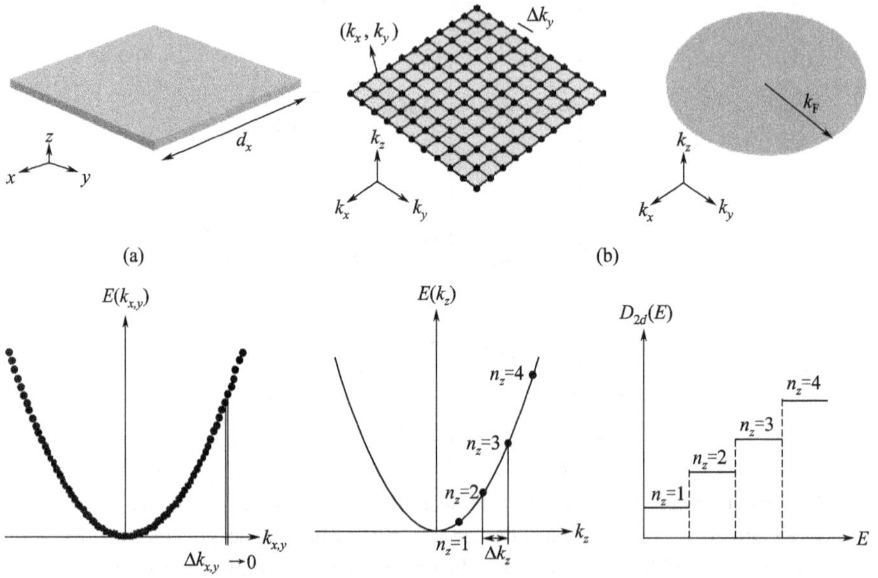

Fig.1.7 Electrons in a two-dimensional system.

The density of electronic states in a two-dimensional solid is therefore remarkably different from the three-dimensional case. The spacing between the allowed energy levels in the bands increases, because fewer levels are now present. In two-dimensional materials the energy spectrum is still quasi-continuous, but the density of states now is a step function ("Staircase") [5, 7].

Fig.1.7 shows electrons in a two-dimensional system. (a) A two-dimensional solid is (almost) infinitely extended in two dimensions (here x, y), but is very thin along the z-direction, which is comparable to the De Broglie wavelength of a free electron ($d_z \rightarrow \lambda$). (b) Electrons can still move freely along the x- and y-directions. The wavefunctions along such directions can be found again by assuming periodic boundary conditions. k_x and k_y states are quasi-continuously distributed in k-space. The movement of electrons in the z-direction is restricted and electrons are confined to a "box". Only certain quantized states are allowed along this direction. For a discrete k_z-state, the distribution of states in three-dimensional k-space can be described as a series of planes parallel to the k_x - and k_y-axes. For each discrete k_z-state, there is a separate plane parallel to the k_x and to the k_y-axes. Here only one of those planes is shown. The k_x-and k_y-states within one plane are quasi-continuous, since $\Delta k_{x,y} = 2\pi/d_{x,y} \rightarrow 0$. The distance between two planes for two separate k_z-states is large, since $\Delta k_z = \pi/d_z \rightarrow >>0$. (c) Free electrons have a parabolic dispersion relation $E(k) \propto k^2$. The energy levels $E(k_x)$ and $E(k_y)$ for the electron motion along the x- and the y-directions are quasi-continuous (they are shown here as circles). The wavefunction $\psi(z)$ at the border of a small "box" must be zero, leading to standing waves inside the box. This constraint causes discrete energy levels $E(k_z)$ for the motion along the z-direction. Electrons can only occupy such discrete states (n_{z1}; n_{z2}; ⋯; shown here as circles). The position of the energy levels now changes with the thickness of the solid in the z-direction, or in other words with the size of the ''box''. (d) It is the density of states for a two-dimensional electron gas. If electrons are confined in one direction (z) but can move freely in the other two directions (x,y), the density of states for a given k_z-state ($n_z = 1, 2, 3\cdots$) does not depend on the energy E.

Quantum confinement naturally restricts the motion of the carriers in some ways. When the electronic motion is confined in one dimension (quantum confinement in one dimension) and the motion is free in the other two dimensions, it results in the creations of quantum wells or quantum films.

The quantum well implies that the electrons drop into a "potential well" as well as they are trapped in the film. 2D systems are usually formed at interfaces between different materials or in layered systems in which some of the layers may be only a few nanometers thick. Layers made by using different semiconductor materials may result in the trapping of carriers (electrons, holes) in a particular layer. The trapped carriers are restricted to movements in certain directions.

The quantum-mechanical behavior of electrons in a two-dimensional solid is the origin of many important physical effects. With recent progress in nanoscience and technology, the fabrication of two-dimensional structures has become routine. 2D systems are usually formed at interfaces between different materials or in layered systems in which some of the layers may be only a few nanometers thick. Structures like this can be grown, for example, by successive deposition of the individual layers by molecular beam epitaxy. In such geometry, charge carriers (electrons and holes) can move freely parallel to the semiconductor layer, but their movement perpendicular to the interface is restricted. The study of these nanostructures led to the discovery of remarkable 2-dimensional quantized effects, such as the Integer and the Fractional Quantum Hall Effect [8~11].

(3) Quantum wires (1D)

Now electrons can only move freely in the x-direction, and their motions along the y- and z-axis are restricted by the borders of the solid. Such a system is called quantum wire. In one-dimensional systems, the electrons are free to move only in one direction with quantum confinement in the other two dimensions.

The states of a one-dimensional solid now can be obtained by methods analogous to those described for the three- and two-dimensional materials. In the x-direction electrons can move freely, and again we can apply the concept of periodic boundary conditions. This yields a quasi-continuous distribution of states parallel to the k_x-axis as well as a quasi-continuous distribution of the corresponding energy levels. Electrons are confined along the remaining directions and their states can be derived from the Schrödinger equation for a particle-in-a-box potential. Again, this yields discrete k_y and k_z-states. We can now visualize all possible states as lines parallel to the k_x-axis. The lines are separated by discrete intervals along k_y and k_z, but within each line the distribution of k_x states is quasi-continuous (Fig.1.8). We can count the number of states along one line by measuring the length of the line. The number of states is therefore proportional to $k = k_x$. Hence, the number of states with wavenumbers in the interval between k and $k + \Delta k$ is proportional to Δk:

$$D_{1d}(k) = \frac{\mathrm{d}N(k)}{\mathrm{d}k} \propto 1 \tag{1.6}$$

In the ground state, all states with $k \leqslant k_F$ are occupied by two electrons. We know that the relation between k and E for free electron is: $E(k) \propto k^2$ and thus $k \propto \sqrt{E}$ and $\mathrm{d}k / \mathrm{d}E \propto 1/\sqrt{E}$. By using Eq. (1.6), we obtain the density of states for a 1-dimensional electron gas:

$$D_{1d}(E) = \frac{\mathrm{d}N(E)}{\mathrm{d}E} = \frac{\mathrm{d}N(k)}{\mathrm{d}k}\frac{\mathrm{d}k}{\mathrm{d}E} \propto 1 \cdot 1/\sqrt{E} \propto 1/\sqrt{E} \tag{1.7}$$

The density of states is depicted in Fig.1.8. In one-dimensional systems the density of states has an $E^{-1/2}$-dependence and thus exhibits singularities near the band edges [5]. Each of the hyperbolas contains a continuous distribution of k_x states but only one discrete k_y- and k_z-state.

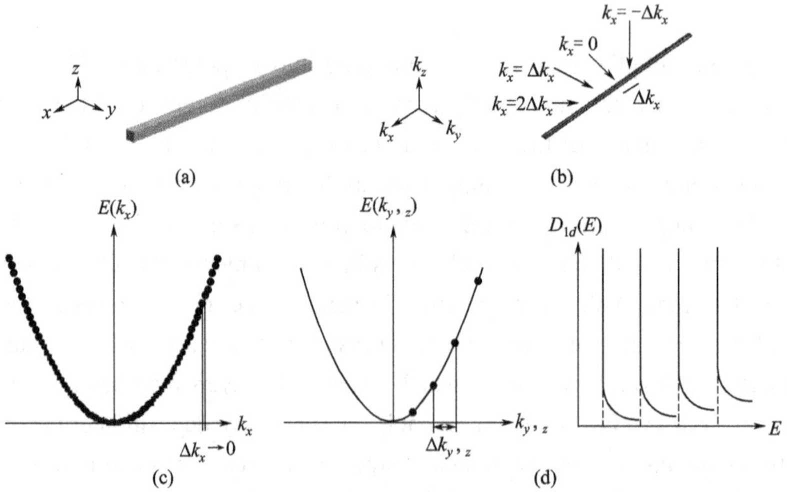

Fig.1.8 Electrons in a one-dimensional system

Fig.1.8 shows electrons in a one-dimensional system. (a) It is one-dimensional solid. (b) The allowed (k_x, k_y, k_z)-states can be visualized as lines parallel to the k_x-axes in the three-dimensional k-space. In this figure only one line is shows as an example. Within each line, the distribution of states is quasi-continuous, since $\Delta k_x \to 0$. The arrangement of the individual lines is discrete, since only certain discrete k_y-and k_z-states are allowed. (c) This can also be seen in the dispersion relations. Along the k_x-axes the energy band $E(k_x, k_y. k_z)$ is quasi-continuous, but along the k_y-and k_z-axes are discrete. (d) The density of states within one line along the k_x-axes is proportional to $E^{-1/2}$. Each of the hyperbolas shown in the D_{1d}-diagram corresponds to an individual (k_y, k_z)-state.

The quantization of states in two dimensions has important consequences for the transport of charges. Electrons can only flow freely along the x-axes but are limited to discrete states in the y- and z-directions. Therefore they are only transported in discrete "conductivity channels". This may be of considerable importance for the microelectronics industry. If the size of electronic circuits is reduced more and more, at one point the diameter of wires will become comparable to the De Broglie wavelength of the electrons. The wire will then exhibit the behavior of a quantum wire. More recent examples of such 1D wires include short organic semiconducting molecules, inorganic semiconductor and metallic nanowires [12~18] or break junctions [19~21]. A particular role is played by carbon nanotubes [22~26], which have been extensively studied both as model systems for one-dimensional confinement and for potential applications, such as electron emitters [27].

(4) Quantum dots (0D)

When electrons are confined in all three dimensions, we get a zero-dimensional system and the system is called a "quantum dot". In a quantum dot, the movement of electrons is confined in all three dimensions and there are only discrete (k_x, k_y, k_z)-states in the k-space. Each individual state in k-space can be represented by a point. The final consequence is that only discrete energy levels are allowed, and they can be seen as discrete peaks in the distribution $D_{0d}(E)$.

Fig.1.9 shows electrons in a zero-dimensional solid. (a) The solid is shrunk in all three dimensions to a thickness that is comparable to the De Broglie wavelength of its electrons. (b) Because

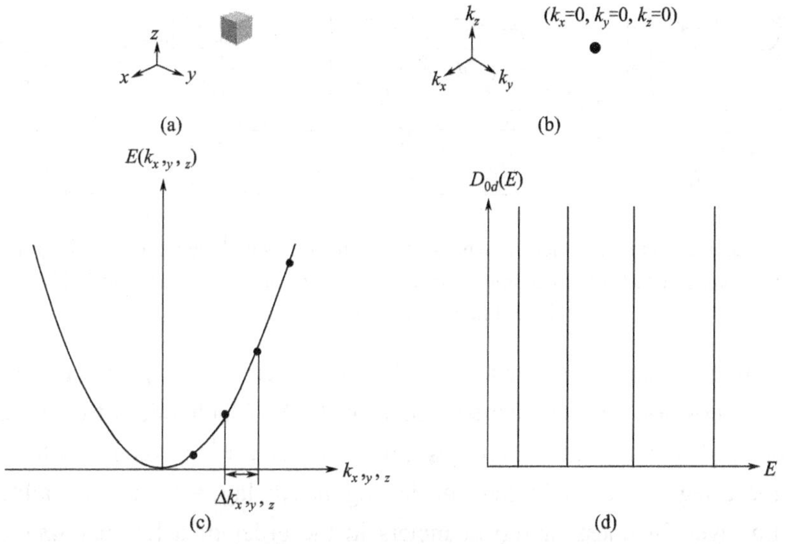

Fig.1.9 Electrons in a zero-dimensional system

of such confinement, all states (k_x, k_y, k_z) are discrete points in the three-dimensional k-space. (c) Only discrete energy levels are allowed. (d) The one-dimensional density of states $D_{0d}(E)$ contains discrete peaks which correspond to the individual states. Electrons can occupy only states with these discrete energies [28].

Quantum dots are also called artificial atoms or boxes of electrons, because of their discrete charge states and the energy-level structures which are similar to atomic systems containing from a few thousands to one electron [29,30].

1.3.3 According to material properties

The thousands of substances that are solids under normal temperature and pressure can be subdivided into metals, ceramics, semiconductors, composites, and polymers. These can be further subdivided into biomaterials, catalytic materials, coatings, glasses, and magnetic and electronic materials. All of these solid substances with their widely variable properties take on another subset of new properties when they were produced in the form of nanoparticles. They can be divided into different forms such as nanocrystal materials, nanoceramic materials, nanocomposite materials, nano high molecular materials and so on.

1.3.4 According to the shape and chemical composition

(1) Nanostructured materials

The term "nanostructured material" (NSM) is often used to describe materials with typical crystallite dimensions on the nanometer scale (Fig.1.10). A nanocrystalline bulk [Fig.1.10(a)] is a body that is macroscopic in three dimensions (3D) but comprised of nanometer scale crystallites. Nanocrystalline materials are produced by a very wide range of methods. A thin film [Fig.1.10(b)] is macroscopic in 2D, but typically has submicron thickness and may have an in-plane grain size that ranges from much smaller, to much greater than the film thickness. Objects that are limited to the nanometer scale in 2 or 3 dimensions are referred to here as lines [1D, Fig.1.10(c)] and dots [0D, Fig.1.10(d)], respectively. Lines and dots are often patterned from thin films [31].

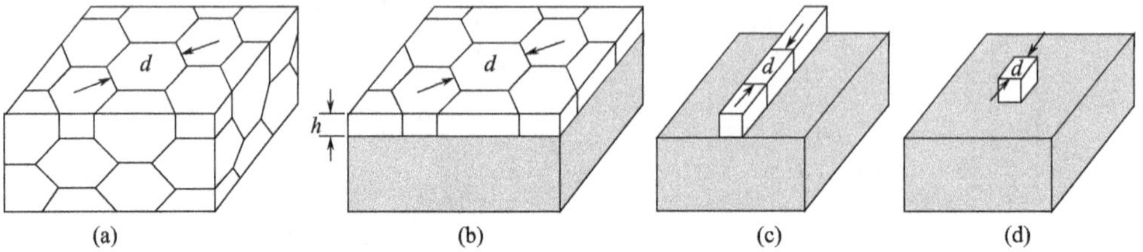

Fig.1.10 Nanostructured materials may be bulk (3D) materials with nanometer scale grain size (a); or may be constrained to the nanometer scale in 1, 2 or 3 dimensions resulting in 2D thin films (b); 1D lines (c); and 0D dots (d) [31]

NsM consisting of nanometer-sized crystallites and interfaces may be classified according to their chemical composition and the shape (dimensionality) of their microstructure constituents of boundary regions and crystallites by Gleiter as shown in Fig.1.11 [32]. According to the shape of the crystallites, three categories of NsM may be distinguished: layer-shaped crystallites, rod-shaped crystallites (with layer thickness or rod diameters in the order of a few nano-meters), and NsM composed of equiaxed nanometer-sized crystallites. Depending on the chemical composition of the crystallites, the three categories of NsM may be grouped into four families.

In the first family shown in Fig.1.11, all crystallites and interfacial regions have the same chemical composition. Examples of this family of NsM are semicrystalline polymers consisting of stacked crystalline lamellae separated by non-crystalline regions as shown in first category in Fig.1.11 or NsM made up of equiaxed nanometer-sized crystals, e.g. of Cu (third category). NsM belonging to the second family consist of crystallites with different chemical compositions (indicated in Fig.1.11 by different thickness of the lines used for hatching). Quantum well (multilayer) structures are probably the most well-known examples of this type (first category).

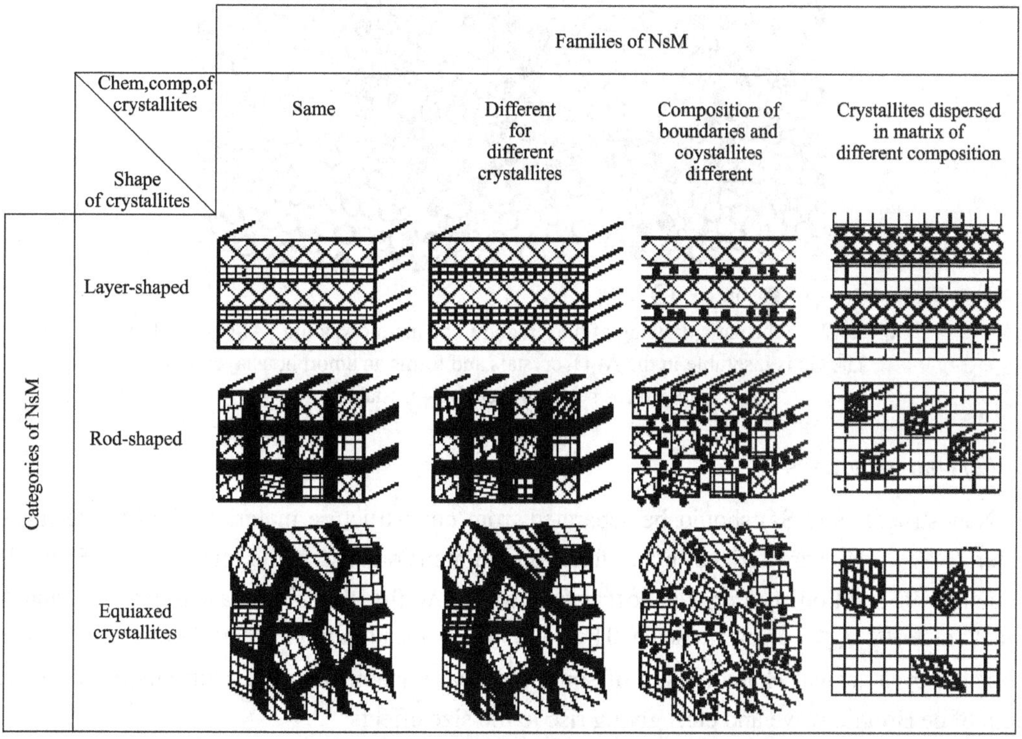

Fig.1.11 Classification schema for NsM according to their chemical compositions and the dimensionality (shape) of the crystallites (structural elements) the NsM [32]

If the compositional variation occurs primarily between crystallites and the interfacial regions, the third family of NsM is obtained. In this case one type of atoms (molecules) segregates preferentially to the interfacial regions so that the structural modulation (crystals/interfaces) is coupled to the local chemical modulation. NsM consisting of nanometer-sized W crystals with Ga atoms segregated to the grain boundaries are an example of this type (third category). An interesting new type of such materials was recently produced by co-milling Al_2O_3 and Ga. It turned out that this procedure resulted in nanometer-sized Al_2O_3 crystals separated by a network of non-crystalline layers of Ga. Depending on the Ga content, the thickness of the Ga boundaries between the Al_2O_3 crystals varies between less than a monolayer and up to about seven layers of Ga as shown in Fig.1.12. The fourth family of NsM is formed by nanometer-sized crystallites (layers, rods or equiaxed crystallites) dispersed in a matrix of different chemical composition. Precipitation-hardened alloys belong in this group of NsM. Nanometer-sized Ni_3Al precipitates dispersed in a Ni matrix-generated by annealing a supersaturated Ni-Al solid solution are an example of such alloys. Most high-temperature materials used in jet engines of modern aircraft are based on precipitation-hardened Ni_3Al/Ni alloys.

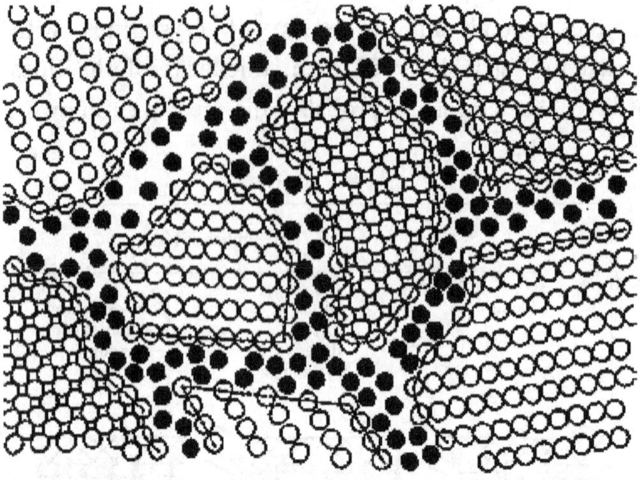

Fig.1.12 Schematic drawing of the structure of a nano-crystalline Al_2O_3-Ga alloy. The Ga is insoluble in the Al_2O_3 crystals and forms an amorphouslayer in the interfaces between the Al_2O_3 crystals [33]

(2) Nanostructures

Nanostructures (NSs) should be separated from nanostructure materials (NSMs) because the former (NSs) are characterized by a form and dimensionality while the latter (NSMs) by a component in addition. Under a nanostructure we know the size d at least is less or equal to a critical value d^*, $d \leqslant d^*$ at nm scale. The value of d^* does not have a certain meaning because it is dictated by a critical characteristic of some physical phenomena (free path length of electrons, length of de Broglie wave and etc.) giving rise to the size effects [34].

According to dimensionality, NSs may be one of the fourth, 0D, 1D, 2D and 3D. All NSs can be built from elementary units (building block) having low dimensionality 0D, 1D, and 2D. The 3D units are excluded because they can't be used to build low dimensional NSs except 3D matrix. However 3D structures can be considered as NSs if they involve the 0D, 1D and 2D NSs. According to Pokropivny, the nanostructures (NSs) can be divided into 36 classes as shown in Fig.1.13 [34].

Let us introduce the notation of NSs, *k*D*lmn,* where *k* is number, *D* is the dimension, *kD* represent a dimensionality of NS as a whole. while the integers *l,m,n* denotes the dimensionality of the NS's building units of different types. Each integer l,m,n refers to different type unit, so the number of these integers must be equal to the number of the different constituting units. From the definition of NSs the condition leads, namely, $k \geqslant l,m,n$, and $k,l,m,n = 0,1,2,3$.

For instances, the name of 3D21 in Fig.1.13 means the three-dimensional NS's is composed of the building units of two-dimension and one-dimension. The term of 2D201 implies that the 2D NS's is made up of the building units of two-dimension, zero-dimension and one-dimension.

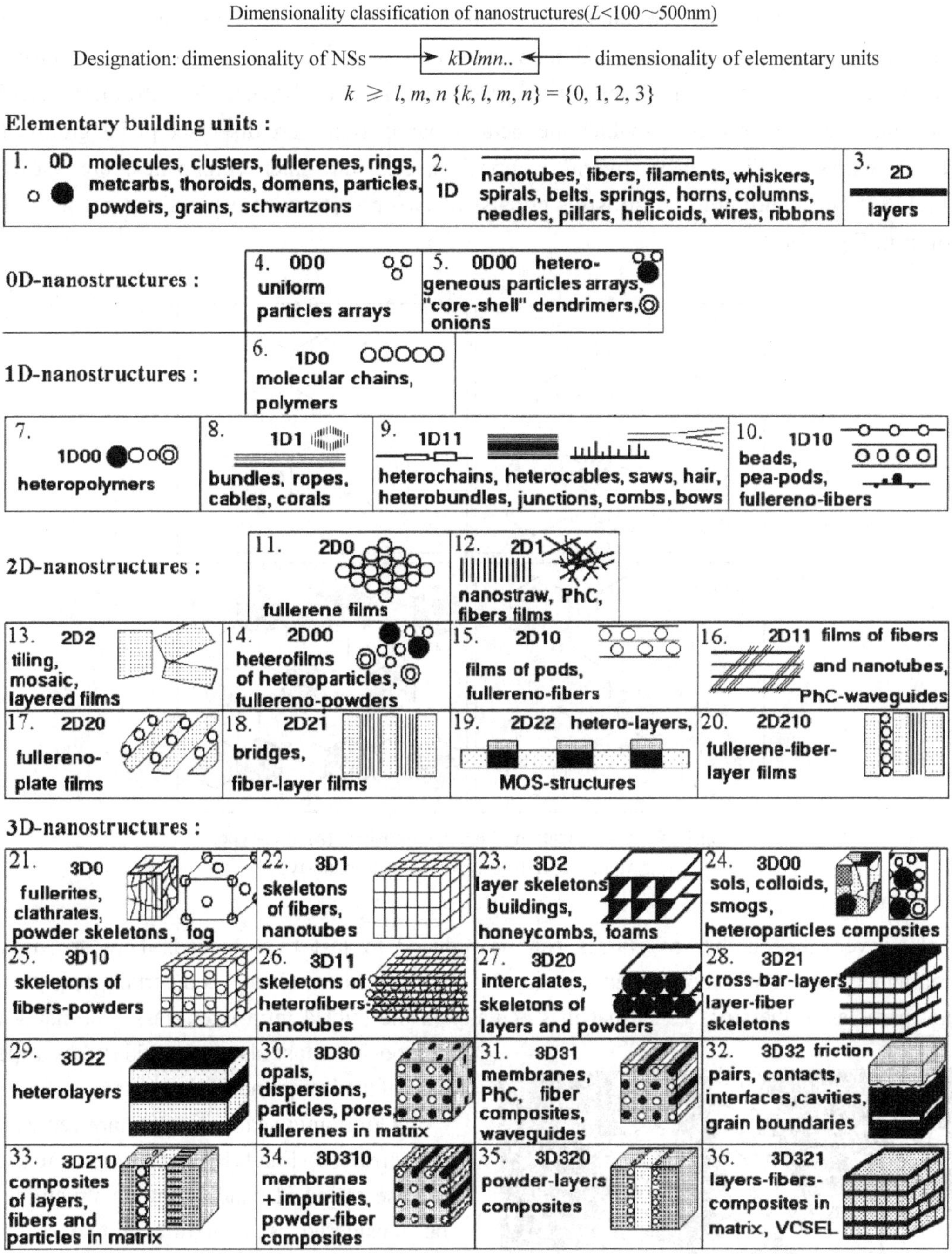

Fig. 1.13 Dimensionality classification of nanostructures [34]

Notices: 1. Interfaces between building units not regarded as additional 2D-NSs;
2. Inverse NSs with cavity building units not regarded as separate ones;
3. The classification may be extended with account of fourfold combinations

(3) Nanocomposites

Nanocomposite refers to composites of more than one solid phase where at least one dimension is in the nanometer range and typically all solid phases are in the range of 1～100nm. The solid phases can be amorphous, semicrystalline, or crystalline, or combinations thereof.

The metal-ceramic and ceramic-ceramic nanocomposites can be divided into four categories: intergranular, intragranular, hybrid, and nano/nano composites[35], intergranular nanocomposites, in which the nanoparticles are distributed at the grain boundaries of the ceramic matrix in Fig.1.14(a), intragranular nanocomposites, in which the metallic nanoparticles are dispersed inside grains of the ceramic matrix in Fig.1.14(b), hybrid composites, i.e. intra-inter composites, are shown in Fig.1.14(c), The nano/nano composites consist of two phases with grain sizes not more than 100nm in Fig.1.14(d).

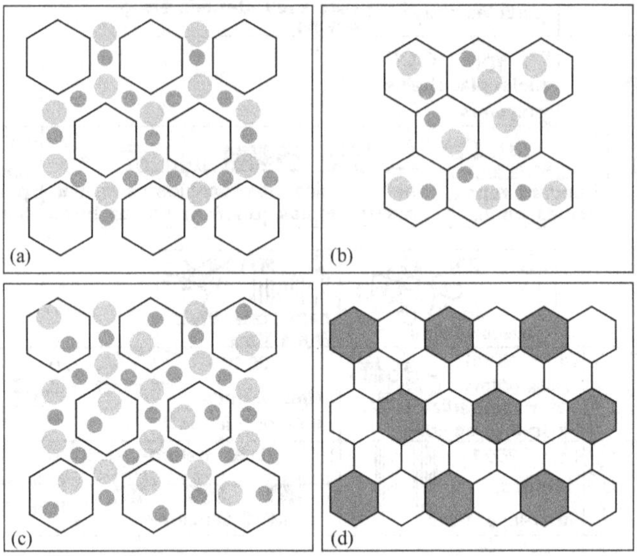

Fig.1.14 Classification of nanocomposites: (a) inter-type; (b) intra-type; (c) hybrid-type; (d) nano/nano-type [35]

However, the number of classes can be reduced to just two. Indeed, there are no pure intragranular nanocomposites, because in any real material, there always is a certain proportion of metal particles at the surface, so that it is actually in the hybrid intra-inter class. The nano-nano composites must also be considered as a special case of the intergranular class[36].

Fig.1.15 Schematic representation of a nanocrystalline structure on atomic scale [37]

On atomic scale, if atoms are represented by circles (Fig.1.15), a two-dimensional representation of nanocrystalline structures can be considered as a mixture of two structural components. Nanograins with a crystal lattice of ordered atoms are denoted by black circles, and intercrystallite boundaries are depicted with white circles in Fig.1.15. The average grain size of the nanocrystalline varies from 2nm to 100nm and can contain from 10^2 to 10^8 atoms. This nanocrystalline structures belong to nanocomposites[37].

1.4 Nanoscale science and technology

Generally, nanoscience is the world of atoms, molecules, macromolecules, quantum dots, and macromolecular assemblies, and is dominated by surface effects such as Van der Waals force attraction, hydrogen bonding, electronic charge, ionic bonding, covalent bonding, hydrophobicity, hydrophilicity, and quantum mechanical tunneling.

Nanoscience is the study of phenomena and manipulation of materials at atomic, molecular, macromolecular scales where properties significantly differ from those at a larger scale [38, 39].

Nanotechnology is the design, characterization, production, and application of structures, devices, and systems by controlling shape and size at the nanometer scale [38, 39].

The term "nanotechnology" was first introduced by a Japanese engineer, Norio Taniguchi [40]. The term originally implied a new technology that went beyond controlling materials and engineering on the micrometer scale, which had dominated the twentieth century.

Generally, nanotechnology should really be called "nanotechnologies": There is no single field of nanotechnology. The term broadly refers to the fields such as biology, physics, chemistry, any other scientific fields, and also deals with controlled manufacturing of nanostructures.

In some senses, nanoscience and nanotechnologies are not new. Chemists have been making polymers which are large molecules made up of nanoscale subunits, for many decades and nanotechnologies have been used to create the tiny features on computer chips for the past 20 years. However, advances in the tools that now allow atoms and molecules to be examined and probed with great precision have enabled the expansion and development of nanoscience and nanotechnologies.

1.5 Driven by industrial revolution

In the field of electronics, for example, since the invention of the transistor by Shockley, Brattain, and Bardeen in 1940s, downsizing of the electronic devices has been continued.

Technology development and industrial competition have been driving the semiconductor industry to produce smaller, faster, and more powerful logic devices. In 1965, Gordon Moore, the cofounder of Intel Co., observed that the number of transistors per square inch on integrated circuits doubled every year since the integrated circuit was invented. Moore predicted that this trend would continue in the future. In the subsequent years, this pace slowed down, but the data density doubled approximately every 18 months. This is the current definition of Moore's Law. Most of experts, including Moore himself, expect Moore's Law to hold for some more time.

However, the exponentially increasing rate of circuit densification has continued into the present as shown in Table 1.2 [41]. The Pentium 4 microprocessor contains 42 million transistors connected to each other on a single piece of silicon. The increases in packing density of the circuitry are achieved by shrinking the line widths of the metal interconnects, by decreasing the size of other features, and by producing thinner layers in the multilevel device structures. These changes are only brought about by the development of new fabrication techniques and materials of construction. As an example, commercial metal interconnect line widths have decreased to 130nm. In 2004, the technology graduated to 90nm, well into the nanotechnology domain (under 100nm) and 50nm technology is being discussed currently.

Table 1.2 Complexity of integrated circuits as seen in the evolution of Intel microprocessors [41]

Name	Year	Transistors	Microns	Clock speed
8080	1974	6000	6	2MHz
8088	1979	29000	3	5MHz
80286	1982	134000	1.5	6MHz
80386	1985	275000	1.5	16MHz
80486	1989	1200000	1	25MHz
Pentium	1993	3100000	0.8	60MHz
Pentium II	1997	7500000	0.35	233MHz
Pentium III	1999	9500000	0.25	450MHz
Pentium 4	2000	42000000	0.18	1.5GHz
Pentium 4 "Prescott"	2004	125000000	0.09	3.6GHz

1.6 Fundamental limitations of present technologies

This top-down method of producing faster and more powerful computer circuitry by shrinking features cannot continue because there are fundamental limitations of present technologies and they can not be overcome by engineering. For instance, charge leakage becomes a big problem when the insulating silicon oxide layers are thinned to about three silicon atoms deep. Moreover, silicon loses its original band structure when it is restricted to very small sizes.

As devices increase in complexity, and it makes nearly every device work perfectly, the defect and contamination control become big problems. A new technology would bring revolutionary progress for the semiconductor industry if it produced faster and smaller logic and memory chips, reduced complexity, saved days to weeks of manufacturing time, and reduced the consumption of natural resources.

1.7 Molecular electronics

How do we overcome the limitations of the present solid-state electronic technology? Molecular electronics is a fairly new and fascinating area of research that is arousing the imagination of scientists [42]. Molecular electronics involves the search for single molecules or small groups of molecules that can be used as the fundamental units for computing, i.e., wires, switches and memory [43]. The goal is to design molecule devices which have the molecules further self-assemble into the higher ordered structural units using manipulating technologies from the bottom-up, instead of present solid state electronic devices that are constructed using lithographic technologies from the top-down.

1.8 Technical challenges in future

The nanotechnology products have stridden over four generations as shown in Fig.1.16 [44, 45]. Each generation of products is marked here by the creation of first commercial prototypes using systematic control of the respective phenomena and manufacturing processing.

The first generation of products (~2001) is "passive nanostructures" and is typically used to tailor macroscale properties and functions of materials. Main purposes are to fabricate nanostructure coatings, dispersion of nanoparticles, nanostructure metals, polymers, and ceramics.

The second generation of products (~2005) is "active nanostructures" for mechanical, electronic, magnetic, photonic, biological, and other effects. It is typically integrated into microscale devices and systems such as new transistors, components of nanoelectronics of metal oxide semiconductor, amplifiers, targeted drugs, chemicals, actuators, artificial "muscles," self-assembling materials and etc.

The third generation (~2010) is 3D nanosystems using various syntheses and assembling techniques such as bio-assembling and robotics. A key challenge is component networking at the nanoscale and hierarchical architectures. The research focus will shift towards heterogeneous nanostructures and supramolecular system engineering. This includes directed multiscale self-assembling, artificial tissues, sensorial systems, quantum interactions within nanoscale systems, processing of information using photons or electron spin, assemblies of nanoscale electromechanical systems and etc.

The fourth generation (2015~2020) will bring "heterogeneous molecular nanosystems", where the components of the nanosystems are reduced to molecules and macromolecules. Designing new atomic and molecular assemblies is expected to increase in importance, including macromolecules "by design", nanoscale machines, directed and multiscale self assembling, exploiting quantum control, nanosystem biology for health care, human-machine interface at the tissue and nervous system level.

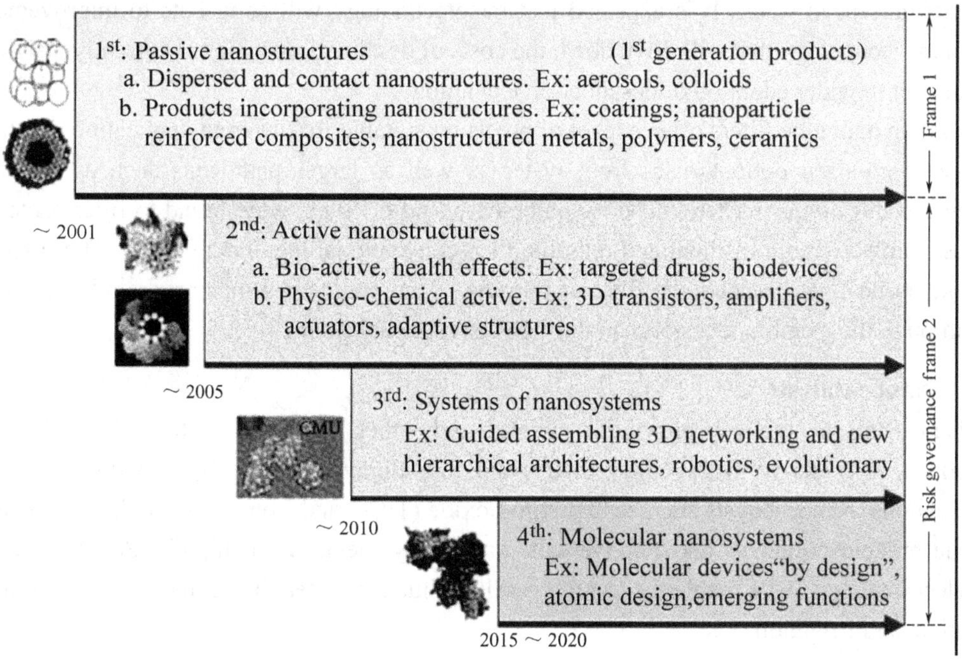

Fig.1.16 Timeline for beginning of industrial prototyping and nanotechnology commercialization: four generations of nanoproducts[44, 45]

Research will include topics such as: atomic manipulation for design of molecules and supramolecular systems, controlled interaction between light and matter with relevance to energy conversion, and exploiting quantum control mechanical-chemical molecular processes.

1.9 Applications of nanomaterials

Nanotechnology encompasses a broad range of tools, techniques, and applications. This technology attempts to manipulate materials at the nanoscale in order to yield novel properties that do not exist at larger scales. These novel properties may enable new or improved solutions to problems that have been challenging to solve with conventional technology. For developing countries, these solutions may include more effective and inexpensive water purification devices, energy sources, medical diagnostic tests and drug delivery systems, durable building materials, and other products. It is quite apparent that there are innumerable potential benefits for society, the environment, and the world. Some of them have been briefly described as below.

1.9.1 Water purification

Nanotechnology for water purification has been identified as a priority area because the water treatment devices that incorporate nanoscale materials are already and the requirements of human development for clean water are pressing. Some water treatment devices manufactured by nanotechnology are already on the market and others are in advanced stages.

Nanofiltration membrane technology is already widely applied for removal of dissolved salts from salty or brackish water, removal of micro pollutants, water softening, and waste water treatment. Nanofiltration membranes are able to selectively remove harmful pollutants and retain nutrients present in water. It is expected that nanotechnology will contribute to improvements in membrane technology that will drive down the costs of desalination, which is currently a significant impediment to wider adoption of desalination technology.

Carbon nanotube filters offer a level of precision suitable for different applications as they can remove 25nm-sized polio viruses from water as well as larger pathogens such as *E. coli* and *Staphylococcus aureus* bacteria. The nanotube-based water filters were found to filter bacteria and viruses, and were more resilient and reusable than conventional membrane filters. The filters were reusable by heating the nanotube filter or purging. Nano-engineered membranes allowed water to flow through the membrane faster than through conventional filters [46].

1.9.2 Nanocatalysts

Nanocatalysts include enzymes, metals, and other materials with enhanced catalytic capabilities that derive from either their nanoscale dimensions or from nanoscale structural modifications. Nanocatalysts such as titanium dioxide (TiO_2) and iron nanoparticles can be used to degrade organic pollutants and remove salts and heavy metals from liquids. People expect that nanoelectrocatalysts will enable the use of heavily polluted and heavily salinated water for drinking, sanitation, and irrigation [47].

1.9.3 Nanosensors

Nanosensors can detect single cell or even atom, making them far more sensitive than counterparts with larger components. Conventional water quality studies rely on a combination of on-site and laboratory analysis that requires trained staff to take water samples and access to a nearby laboratory to conduct chemical and biological analysis. New sensor technology combined with micro- and nanofabrication technologies is expected to lead to small, portable, and highly accurate sensors to detect chemical and biochemical parameters.

1.9.4 Energy

Cheap solar-powered electricity has long been an aspiration for tropical countries, but glass and silicon photovoltaic panels remain too expensive and delicate. Nanotechnology may allow for the production of cheap photovoltaic films that can be unrolled across the roofs of buildings. A novel method is now developed to mass-produce nanostructure for photovoltaic technology.

Nanomaterials manufacturer, Altair Nanotechnologies Co., is developing lithium ion batteries containing nano-titanate material [48]. Because the nano-titanate material does not have to expand or shrink when ions enter and leave the system during charging and discharging, the nano-titanate batteries can be charged over 9000 times while retaining 85% of their charge capacity, while conventional lithium ion batteries typically have a useful life of 750 charges. Besides, the nano-titanate batteries can also be charged to 80% of their capacity in about one minute.

1.9.5 Medical applications

Nanotechnology offers a range of possibilities for health care and medicinal breakthroughs, including targeted drug delivery systems, extended-release vaccines and enhanced diagnostic and imaging technologies.

Nanoporous membranes may help with disease treatments in the developing world. They are a new way of slowly releasing a drug and are important for people far from hospitals. The nanopores are only slightly larger than that of molecules of drugs. Nanoporous membranes can control the rate of diffusion of the drug regardless of the amount of drug remaining inside a capsule. US nano-biopharmaceutical company NanoViricides, Inc. has developed a viral therapy that uses an engineered flexible nanomaterial encapsulated active pharmaceutical ingredients to target specific viruses such as avian flu, common influenza and dismantle the viruses before they can infect cells [49].

Can it be that someday nanorobots will be able to travel through the body searching out and clearing up diseases, such as an arterial atheromatous plaque?

The greatest power of nanomedicine will emerge, perhaps in the 2020s, when we can design and construct complete artificial nanorobots using rigidly diamondoid nanometer-scale parts like molecular gears in Fig.1.17 and bearings[50]. These nanorobots will possess full panoply of autonomous subsystems including onboard sensors, motors, manipulators, power supplies, and molecular computers. But getting all these nanoscale components to spontaneously self-assemble in the right sequence will prove increasingly difficult as machine structures become more complex. Making complex nanorobotic systems requires manufacturing techniques that can build a molecular structure by what is called positional assembly. This will involve picking and placing molecular parts one by one, moving them along controlled trajectories much like the robot arms that manufacture cars on automobile assembly lines. The procedure is then repeated over and over with all the different parts until the final product, such as a medical nanorobot, is fully assembled.

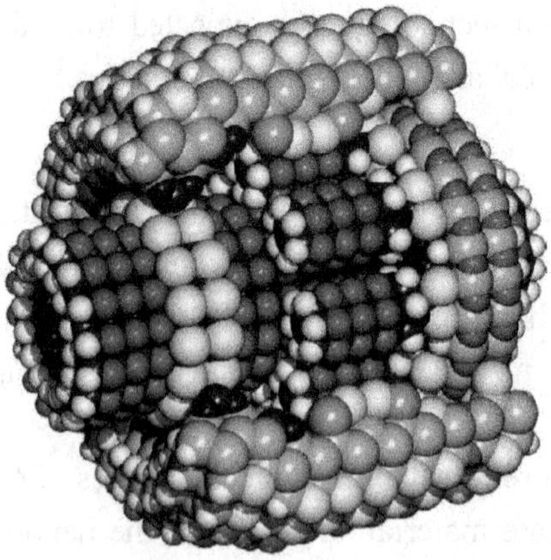

Fig.1.17 A molecular planetary gear is a mechanical component that might be found inside a medical nanorobot. The gear converts shaft power from one angular frequency to another. The casing is a strained silicon shell with predominantly sulfurs termination, with each of the nine planet gears attached to the planet carrier by a carbon-carbon single bond. The planetary gear shown here has not been built experimentally but has been modeled computationally [50]

Future nanorobots equipped with operating instruments and mobility will be able to perform precise and refined intracellular surgeries which are beyond the capabilities of direct manipulation by the human hand. We envision biocompatible surgical nanorobots that can find and eliminate isolated cancerous cells, remove microvascular obstructions and recondition vascular endothelial cells, perform "noninvasive" tissue and organ transplants, conduct molecular repairs on traumatized extracellular and intracellular structures, and even exchange new whole chromosomes for old ones inside individual living human cells[51].

Numerous studies have reported that nanotechnology accelerates various regenerative therapies, such as those for the bone, vascular, heart, cartilage, bladder and brain tissue. Tissue cells elicit specific, distinct reactions depending on implant surface structures [52]. The majority of conventional biomaterials possess micron scale or larger surface features [53]. Since most of the surface features found in and on natural tissues are on the nanometer scale, it has been widely speculated that adding nano-topographies to the surfaces of conventional biomaterials may promote the functions of various cell types. In this light, many nanostructured materials have been called bio-inspired nanomaterials [54]. For example, nanostructured titanium implant surfaces promote bone cell responses leading to accelerated calcium deposition improving integration with surrounding bone compared to conventional titanium surfaces [55]. But bone is not the only application of nanomaterials in regenerative medicine. For cartilage applications, nanostructured polylactic-co-glycolic acid (PLGA) surfaces have been shown to accelerate chondrocyte adhesion and proliferation, as well as extracellular matrix production [56]. Furthermore, vascular graft (PLGA) and stent (titanium) surfaces with nanometer surface roughness values improve endothelial (inner vessel cells) cell functions [57].

Current strategies for the development of improved biomaterials can be grouped into two main categories. The first strategy is through altered chemistry (such as using titanium for orthopedic applications or using controlled drug release form implant surfaces) [58]. The second strategy involves changing physical implant properties such as generating nanometer surface features and, thus, changing surface roughness. For these reasons, biomaterials can be tailored by selectively varying both chemical and physical factors in order to maximize favorable cellular interactions (i.e., increasing functions of tissue-forming cells but decreasing immune cell and bacteria functions).

Various nano-structured polymers and metals (alloys) have been investigated for their bio (and cyto) compatibility properties. Various nanomaterials have greater bone-forming cell (osteoblast) functions, such as nano-hydroxyapatite [59], electro-spun silk [60], anodized titanium [61] and nano-structured titanium surfaces as compared to conventional orthopedic implant materials [62]. As a next step, efforts have focused on understanding the mechanisms of enhanced bone cell functions on these materials and these studies have provided clear evidence that altered amounts and bioactivity of adsorbed proteins (such as vitronectin, fibronectin and collagen) were responsible due to increased surface energy of nanostructured surfaces [63].

References

[1] K.J. Klabunde, Nanoscale Materials in Chemistry, (2001) John Wiley & Sons, Inc., ISBN: 0-471-38395-3 (Hardback), 0-471-22062-0 (Electronic).

[2] E. Serrano, G. Rus, J. Garcia-Martinez, Renewable and Sustainable Energy Reviews, 13 (2009) 2373-2384.

[3] K. Robert, H. Ian, G. Mark, Nanoscale Science and Technology, (2007), ISBN: 978-7-03-018257-9, P4-5.

[4] N.W. Ashcroft, N.D. Mermin, Solid State Physics, Saunders College: Phiadelphia, 1976.

[5] J.H. Davies, The Physics of Low-Dimensional Semiconductors; Cambridge University Press: Cambridge, 1998.

[6] C. Kittel, Einfuhrung in die Festkorperphysik; 8th edn.; R. Oldenbourg Verlag: Munchen, Wien, 1989.

[7] P. Moriarty, Reports on Progress in Physics 64 (2001) 297-381.

[8] N.B. Zhitenev, T.A. Fulton, A.Yacoby, H.F. Hess, L.N. Pfeiffer, K.W. West, Nature 404 (2000) 473-476.

[9] Y.W. Suen, L.W. Engel, M.B. Santos, M. Shayegan, D. C. Tsui, Phys. Rev. Lett. 68 (1992) 1379-1382.

[10] H.L. Stormer, Solid State Commun. 107 (1998) 617-622.

[11] H.L.Stormer, R.R. Du, W. Kang, D.C. Tsui, L.N. Pfeiffer, K.W. Baldwin, K.W.West, Semicond. Sci. Technol. 9 (1994) 1853-1858.

[12] M.P. Anantram, S. Datta, Y.Q. Xue, Phys. Rev. B 61 (2000) 14219-14224.

[13] J.T. Hu, T.W. Odom, C.M. Lieber, Accounts Chem. Res. 32 (1999) 435-445.

[14] Y. Cui, X. Duan, J. Hu, C. M. Lieber, J. Phys. Chem. B 104 (2000) 5213-5216.

[15] V. Rodrigues, T. Fuhrer, D. Ugarte, Phys. Rev. Lett. 85 (2000) 4124-4127.

[16] C.N.R. Rao, G.U. Kulkarni, A. Govindaraj, B.C. Satishkumar, P. Thomas, J. Pure. Appl. Chem. 72 (2000) 21-33.

[17] H. Hakkinen, R.N. Barnett, A.G. Scherbakov, U. Landman, J. Phys. Chem. B 104 (2000) 9063-9066.

[18] Y. Cui, C.M. Lieber, Science 291 (2001) 851-853.

[19] M.A. Reed, C. Zhou, C.J. Muller, T.P. Burgin, J.M. Tour, Science 278 (1997) 252-254.

[20] H.E. Van den Brom, A.I. Yanson, J.M. Ruitenbeek, Physica B 252 (1998) 69-75.

[21] H.X. Xe, C.Z. Li, N.J. Tao, Appl. Phys. Lett. 78 (2001) 811-813.

[22] S.J. Tans, M. H. Devoret, H. Dai, A. Thess, R.E. Smalley, L.J. Geerligs, C. Dekker, Nature 386 (1997) 474-477.

[23] S.Saito, Science 278 (1997) 77-78.

[24] P.L. McEuen, M. Bockrath, D.H. Cobden, J.G. Lu, Microelectronic Eng. 47 (1999) 417-420.

[25] Z. Yao, H.W.C. Postma, L. Balents, C. Dekker, Nature 402 (1999) 273-276.

[26] T.W. Odom, J.-L. Huang, P. Kim, C.M. Lieber, J. Phys. Chem. B 104 (2000) 2794-2809.

[27] W.A. de Heer, E. Al, Science 270 (1995) 1179-1180.

[28] G. Schmid, Nanoparticles: From Theory to Application. (2004) WILEY-VCH Verlag GmbH & Co. KGaA, Weinheim ISBN: 3-527-30507-6.

[29] H.S. Nalwa, Encyclopedia of Nanoscience and Nanotechnology, 9 (1) (2004) American Scientific Publishers, ISBN: 1-58883-057-8, P368.

[30] W.A. Goddard, D.W. Brenner, S.E. Lyshevski, G.J. Lafrate, Handbook of nanoscience, engineering, and tech- nology, Second edition, (2007) Taylor & Francis Group, LLC, ISBN:0-8493-7563-0, P4-5.

[31] S. P. Baker, Plastic deformation and strength of materials in small dimensions, Mater.Sci. Eng. A 319-321 (2001) 16-23.

[32] H.Gleiter, Nanostructured materials :basic concepts and microstructure,Acta mater, 48 (2000) 1-29.

[33] H. Konrad, J. Weissmueller, R. Birringer, C. Kamionik, H.Gleiter, Phys. Rev. B 58 (1998) 2142-2146.

[34] V.V. Pokropivny, V.V. Skorokhod, Mater. Sci. Eng., C. 27 (2007) 990-993.

[35] K. Niihara, J. Ceram. Soc. Jpn., 99 (1991) 974.

[36] C. Brechignac, P. Houdy, M. Lahmani, Nanomaterials and Nanochemistry, (2007)Springer-Verlag Berlin Heidelberg, ISBN:978-3-540-72992-1, P429.

[37] H. Gleiter, Prog. Mater. Sci., 33 (1998) 223.

[38] A. Dowling, R. Clift, N. Grobert, D. Hutton, R. Oliver, O. O'neill, J. Pethica, N. Pidgeon, J. Porritt, J. Ryan, A. Seaton, S. Tendler, M. Welland, R. Whatmore, Nanoscience and Nanotechnologies: Opportunities and Uncertainties, (2004) The Royal Society and the Royal Academy of Engineering, London, UK.

[39] P.J. Borm, D. Robbins, S. Haubold, T. Kuhlbusch, H. Fissan, K. Donaldson, R. Schins, V. Stone, W. Kreyling, J. Lademann, J. Krutmann, D. Warheit, E. Oberdorster, 2006b. The potential risks of nanomaterials: a reviewcarried out for ECETOC. Part. Fibre Toxicol. 3 (11).

[40] N. Taniguchi, On the basic concept of 'nanotechnology'. Proceedings of the International Conference on Production Engineering Tokyo, Part II, Japan Society of Precision Engineering, (1974).

[41] T. Pradeep, Nano: the essentials, (2007) McGraw-Hill Companies, Inc. publisher, ISBN: 0-07-154829-7.

[42] R. Overton, Molecular electronics will change everything, Wired, 8(7) (2000) 242.

[43] J.M. Tour, Molecular electronics, synthesis and testing of components, Acc. Chem. Res., 33(2000) 791.

[44] M.C. Roco, International perspective on government nanotechnology funding in 2005, J. Nanoparticle Res., 7 (6) (2005) 707-712.

[45] W.A. Goddard, D.W. Brenner, S.E. Lyshevski, G.J. Lafrate, Handbook of nanoscience, engineering, and technology, second edition, (2007) Taylor & Francis Group, LLC, ISBN 0-8493-7563-0, P3-18.

[46] A. Srivastava, et al., Carbon nanotube filters, Nature Materials, 3 (2004) 610-614.

[47] UN Millennium Project. 2004. Forging Ahead: Technological Innovation and the Millennium Development Goals. Task Force on Science, Technology, and Innovation. 8 (2004).

[48] Ring, Nano-Titanate Car Batteries. EcoWorld 11 (2006).

[49] T.F. Barker, L. Fatehi, M.T. Lesnick, T.J. Mealey, R.R. Raimond, Nanotechnology & Society: Current and Emerging Ethical Issues, Chapter 13, Nanotechnology and the Poor: Opportunities and Risks for Developing Countries, (2009)Springer Science and Business Media B.V.

[50] K.E. Drexler, Nanosystems: molecular machinery, manufacturing, and computation, (1992) New York: John Wiley & Sons.

[51] Jr.R.A. Freitas Nanomedicine, Vol. IIA: Biocompatibility. Georgetown, (2003) Landes Bioscience.

[52] M.M. Stevens, J.H. George, Exploring and engineering the cell surface interface, Science 310(5751) (2005)1135-1138.

[53] T.J. Webster, Nanophase ceramics as improved bone tissue engineering materials, Am. Ceram. Soc. Bull. 82(6) (2003) 23-28B.

[54] M. Sato, T.J. Webster, Nanobiotechnology: implications for the future of nanotechnology in orthopedic applications, Expert. Rev. Med. Devices 1(1) (2004) 105-114.

[55] C. Ergun, H.N. Liu, T.J. Webster, E. Olcay, S. Yilmaz, F.C. Sahin, Increased osteoblast adhesion on nanoparticulate calcium phosphates with higher ca/p ratios, J. Biomed. Mater. Res. Part A 85A (1) (2008) 236-241.

[56] G.E. Park, M.A. Pattison, K. Park, T.J. Webster, Accelerated chondrocyte functions on naoh-treated plga

scaffolds, Biomaterials 26(16) (2005) 3075-3082.

[57] J. Lu, M.P. Rao, N.C. MacDonald, D. Khang, T.J. Webster, Improved endothelial cell adhesion and proliferation on patterned titanium surfaces with rationally designed, micrometer to nanometer features, Acta Biomaterialia 4(1) (2008) 192-201.

[58] S. Svenson, D.A. Tomalia, Commentary-dendrimers in biomedical applications reflections on the field, Adv. Drug Deliv. Rev. 57 (15) (2005) 2106-2129.

[59] M. Sato, E.B. Slamovich, T.J. Webster, Enhanced osteoblast adhesion on hydrothermally treated hydroxyapatite/titania/poly (lactide-co-glycolide) sol-gel titanium coatings, Biomaterials 26 (12) (2005) 1349-1357.

[60] H.J. Jin, J.S. Chen, V. Karageorgiou, G.H. Altman, D.L. Kaplan, Human bone marrow stromal cell responses on electrospun silk fibroin mats ,Biomaterials 25(6) (2004) 1039-1047.

[61] C. Yao, E.B. Slamovich, T.J. Webster, Enhanced osteoblast functions on anodized titanium with nanotube-like structures, J. Biomed. Mater. Res. Part A 85A(1) (2008) 157-166.

[62] D. Khang, J. Lu, C. Yao, K.M. Haberstroh, T.J. Webster, The role of nanometer and sub-micron surface features on vascular and bone cell adhesion on titanium, Biomaterials 29(8) (2008) 970-983.

[63] D. Khang, S.Y. Kim, P. Liu-Snyder, G.T.R. Palmore, S.M. Durbin, T.J. Webster, Enhanced fibronectin adsorption on carbon nanotube/ poly(carbonate) urethane: independent role of surface nanoroughness and associated surface energy, Biomaterials 28(32) (2007) 4756-4768.

Review questions

1. What is nanometer?
2. What is nanoscience?
3. What is nanotechnology?
4. What are the research objects of the nanoscience and technology?
5. What are nanomaterials?
6. What is different between nanomaterials and nanostructural materials?
7. What are the zero-dimension, one-dimension, two-dimensions and bulk materials?
8. What are the quantum wells, quantum wires and quantum dots?
9. Are there nano objects around you? Please enumerate some examples of applications of nanomaterials.
10. What is quantum confinement?
11. What are the fundamental limitations of present technology?
12. Why do the scientists propose molecular electronics?

Vocabulary

activate ['æktɪveɪt] vt. 刺激；使活泼；vi. 有活力；激活

actuator ['æktjʊeɪtə] n. 致动器，驱动器

adaptive [ə'dæptɪv] adj. 适应的，适合的

adhesion [əd'hiːʒ(ə)n] n. 黏附；支持；固守

adjacent [ə'dʒeɪs(ə)nt] adj. 邻近的，毗连的

aerosol ['eərəsɒl] n. 喷雾器，气溶胶；adj 喷雾的

agglomerate [ə'glɒməreɪt] adj. 凝聚的；成团的，结块的；n. 团块；[地] 集块岩；[化] 附聚物；vt. 使结块；使成团

alloy ['ælɒɪ] vt. 使减低成色；使成合金；n. 合金

alter ['ɔːltə; 'ɒl-] vt. 改变，更改；vi. 改变；修改

amplifier ['æmplɪfaɪə] n. 放大器，扩大器；扩音器

analogous [ə'næləgəs] adj. 类似的；[昆] 同功的；可比拟的

angular ['æŋgjʊlə] adj. 有角的；生硬的，笨拙的；瘦削的

anodize ['ænədaɪz] vt. 阳极电镀；作阳极化处理（等

于 anodise）
antimicrobial [ˌæntɪmaɪ'krəʊbɪəl] n. 抗菌剂；杀菌剂；adj. 抗菌的
application [ˌæplɪ'keɪʃ(ə)n] n. 应用；申请；敷用；应用程序
artificial [ɑːtɪ'fɪʃ(ə)l] adj. 人造的；仿造的；虚伪的；非原产地的；武断的
aspiration [æspə'reiʃ(ə)n] n. 渴望；抱负；呼气；吸引术
assemble [ə'semb(ə)l] vt. 收集；装配；vi. 集合，聚集
assumption [ə'sʌm(p)ʃ(ə)n] n. 假定；设想；担任；采取
atheromatous [ˌæθə'rɑmətəs] adj. 动脉粥样化的
aureus ['ɔːriəs] n. 奥里斯（古罗马的金质货币），金黄色
available [ə'veɪləb(ə)l] adj. 可利用的；有效的，可得的；空闲的
avian ['eɪvɪən] adj. 鸟的；鸟类的；n. 鸟，禽
bacteria [bæk'tɪərɪə] n. 细菌
band edge 带边沿
bearing ['beərɪŋ] n. 轴承；关系；方位；举止；v. 忍受
biocompatible [ˌbaiəukəm'pætəbl] adj. 生物适合的；不会引起排斥的
bio-inspired n. 仿生的
biomedical [baɪə(ʊ)'medɪk(ə)l] adj. 生物医学的
biopharmaceutical ['baiəuˌfɑːmə'sjuːtikəl] adj. 生物制药学
bladder ['blædə] n. 膀胱；囊状物，可充气的囊袋
block [blɒk] n. 块；街区；vt. 限制；阻止
Bohr model 玻尔模型
boundary ['baʊnd(ə)rɪ] n. 边界；范围；分界线
breakthrough ['breɪkθruː] n. 突破；突破性进展
bridge [brɪdʒ] n. 桥；桥牌；桥接器；船桥；vt. 架桥；渡过
bulk [bʌlk] n. 大块；体积，容量；vt. 使扩大，使形成大量；使显得重要
cancerous ['kænsərəs] adj. 癌的；生癌的；像癌的
carbon ['kɑːbən] n. 碳；碳棒；adj. 碳的
Cartesian coordinate 笛卡尔坐标
cartilage ['kɑːt(ɪ)lɪdʒ] n. 软骨
category ['kætigəri] n. 种类，分类；范畴
cell [sel] n. 电池；细胞；vi. 住在牢房或小室中
cellular ['seljʊlə] adj. 细胞的；由细胞组成的；n. 移动电话；单元
ceramic [si'ræmik] adj. 陶瓷的；陶器的；n. 陶瓷制品；陶瓷
cermet ['sɜːmet] n. 金属陶瓷；金属陶瓷材料
channel ['tʃæn(ə)l] vt. 引导，开导；形成河道；n. 通道；频道；海峡
characteristic [kærəktə'ristik] adj. 典型的；特有的；n. 特征；特性；特色
charge [tʃɑːdʒ] n. 命令；掌管；vt. 指责；使充电
charge carrier n. [电子] 电荷载子
chondrocyte [kʌdrɒsɪt] n. [组织] 软骨细胞
chromosome ['krəʊməsəʊm] n. 染色体
circuit ['səːkit] n. 电路，回路，巡回；环道；vi. 环行；vt. 绕回…环行
claim [kleim] vi. 提出要求；vt. 需要；声称
classification [ˌklæsifi'keiʃ(ə)n] n. 分类；类别，等级
clay [klei] n. 肉体；黏土；vt. 用黏土处理
cluster ['klʌstə] n. 群；串；vi. 丛生；群聚
coating ['kəʊtiŋ] n. 涂层
collagen ['kɑlədʒ(ə)n] n. [生化] 胶原，胶原质
colloidal [kə'lɒɪdəl] adj. 胶质的；胶体的；胶状的
compatibility [kəmˌpætɪ'bɪlɪtɪ] n. [计] 兼容性
component [kəm'pəʊnənt] adj. 组成的，构成的；n. 成分；组件；[电子] 元件
computational [kɒmpjʊ'teiʃənl] adj. 计算的
conduct ['kɒndʌkt] vi. 导电；带领；vt. 管理；引导；表现；n. 进行；行为；实施
conductivity [kɒndʌk'tɪvətɪ] n. 导电性；[物][生理] 传导性
confine [kən'faɪn] n. 界限，边界;约束；限制；vt. 限制；禁闭
confinement [kən'faɪnmənt] n. 限制；监禁；分娩
considerable [kən'sɪd(ə)rəb(ə)l] adj. 相当大的；重要的，值得考虑的
consolidation [kənˌsɒli'deiʃən] n. 巩固；合并；团结，固化
constraint [kən'streɪnt] n. [数] 约束；局促，态度不自然；强制
content ['kəntent] n. 内容，目录；满足；容量；adj. 满意的；vt. 使满足
converge [kən'vɜːdʒ] vi. 集中于一点；聚合；vt. 使聚集；使向一点会合；会聚
counterpart ['kaʊntəpɑːt] n. 副本；配对物；极相似的人或物
couple ['kʌp(ə)l] n. 对；夫妇；vi. 结合；成婚
covalent [kəʊ'veɪl(ə)nt] a. 共有原子价，共价
crystal ['krɪst(ə)l] n. 水晶；晶体；adj. 水晶的；透明的，清澈的
crystallite ['krɪstəlaɪt] n. 微晶；雏晶
crystallographic [ˌkrɪstəlɑ'græfɪk] adj. [建] 结晶的
cyto ['sɪtəʊ] n. 细胞
De Broglie wavelength [量子] 德布罗意波长

degrade [dɪ'greɪd] vt. 贬低;使……降级；vi. 退化；降级，降低
delivery [dɪ'lɪv(ə)rɪ] n. 交付；分娩；递送
densification [ˌdensifi'keiʃən] n. 密实化；封严；稠化
density ['densɪtɪ] n. 密度
depict [dɪ'pɪkt] vt. 描述；描画
desalination [di: ˌsælɪ'neɪʃən] n. 脱盐作用；减少盐分
detection [dɪ'tekʃ(ə)n] n. 察觉；侦查，探测；发觉，发现
diagnosis [ˌdaɪəg'nəʊsɪs] n. 诊断
diagnostic [daɪəg'nɒstɪk] a. 特征的，诊断的
diameter [daɪ'æmɪtə] n. 直径
diamondoid ['daɪəmәndɒɪd] adj. 菱形的；钻石形的
dielectric [ˌdaɪɪ'lektrɪk] adj. 非传导性的；诱电性的；n. 电介质；绝缘体
dimension [dɪ'mɛnʃ(ə)n; daɪ-] n. 方面；[数] 维；尺寸；vt. 标出尺寸
dimensionality [dɪˌmenʃə'nælətɪ] n. 维度；幅员；广延
discrete [dɪ'skri:t] adj. 离散的，不连续的；n. 分立元件；独立部件
discrete value 不连续值，[数] 离散值
dismantle [dɪs'mænt(ə)l] vt. 拆除；取消；解散；除掉…的覆盖物；vi. 可拆卸
disperse [dɪ'spɜ:s] vt. 传播；分散；vi. 分散
dispersion relation [物][核] 色散关系；分散关系
distribution [dɪstrɪ'bju:ʃ(ə)n] n. 分配；分布
driven ['drɪvn] a. 受到驱策的；v. 驱使，驾车
dwarf [dwɔ:f] vi. 变矮小；n. 侏儒，矮子；vt. 使矮小；adj. 矮小的
electronics [ɪlek'trɒnɪks; el-] n. 电子学；电子工业
elementary [elɪ'ment(ə)rɪ] adj. 基本的；初级的；元素的
elicit [ɪ'lɪsɪt] vt. 抽出，引出；引起
emitters [ɪ'mɪtə] n. 发射器，发射体
encapsulate [ɪn'kæpsjʊleɪt; en-] vt. 压缩；将…装入胶囊；将…封进内部；vi. 形成胶囊
encompass [ɪn'kʌmpəs; en-] vt. 包含；包围，环绕；完成
endothelial [ˌendəʊ'θi:lɪəl] adj. 内皮的
energy spectrum [物] 能谱
envision [en'vɪʒ(ə)n] vt. 预想；想象
enzyme ['enzaɪm] n. 酶
equiaxed ['i:kwiækst] adj. 各向等大的，等轴的
essentially [ɪ'senʃ(ə)lɪ] ad. 本质上，本来
exclusion [ɪk'sklu:ʒ(ə)n; ek-] n. 排斥；排除；驱逐；被排除在外的事物
explosion [ɪk'spləʊʒ(ə)n; ek-] n. 爆炸；爆发；激增
exponential [ˌekspə'nenʃ(ə)l] adj. 指数的；n. 指数
extracellular [ekstrə'seljʊlə] adj. （位于或发生于）[生物] 细胞外的
fabrication [fæbrɪ'keɪʃ(ə)n] n. 制造，建造；装配；伪造物
feature ['fi:tʃə] n. 容貌;特色，特征；vi. 起重要作用
Fermi energy 费米能级；[统物] 费米能量
fertile ['fɜ:taɪl] adj. 富饶的，肥沃的；能生育的；n. 肥沃，多产
fiber ['faɪbə] n. 纤维；光纤（等于 fibre）
fibronectin ['faibrəu'nektin] n. [生化] 纤连蛋白
filament ['fɪləm(ə)nt] n. 细丝；灯丝；细线；单纤维
film [fɪlm] n. 膜，电影；vt. 在…上覆以薄膜
filtration [fɪl'treɪʃn] n. 过滤
fine [faɪn] adj. 好的；优良的；细小的，精美的；健康的；晴朗的；n. 罚款；vt. 罚款；澄清；adv. 很好地；精巧地
foundation [faʊn'deɪʃ(ə)n] n. 基础；地基；基金会；根据；创立
fragment ['frægm(ə)nt] n. 碎片；片断或不完整部分；vt. 使成碎片；vi. 破碎或裂开
frequency ['fri:kw(ə)nsɪ] n. 频率；频繁
frication [fri'keiʃən] n. 摩擦
generate ['dʒenәreɪt] vt. 使形成；生殖；发生
geometry [dʒɪ'ɒmɪtrɪ] n. 几何学，几何结构
gradient ['greɪdɪәnt] n. 坡度；倾斜度；adj. 倾斜的；步行的
graft [grɑ:ft] vi. 移植；嫁接；贪污；vt. 移植；嫁接；贪污；n. 移植；嫁接；渎职
grain [greɪn] n. 粮食；颗粒；谷物；晶粒
granular ['grænjʊlə] a. 由小粒而成的，粒状的
graphite ['græfaɪt] n. 石墨；黑铅;vt. 用石墨涂（或掺入等）
ground state n. [物] 基态…
hazard ['hæzəd] vt. 赌运气；冒…的危险，使遭受危险；n. 危险，冒险；冒险的事
heterogeneous [ˌhet(ə)rə(ʊ)'dʒi:nɪəs; -'dʒen-] adj. [化] 多相的；异种的；不均匀的
hierarchical [haɪə'rɑ:kɪk(ə)l] adj. 分层的；等级体系的
homogeneously [ˌhomə'dʒinɪrslɪ] adv. 同样地
hybrid ['haɪbrɪd] n. 混合物;杂种，混血儿；adj. 混合的，杂化的
hydrogen ['haɪdrədʒ(ə)n] n. 氢
hydrophilic [ˌhaɪdrə(ʊ)'fɪlɪk] n. 软性接触透镜吸水隐形眼镜片；adj. [化] 亲水的
hydrophilicity [-'haidrəfilisiti] n. 亲水性，亲和性
hydrophobic [haɪdrə(ʊ)'fəʊbɪk] adj. 狂犬病的；恐水病的；疏水的
hydrophobicity [haɪdrəfəʊ'bɪsɪtɪ] n. [化] 疏水性

hydroxyapatite [haɪˌdroksi'æpɪˌtaɪt] n. [矿物] 羟磷灰石
hyperbola [haɪ'pɜːbələ] n. [数] 双曲线
identical [aɪ'dentɪk(ə)l] adj. 同一的；完全相同的；n. 完全相同的事物
illustrate ['ɪləstreɪt] vt. 阐明，举例说明；图解；vi. 举例
illustration [ɪlə'streɪʃ(ə)n] n. 说明；插图；例证；图解
impediment [ɪm'pedɪm(ə)nt] n. 妨碍；阻止；口吃
implant [ɪm'plɑːnt] vt. 种植；灌输；嵌入；n. [医] 植入物；植入管；vi. 被移植
impose [ɪm'pəʊz] vi. 利用；欺骗；施加影响；vt. 强加；征税
incoherent [ɪnkə(ʊ)'hɪər(ə)nt] adj. 语无伦次的；不连贯的；不合逻辑的
incorporate [in'kɔːpəreit] vt. 包含，吸收；体现；把……合并；vi. 合并；混合；组成公司；adj. 合并的；一体化的；
incredible [ɪn'kredɪb(ə)l] adj. 难以置信的，惊人的
inertia [ɪ'nɜːʃə] n. 惯性；惰性，迟钝；不活动
infinite ['ɪnfɪnɪt] adj. 无限的，无穷的；无数的；极大的；n. 无限
influenza [ɪnflʊ'enzə] n. [医] 流行性感冒（简写 flu）；[兽医] 家畜流行性感冒
ingredient [ɪn'griːdɪənt] n. 原料；要素；adj. 构成组成部分的
innumerable [ɪ'njuːm(ə)rəb(ə)l] adj. 无数的，数不清的
insulate ['ɪnsjʊleɪt] vt. 隔离，使孤立；[物] 使绝缘，使隔热
integer ['ɪntɪdʒə] n. [数] 整数；整体；完整的事物
integrate ['ɪntɪgreɪt] n. 一体化；集成体；vi. 求积分；取消隔离；成为一体；adj. 整合的；完全的；vt. 使…完整
interconnect [ɪntəkə'nekt] vt. 使互相连接；vi. 互相联系
integration [ɪntɪ'greɪʃ(ə)n] n. 集成；综合；[数] 积分法
interface ['ɪntəfeɪs] n. 接口；界面；接触面；vi. 相互作用
interval ['ɪntəv(ə)l] n. 间隔；间距；幕间休息
intracellular [ˌɪntrə'seljʊlə] adj. 细胞内的
irrigation [ɪrɪ'geɪʃən] n. 灌溉；冲洗；[医] 冲洗法
laminate ['læmɪneɪt] vt. 将锻压成薄片；分成薄片；n. 薄片制品；层压制件
langmuir ['læŋmjuə] n. [姓氏] 朗缪尔(美国化学家，1881～1957)
leakage ['liːkɪdʒ] n. 泄漏；渗漏物；漏出量
limitation [lɪmɪ'teɪʃ(ə)n] n. 限制；限度；极限；[律] 追诉时效；有效期限
linear ['lɪnɪə] adj. 线的，线型的；直线的，线状的；长度的
literature ['lɪt(ə)rətʃə] n. 文学；文艺；文献；著作
lithium ['lɪθɪəm] n. 锂（符号 Li）
lithographic [ˌlɪθə'græfɪk] adj. 平版的；平版印刷的
logic ['lɒdʒɪk] n. 逻辑；逻辑性；adj. [计算机] 逻辑的
macroporous ['mækrə'pɔːrəs] adj. 大孔的；大孔隙的；多孔的
macroscopic [ˌmækrə(ʊ)'skɒpɪk] adj. 宏观的；肉眼可见的
magical ['mædʒɪk(ə)l] adj. 魔术的；有魔力的
magnetic [mæg'netɪk] adj. 有吸引力的；有磁性的；地磁的
mania ['meɪnɪə] n. 狂热；狂躁；热衷
manipulate [mə'nɪpjʊleɪt] vt. 操纵；操作；巧妙地处理；篡改
manipulation [məˌnɪpjʊ'leɪʃ(ə)n] n. 操纵；操作；处理；篡改
manufacture [mænjʊ'fæktʃə] n. 产品；制造业；vt. 捏造；加工
matrix ['meɪtrɪks] n. [数] 矩阵；模型；[生物][地质] 基质；母体；子宫；[地质] 脉石
membrane ['membreɪn] n. 膜；薄膜；羊皮纸
mesopore ['mesəupɔː] n. 中孔；间隙孔
micron ['maɪkrɒn] n. 微米
microporous [ˌmaikrəu'pɔːrəs] adj. 微孔的
microvascular [ˌmaɪkrəʊ'væskjʊlə] adj. [解] 微脉管的
mobility [məʊ'bɪlətɪ] n. 移动性；机动性；迁移率
modification [ˌmɒdɪfɪ'keɪʃ(ə)n] n. 修改，修正；改变
molecular [mə'lekjʊlə] adj. 分子的
molecular beam epitaxy [物] 分子束外延
molecule ['mɒlɪkjuːl] n. 分子
momentum [mə'mentəm] n. 势头；[物] 动量；动力；冲力
monolayer ['mɒnəleɪə] n. 单层；adj. 单层的
motion ['məʊʃ(ə)n] n. 意向；请求；vi. 运动；打手势
motor ['məʊtə] n. 发动机，马达；汽车；adj. 汽车的；机动的； vt. 以汽车载运
multilevel ['mʌltəˌlevəl] adj. 多级的；多层的
nano ['nænəʊ] n. 纳；毫微
nanoscale ['nænoskel] n. 纳米级
nature ['neɪtʃə] n. 自然；性质；种类；本性
nervous ['nɜːvəs] adj. 神经的；紧张不安的；强健有力的
nitride ['naɪtraɪd] n. [化] 氮化物；vt. 使氮化；[冶] 渗氮于
noninvasive [ˌnɒnɪn'veɪsɪv] adj. 非侵袭的；非侵害的

novel ['nɒv(ə)l] adj. 新奇的；异常的；n. 小说
nutrient ['nju:trɪənt] n. 营养物；滋养物；adj. 营养的；滋养的
object ['ɒbdʒɪkt; -dʒekt] n. 物体;客体；vt. 提出…作为反对的理由
obstruction [əb'strʌkʃ(ə)n] n. 障碍；阻碍；妨碍
orbital ['ɔ:bɪt(ə)l] adj. 轨道的；眼窝的
orientation [ɔ:rien'teiʃən] n. 方向；定向；向东方；适应；情况介绍
orthopedic [ɔ:θəʊ'pi:dɪk] adj. 整形外科的（等于orthopaedic）
osteoblast ['ɒstɪə(ʊ)blæst] n. [基医] 成骨细胞
overlap [əʊvə'læp] n. 重叠，重复；vi. 部分重叠，部分的同时发生
pack [pæk] n. 背包；一副；vt. 塞满；压紧
panoply ['pænəplɪ] n. 华丽服饰；全套甲胄，全副盔甲
parabola [pə'ræb(ə)lə] n. 抛物线
parallel ['pærəlel] n. 平行线；对比；vt. 使…与…平行；adj. 平行的；类似的
parameter [pə'ræmɪtə] n. 参数；参量；系数
particle ['pɑ:tɪk(ə)l] n. 颗粒；质点；极小量；小品词
pathogen ['pæθədʒ(ə)n] n. 病原体
patterned ['pætənd] adj. 有图案的，组成图案的
Pauli's exclusion principle 泡利不相容原理
perpendicular [ˌpɜ:p(ə)n'dɪkjʊlə] adj. 垂直的，正交的，陡峭的；n. 垂线，垂直的位置
pharmaceutical [ˌfɑ:mə'su:tɪk(ə)l; -'sju:-] adj. 制药（学）的 n. 药物
phase [feɪz] n. 相；阶段；位相；vt. 使定相；逐步执行；vi. 逐步前进
photonic [ˌfoʊ'tɑnɪk] adj. 光激性的
photovoltaic [ˌfəʊtəʊvɒl'teɪɪk] adj. 光伏的，光电的
planetary ['plænɪt(ə)rɪ] adj. 行星的
plaque [plæk; plɑ:k] n. 匾；[医] 血小板；饰板
platelet ['pleɪtlɪt] n. 血小板；薄片
PLGA (polylactic-co-glycolic acid) 乳酸-羟基乙酸共聚物
polio ['pəʊlɪəʊ] n. [医] 脊髓灰质炎，小儿麻痹症
polycrystalline [pɒlɪ'krɪstəlaɪn] adj. [物] 多晶的
polymer ['pɒlɪmə] n. [化] 聚合物
porous ['pɔ:rəs] adj. 多孔渗水的；能渗透的；有气孔的
portable ['pɔ:təb(ə)l] adj. 手提的，便携式的；轻便的；n. 手提式打字机
potential well 势阱
precaution [prɪ'kɔ:ʃ(ə)n] n. 预防，警惕；预防措施；vt. 预先警告；警惕
precipitation [prɪˌsɪpɪ'teɪʃ(ə)n] n. 坠落；沉淀，沉淀物；鲁莽；冰雹
prefix ['pri:fɪks] n. 前缀；vt. 加前缀；将某事物加在前面
primitive ['prɪmɪtɪv] adj. 原始的，远古的；简单的，粗糙的；n. 原始人
probe [prəʊb] n. 探针；调查；vi. 调查；探测
proliferation [prəˌlɪfə'reʃən] n. 增殖，扩散；分芽繁殖
promise ['prɒmɪs] n. 许诺，允诺；希望；vt. 允诺，许诺；vi. 许诺；有指望，有前途
propagate ['prɒpəgeɪt] vt. 传播；传送；繁殖；宣传；vi. 繁殖；增殖
property ['prɒpətɪ] n. 财产；性质，性能；所有权
prophetic [prə'fetɪk] adj. 预言的，预示的；先知的
proportional [prə'pɔ:ʃ(ə)n(ə)l] adj. 比例的，成比例的；相称的，均衡的；n. [数] 比例项
prototype ['prəʊtətaɪp] n. 原型；标准，模范
purge [pɜ:dʒ] vi. 净化；通便；vt. 通便；清洗
purification [ˌpjʊərɪfɪ'keɪʃən] n. 净化；提纯
quantize ['kwɒntaɪz] vt. 使量子化；数字转换
quantum['kwɑntəm] n. 量子论
quantum dot n. [物] 量子点
quantum well [物] 量子阱
quasi ['kweɪzaɪ] adj. 准的；类似的；外表的；adv. 似是；有如
radius ['reɪdɪəs] n. 半径，半径范围；[解剖] 桡骨；辐射光线；有效航程
regenerative [rɪ'dʒen(ə)rətɪv] adj. 再生的，更生的；更新的
region ['ri:dʒ(ə)n] n. 地区；范围；部位
reinforce [ri:ɪn'fɔ:s] vt. 加强，加固；强化；补充；n. 加强；加固物；加固材料
residual [rɪ'zɪdjʊəl] n. 剩余；残渣；adj. 剩余的；残留的
resilient [rɪ'zɪlɪənt] adj. 弹回的，有弹力的
respectively [rɪ'spektɪvlɪ] adv. 分别地；各自地，独自地
retention [rɪ'tenʃ(ə)n] n. 扣留，滞留；保留；记忆力；[医] 闭尿
revolutionize [ˌrevə'lu:ʃənaɪz] vt. 发动革命；彻底改革；宣传革命；vi. 革命化
robot ['rəʊbɒt] n. 机器人；遥控设备，自动机械；机械般工作的人
rod [rɒd] n. 棒；枝条；惩罚；权力
routine [ru:'ti:n] n. [计] 程序；日常工作；例行公事；adj. 日常的；例行的
sanitation [sænɪ'teɪʃ(ə)n] n. 环境卫生；卫生设备；下水道设施

scale [skeɪl] n. 规模；比例；鳞；刻度；天平；数值范围
schema ['skiːmə] n. 图解，计划，模式
schematic [skiː'mætɪk; skɪ-] adj. 图解的；概要的；n. 图解视图；原理图
schematically [ski'mætikli] adv. 计划性地；按照图式
SchrÖdinger equation 薛定谔方程式
selective [sɪ'lektɪv] adj. 选择性的
semiconductor [ˌsemɪkən'dʌktə] n. 半导体
sensor ['sensə] n. 传感器
sensorial [sen'sɔːrɪəl] adj. 知觉的；感觉的
shrink [ʃrɪŋk] n. 收缩；畏缩；vt. 使缩小，使收缩；vi. 收缩；畏缩
silicon ['sɪlɪk(ə)n] n. 硅；硅元素
singularity [sɪŋgjʊ'lærɪtɪ] n. 奇异；奇点；突出；稀有
spatial ['speɪʃ(ə)l] adj. 空间的；存在于空间的；受空间条件限制的
speculate ['spekjʊleɪt] vi. 推测；投机；思索；vt. 推断
sphere [sfɪə] n. 范围；球体；vt. 包围；使…成球形；adj. 球体的
spherical shell [天] 球壳
spin [spɪn] vi. 旋转；纺纱；晕眩；vt. 使旋转；纺纱；编造；结网；n. 旋转
spontaneous [spɒn'teɪnɪəs] adj. 自发的；自然的；无意识的
spun [spʌn] adj. 纺成的
square root 平方根；二次根
staircase ['steəkeɪs] n. 楼梯，阶梯
staphylococcus [ˌstæfɪlə(ʊ)'kɒkəs] n. 葡萄球菌
stent [stent] adj. 扩张的
step function 阶梯函数
subdivide [sʌbdɪ'vaɪd] vi. 细分，再分；vt. 把……再分，把……细分
submicron [ˌsʌb'maɪkrɒn] n. 亚微细米；adj. 亚微细粒的
subset ['sʌbset] n. 子集；子设备；小团体
substance ['sʌbst(ə)ns] n. 物质；实质；资产；主旨
subunit ['sʌbˌjʊnɪt] n. 副族(亚组)，亚单位
successive [sək'sesɪv] adj. 连续的；继承的；依次的；接替的
sulfur ['sʌlfə] vt. 用硫磺处理；n. 硫磺
superlattice [suːpə'lætɪs; sjuː-] n. 超点阵；超结晶格子
supramolecular [ˌs(j)uːprəmə'lekjʊlə] adj. 超分子的（由许多分子组成的）
surgery ['sɜːdʒ(ə)rɪ] n. 外科；外科手术；手术室；诊疗室
sweep [swiːp] vt. 扫除；猛拉；掸去；vi. 扫，打扫；席卷；扫视；袭击；n. 打扫，扫除
synthesis ['sɪnθɪsɪs] n. 综合，合成；综合体
systematic [sɪstə'mætɪk] adj. 有系统的；系统的；体系的；分类的
tailored ['teɪləd] adj. 定做的；裁缝的
term [tɜːm] n. 期限，术语；条款；vt. 把…叫做
termination [tɜːmɪ'neɪʃ(ə)n] n. 结束，终止
theoretical [θɪə'retɪk(ə)l] adj. 理论的；假设的；理论上的；推理的
thickness ['θɪknɪs] n. 厚度；层；浓度；含混不清
tissue ['tɪʃuː; 'tɪsjuː] n. 纸巾；薄纱；组织；vt. 饰以薄纱
titanium [taɪ'teɪnɪəm; tɪ-] n. [化] 钛（金属元素）
tolerance ['tɒl(ə)r(ə)ns] n. 公差；宽容；容忍；公差
topography [tə'pɒgrəfɪ] n. 地势；地形学；地志
trajectory [trə'dʒekt(ə)rɪ; 'trædʒɪkt(ə)rɪ] n. 轨道，轨线；弹道
transistor [træn'zɪstə; trɑːn-; -'sɪ-] n. [电子] 晶体管
trapped ['træpɪd] adj. 捕获的；收集的；受到限制的；vt. 诱捕（trap 的过去分词）
traumatize ['trɔːmətaɪz] vt. 使……受损伤；使……受精神创伤
tremendous [trɪ'mendəs] adj. 极大的，巨大的；惊人的
tropical ['trɒpɪk(ə)l] adj. 热带的；酷热的；热情的
tube [tjuːb] n. 管；电子管；隧道；电视机；vt. 使成管状；把…装管；用管输送
turbulence ['tɜːbjʊl(ə)ns] n. 骚乱，动荡；狂暴；湍流
ultra ['ʌltrə] a. 过度的，过激的，极端的；n. 过激论者，急进论者
ultrafine [ˌʌltrə'fain] adj. 非常细微的
unroll [ʌn'rəʊl] vt. 展开，铺开；显示；vi. 显露，展现
vaccine ['væksiːn; -ɪn] n. 疫苗，牛痘苗；adj. 疫苗的，牛痘的
valve [vælv] n. [机] 阀，[机] 活门；vt. 装阀于，以活门调节
variation [veərɪ'eɪʃ(ə)n] n. 变化；变异，变种
vascular ['væskjʊlə] adj. [生物] 血管的
viral ['vaɪr(ə)l] adj. 滤过性毒菌引起的；滤过性毒菌的
viricide ['vaɪərɪsaɪd] n. 杀病毒剂
virtually ['vɜːtjʊəlɪ] adv. 事实上，几乎；实质上
virus ['vaɪrəs] n. 病毒；恶毒；毒害
vitronectin n. 玻连蛋白；玻璃体结合蛋白
volume ['vɒljuːm] n. 量；体积；卷；音量；大量；册；adj. 大量的；vi. 成团卷起
wall [wɔːl] n. 墙壁，围墙；似墙之物；vt. 用墙围住，

围以墙；adj. 墙壁的
wavevector 波矢；波矢量
wire [waɪə] n. 电线；金属丝；电报；vt. 拍电报；给……装电线；vi. 打电报
yield [ji:ld] vt. 屈服；出产；放弃；vi. 屈服，投降；n. 产量，收益

1. 纳米材料概论

1.1 纳米世界概述

至少在一维尺度上处于纳米量级的纳米材料是连接孤立的原子或小分子与块体材料之间的桥梁，因而被称为介观尺度材料。作为纳米科技基础的纳米材料，近年来已成为最热门的研究领域之一。纳米技术和纳米材料的热点集中在能对经济、技术及科学产生巨大影响的若干领域：①随着半导体芯片的容量和传输速度呈指数增长，实际上所有现代科技制作的关键元器件正在迅速地逼近它们的工作极限值，这就迫使新技术和新材料的问世；②新的纳米材料和器件在提高能源的利用效率，更有效地处理环境公害，快速而精确地检测和诊断人类疾病的一些领域，如能源、环境保护、生物医学及保健科学，具有很大的发展潜力；③当一种材料的尺寸缩减到纳米量级时，即使其组成与可以看得见和触摸到的块体材料是完全相同的，但材料的性能却有着本质的区别。因此，纳米材料对于重大的科学发现和探索的确是一片沃土。

有人称 1 纳米是“长度单位上一个神奇的点”，因为这是最小的人造器件能够达到的自然界中原子和分子尺度的一个点[1]。

的确，纳米科技在过去的数年中得到了迅猛发展。“纳米热”正在席卷整个科学与工程领域，化学家和诺贝尔奖得主 Richard Smally 说过“等着吧，下个世纪将会是一个惊人的世纪。我们将能够制造出在尽可能小的尺度上由一个个原子组成的器件，这些小的纳米器件将给我们的工业和生活带来一场革命性的变革[1]。”公众对他这一言论有了越来越多的认识。

1.2 纳米材料的定义

1.2.1 纳米

前缀“nano”来源于希腊语“nanos”，是“矮小”之意。纳米是一个长度单位。1 纳米（nm）等于十亿分之一米（$1nm = 1 \times 10^{-9}m$）。

图 1.1 是一些合成材料和一些生物的长度。从小尺度开始，金原子的直径为 0.1nm，碳纳米管的直径为 1～2nm，双螺旋 DNA 的直径约为 3nm。艾滋病毒（HIV）的直径约为 100nm 等[2]。如图 1.2 所示，单个原子的直径为 0.1～0.2nm，8～10 个原子的长度约为 1nm。

1.2.2 纳米材料的定义

纳米材料是指三维空间中至少有一维尺寸处于纳米量级（1～100nm）的材料。

图 1.3 给出了单壁碳纳米管与人的头发的比较，头发的直径约为 80000nm，而单壁碳纳米管的直径约是头发直径的 1/1000 倍。

1.3 纳米材料的分类

纳米材料可简单分为离散的纳米材料和纳米结构材料，但也存在其它的分类方法。

离散的纳米材料是指至少在一个维度上具有纳米量级的外观特征的材料，如纳米粒子、纳米纤维、纳米管和薄膜。

纳米结构材料是指材料具有块体材料的外观特征，但是，它可以由离散的纳米颗粒构成，如：通过固结纳米粉形成的块体材料；或是由连续的纳米结构单元所组成的材料，如：多孔材料，包括微孔（<2nm）材料、介孔（2～50nm）材料和大孔（>50nm）材料，还有纳米相材料和多晶材料等。

固结纳米粉体技术是制备块体纳米结构材料的一种方法，然而，由于粉末颗粒尺寸很小，必须特别地谨慎，以减小颗粒间的摩擦，把爆炸或着火的可能降至最低。基于粉体颗粒的制备方法不同，粉体可能具有微尺寸的平均粒径，或者是真正的纳米粉体颗粒。这些粉体颗粒可在低温或中温下压实制备出所谓的“坯体”，其密度可超过理论最大值的90%。其中残余的孔隙率在材料中平均分布，在尺度上，孔径小并具有窄的粒径分布。晶粒尺寸在100nm～1μm之间的多晶材料由许多纳米晶组成，习惯上可称为超细粒子。

1.3.1 依据材料的空间维度分类

当材料结构中的空间维度减小，或纳米粒子在一个特定的晶体方向上受到限制时，通常会导致材料在该晶向上的物理性能发生变化。因而，可依据材料处于纳米范围内的空间维度的数量，对纳米材料和纳米体系进行分类，表1.1及图1.4给出了不同维度系统的范例[1,3]。

（1）零维（0D）材料

零维材料是指材料的空间三维结构尺度均处于纳米量级，也就是说，材料的三维尺寸均被限制在纳米尺度范围内（即材料在三个方向上均不占维度）。该系统的材料包括纳米粒子、纳米晶等。

（2）一维（1D）材料

一维材料是指材料的空间结构尺度中，有二维处于纳米量级，也就是说，其三维中的二维被限制在纳米量级内（即材料在另外两个方向上均不占维度），该体系包括纳米线、纳米棒、纳米丝、纳米管、纳米带等。在这些结构中，长度与直径的比率称为长径比。通常纳米棒的长径比位于10～100范围内。如果纳米棒的长径比大于100，则称为纳米线，因此，纳米线是更长的纳米棒。另一方面，纳米管是中空的纳米棒。

（3）二维（2D）材料

二维材料是指在材料的空间结构尺度中，有一维处于纳米量级，也就是说，其三维中的一维被限制在纳米量级内，该体系包括超薄膜、多层膜、薄膜、表面涂层、超晶格等。

1.3.2 依据材料的量子性质分类

（1）块体材料

材料的电子结构与其性质紧密相关。我们现在可以认为，对于一个三维固体材料，在 d_x，d_y 和 d_z 三个空间方向上包含大量的“自由”电子。这里的“自由”是指这些电子是离域化的，而不受原子核的束缚。

在自由电子模型中，固体中的每个电子以 $\vec{v}=(v_x,v_y,v_z)$ 速度运动着，单个电子的能量恰好为其动能：

$$E=\frac{1}{2}m\vec{v}^2=\frac{1}{2}m(v_x^2+v_y^2+v_z^2) \tag{1.1}$$

根据 Pauli 不相容原理，每个电子都占据一个独立的量子态。由于电子具有两个自旋方向 ($m_s=+1/2$) 和 ($m_s=-1/2$)，只有当两个电子自旋方向相反时才会有相同的速度 $\vec{v}$。这种情况类似于玻尔原子模型，每一个轨道最多只能填充两个自旋方向相反的电子。在固态物理学中，更多的是用粒子的波矢 $\vec{k}=(k_x,k_y,k_z)$ 来描述粒子的状态，而不是用速度。波矢的绝对值 $k=|\vec{k}|$ 是波数，波矢 $\vec{k}$ 与线动量($\vec{p}$)和电子的速度($\vec{v}$)成正比。

块状晶体的能态计算是基于周期性边界条件的假设。周期性边界条件是用一种数学方法来模拟一种无限大的 $(d=\infty)$ 固体材料，这就意味着对固体的边界同样适应。依据这种假设，靠近于边界的一个电子不会真正“感觉”到边界。换句话说，边界处电子的行为就像它们在块体里面一样。通过对电子的波函数施加以下条件：$\psi(x,y,z)=\psi(x+d_x,y,z)$, $\psi(x,y,z)=\psi(x,y+d_y,z)$, 和 $\psi(x,y,z)=\psi(x,y,z+d_z)$, 这种情况在数学上是可以实现的。换言之，波函数的周期性必须与固体中晶格的周期性一致[4,5]。每个函数描述一个自由电子沿一个笛卡儿坐标的运动情况。在这些函数中，$k_{x,y,z}$ 等于 $\pm n\Delta k=\pm n2\pi/d_{x,y,z}$，$n$ 取整数[4~6]。这些解是分别沿正负两个方向传播的波 $k_{x,y,z}>0$ 和 $k_{x,y,z}<0$ 的解。这种周期性边界条件的一个重要结果是 $\vec{k}$ 空间中所有可能的电子态都是均匀分布。不难想象在理想情况下，一维自由电子模型中存在这样的态分布：两个电子 $(m_s=\pm 1/2)$ 时，可以占据 $k_x=0$ $(v_x=0)$ 的态分布；同时亦存在 $k_x=+\Delta k$ $(v_x=+\Delta v)$ 的态分布；$k_x=-\Delta k$ $(v_x=\Delta v)$ 的态分布以及 $k_x=+2\Delta k$ $(v_x=+2\Delta v)$ 的态分布等。

对于一个三维的块体材料而言，与一维体系类似，两个电子 $(m_s=\pm 1/2)$ 可以占据 $(k_x,k_y,k_z)=(\pm n_x\Delta k,\pm n_y\Delta k,\pm n_z\Delta k)$ 的各个状态分布，$n_{x,y,z}$ 为整数，如图 1.5 所示。不难想象在 $\vec{k}$ 空间的电子占据态中，所有这些电子态都被容纳在以电子占据的最高能量波数为半径的一个球体之中。在 0 K 的基态下，球的半径是费米波数 k_F（费米速度 v_F）。费米能 $E_F\propto k_F^2$ 是最低电子占据态的能量。能量 $E\leqslant E_F$ 的所有能态都被电子所占据，而能量 $E>E_F$ 的较高的能态都是空的。在固体中，波数是被一个个的 $\Delta k=\pm n2\pi/d_{x,y,z}$ 所分隔。在块体材料中，$d_{x,y,z}$ 值较大，所以 Δk 很小。因此球体里被准连续态所充满[4]。

现在需要介绍一下态密度 $D_{3d}(k)$ 的概念，$D_{3d}(k)$ 是指每单位间隔波数中所含有的电子态的数量。根据这个定义，$D_{3d}(k)\Delta k$ 是指在固体中波数介于 k 和 $k+\Delta k$ 之间的电子数。如果知道一个固体的态密度，就可以计算电子数。例如，具有波数的总电子数小于一个给定的 k_{max}，我们将其称为 $N(k_{max})$。显然，$N(k_{max})$ 等于 $\int_0^{k_{max}} D_{3d}(k)\mathrm{d}k$。当固体在基态时，所有电子的波数 $k\leqslant k_F$，其中 k_F 为费米波数。因为在块状固体中，各个态均匀分布在 $\vec{k}$ 空间，我们知道，在 k 与 $k+\Delta k$ 之间的态数量与 $k^2\Delta k$ 成正比（图 1.5）。对此，可以进一步理解，三维 k 空间的体积随着 k^3 而变化。如果想要计算出在 k 与 $k+\Delta k$ 之间的一个波数中电子的数目，就需要确定一个半径为 k 和厚度为 Δk 的球壳的体积。这个体积是和球的表面积（随 k^2 变化）、球壳厚度（Δk）成正比的。因此，$D_{3d}(k)\Delta k$ 正比于 $k^2\Delta k$，并且当 Δk 的极限接近零时，可以写成：

$$D_{3d}(k) = \frac{dN(k)}{dk} \propto k^2 \quad (1.2)$$

了解能量介于 E 和 $E + \Delta E$ 之间的电子数量比知道一个已知波数间隔的态密度更有用。我们知道 $E(k)$ 与 k^2 成正比，因此，$k \propto \sqrt{E}$；所以，$dk / dE \propto 1/\sqrt{E}$。利用公式(1.2)，可以得到一个三维电子模型的态密度[5]：

$$D_{3d}(E) = \frac{dN(E)}{dE} = \frac{dN(k)}{dk}\frac{dk}{dE} \propto E\frac{1}{\sqrt{E}} \propto \sqrt{E} \quad (1.3)$$

利用公式(1.3) 可以得出对块状固体的简单描述。从图 1.5 可以看出，电子能态是准连续的，其电子的能态密度随能量的平方根而变化。

图 1.5 呈现的是三维块状固体中的电子。（a）这种固体可以看作是一种沿着 x，y，z 三个方向伸展的一个无限晶体模型。（b）周期性边界条件的假设产生的驻波是自由电子的薛定谔方程的解。相关的波数（k_x，k_y，k_z）在 k 空间周期性地分布[5]。图中所示的每个点（k_x，k_y，k_z）代表一个可能的电子态，在 k 空间的每个态只能被两个电子所占据。在一个大的固体中，每个电子态之间的空间$\Delta k_{x,y,z}$是非常小的，因此，k 空间是处于准连续的满态。一个半径为 k_F 的球形空间包括了 $k = (k_x^2 + k_y^2 + k_z^2)^{1/2} < k_F$ 所有态。在温度为 0K 的基态时，$k<k_F$ 的每个能态都被两个电子所占据，其余态是空的。因为 k-空间是被电子态均匀地填满，在一定体积内电子态的数量随 k^3 而变化。（c）描述的是一个三维固体中自由电子的分布关系。自由电子的能量随波数的平方而变化，能量与波数（k）的关系呈抛物线。对于一个块状固体，其电子填充态是准连续分布的，并且在 k-空间内，两个相邻态（图中以黑点表示）之间的距离非常小。（d）显示的是在三维体系中，自由电子的态密度（D_{3d}）与能量的关系图。电子填充态的能量是准连续的，电子的态密度随能量的平方根 $E^{1/2}$ 而变化[1]。

（2）量子阱（2D）

当一个固体沿着 x-和 y-方向上任意延伸，而沿着 z-方向仅有数纳米的厚度，电子可在 x-和 y-方向上自由运动，但电子在 z-方向上的运动受到限制。这样一个系统被称为二维（2D）系统，亦称为量子阱。

当一个固体在一个维度或多个维度变得比自由电荷载流子的德布罗意波的波长还要小时，限制沿着该维度载流子运动分量所需的能量是必不可少的。亦可以说，电子沿着该维度方向（假如 z 轴方向）的运动是量子化的，这种情形如图 1.6 所示。电子不可能离开固体，而是在 z-方向上运动的电子被诱捕在一个“盒子”里。在数学上，可以通过边界 $z = \pm\frac{1}{2}d_z$ 处无限高的势能阱 $[V(z) = \infty]$ 来描述。

通过对在盒子里电势 $V(z)$为零，在边界处电势为无穷大的一个电子的一维薛定谔方程进行求解，可以获得盒子里粒子情形的解。从图 1.6 可以看出，这些解是能量 $E_{nz} = \hbar^2 k_z^2 / (2m) = h^2 k_z^2 / (8\pi^2 m) = h^2 n_z^2 / (8md_z^2), n_z = 1,2,3 \cdots$ 的驻波。这与 $\Delta k_z = \pi / d_z$ 时的 $k_z = n_z \Delta k_z$ 态相似。此外，这些态中的每个态可以最大限度地被两个电子所占据。

对于一个二维固体，k 空间中的态在 x-y 平面上可以延伸，而其 k_z 的值是离散的。固体在 z-方向愈薄，电子填充态之间的空间Δk_z就愈大。另一方面，k_x-k_y 平面的态分布是保持准连续的。因此，k-空间中可能的态是平行于 k_x-和 k_y-轴的平面，在 k_z-方向该平面之间的间隔为Δk_z。我们可以用 n_z 标明每一个面。因为在一个平面内，态的数量是准连续的，同时，态的数量与平面的面积成正比。这就意味着态的数量正比于 $k^2 = k_x^2 + k_y^2$。因此，在一个半径为 k，厚度为Δk的空间中的态数量与$k\Delta k$成正比，通过对所有的环进行积分就可以获得k-空间中所有平面的总面积。与一个三维固体的情形相比，态密度与k呈线性变化：

$$D_{2d}(k)=\frac{\mathrm{d}N(k)}{\mathrm{d}k}\propto k \tag{1.4}$$

在基态时，$k \leqslant k_F$的每个态都是被两个电子所占据。我们现在欲知道，能量在E和$E+\Delta E$之间的电子有多少个态存在。我们知道k和E的关系：$E(k)\propto k^2$，则$k\propto\sqrt{E}$，并且$\mathrm{d}k/\mathrm{d}E\propto 1/\sqrt{E}$。利用式（1.4），我们就能得到一个二维电子云的态密度，参见图1.7。

$$D_{2d}(E)=\frac{\mathrm{d}N(E)}{\mathrm{d}E}=\frac{\mathrm{d}N(k)}{\mathrm{d}k}\frac{\mathrm{d}k}{\mathrm{d}E}\propto\sqrt{E}\frac{1}{\sqrt{E}}\propto 1 \tag{1.5}$$

因此，一个二维固体中的电子态密度与三维固体中的电子态密度有显著的不同。因为存在较少的能级，在能带中，电子填充的能级之间的间距增大了。在二维材料中，能带依然是准连续的，但是，态密度是一个阶梯函数[5,7]。

图1.7显示了二维系统中的电子。（a）二维固体在x,y两个维度上几乎是可以无限延伸的，但沿着z-方向是非常薄的，相当于一个自由电子的德布罗意波长（$d_z\rightarrow\lambda$）。（b）沿着x-轴和y-轴方向电子是自由运动的。沿着x-轴和y-轴方向，通过假设周期性边界条件，沿着该方向就能够找到其波函数。在k空间，k_x和k_y态是准连续地分布。在z-轴方向上，电子的运动受到了限制，并局限在一个“盒子”里，即沿着z-轴方向，电子的能态是量子化的。对于一个离散的k_z态，在三维k空间中的态分布可以描述成一系列平行于k_x轴和k_y轴的平面。对于每个离散的k_z态，存在一个平行于k_x轴和k_y轴的分隔的平面。图1.7（b）仅展示了其中的一个面。由于$\Delta k_{x,y}=2\pi/d_{x,y}\rightarrow 0$，在一个面内的$k_x$和$k_y$态是准连续的。因为被$k_z$态分隔的两个平面之间的距离是大的，所以$\Delta k_z=\pi/d_z>>0$。（c）自由电子呈现出$E(k)\propto k^2$的抛物线分布关系。对于沿着$x$-轴和$y$-轴方向电子运动的能级$E(k_x)$和$E(k_y)$是准连续的（用黑点表示）。在小“盒子”边界处的波函数$\psi(z)$等于零，导致在盒子内部出现驻波。这种限制造成了沿着z-轴方向运动的电子具有离散的能级$E(k_z)$。电子只能占据这种离散态（n_{z1}；n_{z2}；…；以黑点表示）。能级的位置随着z-方向上固体的厚度而变化，换言之，就是随着“盒子”的尺寸而变化。（d）描述的是二维电子云的态密度，当电子的运动在z-轴方向上受到限制，但电子在另外两个方向（x,y）上是自由运动的，对于给定的k_z态（$n_z=1, 2, 3\cdots\cdots$）的态密度与能量E无关。

量子限域是指载流子的运动在某一维度上受到限制。当材料中的电子运动方向在一个空间维度上受到限制（一维量子限域），而在其它两个空间维度上，电子的运动是自由时，会产生量子阱或量子膜。

量子阱意味着电子掉进一个“势阱”中，或这些电子被捕获在一种膜中。2D系统通常是在不同材料的界面处，或在仅为几个纳米厚度的层状系统中形成。由不同半导体材料制成的层状材料可以将载流子（电子、空穴）俘获在一个特殊的层中。这些俘获的载流子被限制在某个方向上运动。

二维固体中电子的量子力学行为源于许多重要的物理效应。随着纳米科技的最新发展，二维结构材料的制造已成为一种常态。2D系统通常在不同材料的界面处形成，或在一些层只有几个纳米层厚的层状系统的界面处形成。例如，采用分子束外延的方法逐层的连续沉积就能生长出这样的结构。在这种几何结构中，载流子(电子和空穴)可以在平行于半导体层的方向上自由运动，但垂直于界面的运动受到限制。这些纳米结构的研究导致了显著的二维量子效应的发现，如整数和分数量子霍尔效应[8~11]。

（3）量子线（1D）

电子仅在 x-方向上自由运动，而沿 y-轴和 z-轴上的运动则受到固体边界的限制。这样的系统称为量子线。在一维系统中，电子只能在一个方向上运动，在其它二维空间的运动受到限制。

采用对二维以及三维材料描述的类似的方法，可以获得一维固体中的电子态分布。在 x-方向上，电子可以自由的运动，可以再次应用周期性边界条件的概念。这样就可以得到一个平行于 k_x-轴的电子态的准连续分布，以及相应能级的准连续分布。电子在其余方向上的运动受到限制，对于一个盒子里的一个粒子的势阱，可以通过对解薛定谔方程进行求解而获得电子态，得到是离散的 k_y 和 k_z 电子态。我们现在能够想象到所有可能的电子态是平行于 k_x-轴的线。这些线在 k_y 和 k_z 方向是被不连续的间隔所分开，但在每条线内，k_x 态是准连续分布的（图 1.8）。沿着该条线，通过测量线的长度，我们可以计算其电子态的数目，态的数目与 $k = k_x$ 成正比。因此，在 k 与 $k+\Delta k$ 间隔内，具有一定波数的电子态的数目与Δk 成正比：

$$D_{1d}(k) = \frac{\mathrm{d}N(k)}{\mathrm{d}k} \propto 1 \tag{1.6}$$

在基态时，$k \leqslant k_\mathrm{F}$ 的每个态都被两个电子所占据。我们知道，对于自由电子而言，k 和 E 之间有以下关系：$E(k) \propto k^2$ ，则 $k \propto \sqrt{E}$ ，从而得出，$\mathrm{d}k / \mathrm{d}E \propto 1/\sqrt{E}$ 。利用公式（1.6），可以获得一维电子云的态密度：

$$D_{1d}(E) = \frac{\mathrm{d}N(E)}{\mathrm{d}E} = \frac{\mathrm{d}N(k)}{\mathrm{d}k}\frac{\mathrm{d}k}{\mathrm{d}E} \propto 1 \times 1/\sqrt{E} \propto 1/\sqrt{E} \tag{1.7}$$

关于态密度的描述如图 1.8 所示。在一维体系中，态密度与能量 $E^{-1/2}$ 相关，因此，在态密度与能量的关系图［图 1.8(d)］中，在带边附近会出现奇异点[5]，每个双曲线中，k_x 态是连续分布的，而只有 k_y 和 k_z 态是离散的。

图 1.8 为一维系统中的电子。(a）是一维固体。(b）在三维 k 空间中，填充的(k_x, k_y, k_z)-态可以被视为平行于 k_x-轴的多条线。图中只显示了一条线作为例子，因为$\Delta k_x \to 0$，所以在每条线内，态分布是准连续的。因为只有当 k_y 和 k_z 态是不连续，各个独立的线是分立的。(c）从分散关系中也可以看出，沿着 k_x-轴的能级 $E(k_x, k_y, k_z)$是准连续的，而沿 k_y-和 k_z-轴方向能级是离散的。(d）在一条线内，沿着 k_x-轴方向的态密度与 $E^{-1/2}$ 成正比。在 D_{1d}-图中的每条双曲线对应于一个单独的 (k_y, k_z) 态。

这种二维的量子态对于电荷的传输具有重要影响。电子沿 x 轴方向可以自由地流动，而沿 y-和 z-轴方向，电子的运动被限制到离散态。因此，电子只能在离散的“导电通道”内被传输。这对微电子工业是相当重要的。如果电路的尺寸被缩减得越来越小，在某个点，导线的直径将与电子的德布罗意波长相当，那么，导线将会表现出量子线的特性。最近有许多这种 1D 线的例子，包括短的有机半导体分子，无机半导体和金属纳米线[12~18]或者断裂结[19~21]。碳纳米管担负着重要的角色[22~26]，将碳纳米管作为一维限制的模型系统，比如作为电子发射器等一些潜在的应用[27]，已经得到了广泛的研究。

（4）量子点（0D）

当电子的运动在三个维度上都受到限制时，就得到一个零维系统，该系统被称为“量子点”。在量子点中，电子的运动在三个维度上都受到限制，所以，在 k 空间，仅存在离散的 (k_x, k_y, k_z) 态。在 k 空间中的每个独立态可以用一个点表示。其结果，只有离散的能级存在，电子在零维的态密度 D_{0d} 与能量 E 的分布图中，表现为一个个分立的峰。

图 1.9 展示了在零维固体中的电子。(a) 固体在所有三个维度收缩到与电子的德布罗意波

长的尺度相当。(b) 因为这种限域，在三维 k-空间中，(k_x, k_y, k_z) 的所有态都是离散的点。(c) 三维 (k_x, k_y, k_z) 的能带显示出分立的能级。(d) 一维的态密度 D_{0d} 与能量 E 的关系图中，每个分立的峰对应于每个单独的态。电子只能占据这些分立能级的态 [28]。

由于它们离散的电荷态密度和能级结构，与包含有数千个电子，甚至与只包含有一个电子的原子体系相似，因而，量子点也被称为人造原子或电子箱[29,30]。

1.3.3 依据材料的性能分类

在常温常压下，成千上万种固体物质可细分为金属、陶瓷、半导体、复合材料和高分子材料。它们还可以进一步细分为生物材料、催化材料、涂料、玻璃、磁性材料及电子材料。当所有这些固体物质以纳米粒子的形式存在时，这些固体物质的性能将会发生很大的变化，并呈现出另外一些新的特性，可将其分为纳米晶体材料、纳米陶瓷材料、纳米复合材料、纳米高分子材料等。

1.3.4 依据形态和化学组成分类

（1）纳米结构材料

“纳米结构材料”一词常用来描述具有典型的晶体尺度处于纳米量级的材料（图 1.10）。如图 1.10(a) 所示的纳米晶块体材料是由一个个纳米晶体组成的一种三维尺度的宏观材料，有许多方法可以用来制备纳米晶体材料。薄膜是一种二维尺度的宏观材料［图 1.10(b)］，通常具有亚微米量级的厚度，膜平面内的晶粒分布范围是由比薄膜厚度尺寸小得多的小晶粒和比薄膜厚度尺寸大得多的晶粒所组成。两个维度处于纳米量级的材料称为纳米线的一维材料［(1D)，图 1.10(c)］，三个维度处于纳米量级的材料称为量子点（纳米颗粒）的零维材料［(0D)，图 1.10(d)］，由纳米线和纳米颗粒可以组合成薄膜材料[31]。

基于晶体和晶界微观结构的化学组成和形状（维度），Gleiter 对纳米晶与界面组成的纳米结构材料（NsM）进行了分类，如图 1.11 所示 [32]。依据晶体的形状，把纳米结构材料（NsM）分为三类：层的厚度为数纳米的层状纳米晶、棒的直径为数纳米的棒状纳米晶以及等轴状纳米晶体。依据晶体的化学组成，可将三类纳米结构材料（NsM）分为四族，如图 1.11 所示[32]。

在图 1.11 所示的第一族中，所有的晶体和晶界区域具有相同的化学组成。与第一族相对应的第一类的被非晶体区域所分隔的片晶堆垛组成的层状半晶体聚合物，就属于第一族纳米结构材料（NsM）的例子，与第一族相对应的第三类中的等轴状纳米晶体（如：Cu 等轴状纳米晶体）是第一族的另一个例子。第二族的纳米结构材料（NsM）是由具有不同化学成分的纳米晶所组成。以影线的宽度表示其化学成分的不同，与第二族相对应的第一类的层状纳米结构形成的量子阱（多层膜）结构可能是最熟知的一个实例。

如果化学成分的变化主要发生在晶体与晶体的界面区域，就诞生了第三族纳米结构材料（NsM）。在第三族纳米结构材料（NsM）中，原子（分子）被优先分散在界面区域，以至于形成结构调变（晶体/界面）与局部化学调变相结合的纳米结构材料（NsM）。镓(Ga)原子分散在钨（W）纳米晶的边界处是一个典型的例子（与第三族相对应的第三类）（图 1.11）。通过将 Al_2O_3 和 Ga 共粉磨可以制备出一种新型材料，结果表明，形成的新材料的特点是 Al_2O_3 纳米晶分散在 Ga 的非晶体结构的层间，在三氧化二铝（Al_2O_3）纳米晶粒之间的镓(Ga)非晶体边界层厚度取决于 Ga 的含量，Ga 层的厚度介于不足一个单分子到 7 个 Ga 原子的层厚，如图 1.12 所示 [33]。

将纳米晶（层状的、棒状的和等轴状的纳米晶）分散于与该纳米晶化学组成不同的基质（矩阵）中，形成了第四族纳米结构材料（NsM）。析出硬化合金属于这一族纳米结构材料，通过对一种过饱和的 Ni-Al 固溶体进行热处理，可将纳米尺寸的 Ni_3Al 分散在 Ni 基质中是这种合金的一个实例。用于现代飞机发动机中的大多数耐高温材料是基于析出-硬化工艺的 Ni_3Al/Ni 合金（图 1.11）。

（2）纳米结构

纳米结构（NSs）与纳米结构材料（NSMs）有所区别，因为纳米结构是以形状和维度为特征，而纳米结构材料是以赋予纳米结构一种化学成分为特征。在纳米结构中，我们知道，在纳米尺度上，d 的尺寸必须要小于或等于一个临界值 d^* (在纳米尺度范围内), $d \leqslant d^*$。d^* 没有确定的含义，它是由能够引起纳米效应的一些物理现象（电子的自由程长度、德布罗意波长等）的重要特性所决定的[7]。

依据空间维度，纳米结构可能是 0D、1D、2D 和 3D 四类中的任何一类。所有的纳米结构可由具有低维度的 0D、1D 和 2D 的基本构建单元所组成。而 3D 构建单元是除外的，这是因为它不能用于构建除 3D 矩阵以外的低维纳米结构。然而，如果 3D 结构是由 0D、1D、2D 基本单元构成的三维结构，则可被认为是纳米结构。根据 Pokropivny 的分类方法，纳米结构可被分为 36 个类别，如图 1.13 所示[34]。

这里介绍一下纳米结构中 *kDlmn* 符号的含义，k 是维度的数目，D 代表维度，kD 代表整个纳米结构的维度；整数 lmn 分别代表纳米结构中不同类型构建单元的维度，整数 lmn 涉及不同的构建单元，因此 lmn 的数量必须等于不同组建单元的数量；从纳米结构条件得到的定义，即

$$k \geqslant l,m,n;\quad k,l,m,n = 0,1,2,3$$

例如：纳米结构 36 种分类中的 3D21 符号代表三维纳米结构是由二维的基本构建单元与一维的基本构建单元所组成。2D201 意味着二维纳米结构是由二维的构建单元、零维的构建单元和一维的构建单元所组成。

（3）纳米复合材料

纳米复合材料是指两个或两个以上的，至少有一维处于纳米量级，且所有的固相尺寸在 1～100nm 范围之内的固相复合物。固相可以是无定形的、半晶型的或晶体的，或者是它们之间的混合物。

金属-陶瓷、陶瓷-陶瓷纳米复合材料可分为四类：晶间型，晶内型，混合型，纳米/纳米型[8]，晶间型纳米复合材料是指纳米粒子分布在陶瓷基质的晶界处，如图 1.14（a）所示；晶内型纳米复合材料是指金属纳米粒子分散在陶瓷晶粒内部，如图 1.14（b）所示；混合型复合材料如晶内-晶间复合，如图 1.14（c）所示；纳米/纳米复合材料是由晶粒尺寸均小于 100nm 的两相材料复合而成，如图 1.14（d）所示。

然而，该分类的数量可减少为两种。实际上，不存在单纯的晶内型纳米复合材料，因为在任何现实的材料中，在材料的表面上总是存在一定比例的金属粒子，因此，该复合材料实际上为晶内-晶间型的纳米复合材料；纳米/纳米复合材料也可被认为是晶间型纳米复合材料的特例[36]。

在原子尺度，若以圆圈代表原子（图 1.15），一种纳米晶的二维结构可认为是两种纳米结构成分的一种混合物。在图 1.15 中，原子有序排列的晶格点阵的纳米晶以黑圈表示，晶界以白圈表示。纳米晶的平均粒径在 2～100nm 之间，包含 10^2～10^8 个原子。这样的纳米晶结构属于纳米复合材料[37]。

1.4 纳米科学与技术

一般认为，纳米科学属于原子、分子、大分子、量子点和大分子组装的领域，并受表面

效应所支配，如：分子间的范德华力，氢键，电荷，离子键，共价键，疏水性，亲水性和量子力学隧穿等。

纳米科学是研究与较大尺寸的材料性能有本质区别的原子、分子以及大分子材料的现象和操纵的一门科学[38, 39]。

纳米技术是指在纳米尺寸上，通过控制形状和大小，对材料的结构、器件和系统进行设计、表征、制造及其应用的一门技术[38, 39]。

“纳米技术”这个术语最初是由一名日本的工程师 Norio Taniguchi 提出的[40]，而这个术语当初意味着，在 20 世纪占主导地位的微尺度上的材料控制与工程技术之外的一种新技术。

一般而言，纳米技术应该被称为“纳米综合技术”，因为不存在单一的纳米技术领域。该术语广义上是指生物、物理、化学及任何其它的科学领域，以及纳米结构的可控性制造领域。

在某种意义上，纳米科学和纳米技术并不是新鲜事物。数十年来，化学家们已经制备出了由纳米基本单元组成的高分子材料，在过去的 20 年中，已经运用纳米技术在计算机芯片上制造出微小的元件。如今，对原子和分子进行精确检测的工具的进步，促进了纳米科学和技术的迅猛发展。

1.5 工业革命的驱动

在电子学领域，自从 Shockley、Brattain 和 Bardeen 在 1940 年发明了晶体管以来，电子器件的小型化从来就没有停止过。

技术的发展和工业竞争已经驱使半导体工业制备出尺寸更小、运算速度更快、功能更强大的逻辑器件。1965 年，Intel 公司的创始人之一 Gordon Moore（摩尔）观察到，自从集成电路发明以来，每平方英寸集成电路中晶体管的数目每年增加一倍。Moore 当时预言这种倾向在未来将会延续。在随后的时间里，这种增加的步伐有所减缓，其数据密度大约每 18 个月翻一番，这也就是当前摩尔定律的含义。大多数专家，包括摩尔本人在内，都认为摩尔定律将持续更长一段时间。

如表 1.2 所示，直至目前，电路的致密化程度也一直呈现指数式增长[41]。Pentium 4 微处理器中每个硅片上拥有 42000000 个相互连接的晶体管。然而，通过缩减金属连线的线宽、减小其它元件的尺寸、在多层器件结构中组装更薄的层状结构，使电路的组装密度增加成为可能，新的制备技术的发展带来了这些变化。例如，商业金属连接线的线宽已经缩小至 130nm，在 2004 年，已经缩小至 90nm，已经进入了纳米领域（小于 100nm）。当前，50nm 的技术亦在研究之中。

1.6 目前技术的基础性缺陷

通过不断缩小元件的尺寸，采用“自上而下”的方法生产出运行更快、功能更强大的计算机电路系统是不可能的，因为当前的技术存在着基础性的缺陷。而且，这种基础性的缺陷仅靠工程设计的方式是无法克服的。比如：当氧化硅的绝缘层厚度被减薄至大约 3 个硅原子的厚度时，漏电就成为一个大问题。加之，当硅被限制在很小的尺寸时，将会失去它固有的带结构。

由于更多的器件增加了体系的复杂性，为保障每一个器件都能正常工作，缺陷和污染控制就显得尤为重要。如果一个新的技术可以制备出运算更快，且更小的逻辑存储芯片，并能降低体系的复杂性，不仅可缩短数天乃至数周的制造时间，而且可减少自然资源的消耗量，那么这种新技术对半导体工业必将带来革命性的进步。

1.7　分子电子学

我们如何才能克服当前固态电子学技术中的局限性呢？分子电子学是一个崭新的和诱人的研究领域，该研究领域正在唤起科学家的想象力[42]。分子电子学涉及寻找具有运算基本单元的导线、开关和存储功能的单分子或者分子的组合体[43]。该研究的目标是采用“自下而上”的操控技术，设计能够使分子进一步自组装，形成有序结构单元的分子器件，以替代目前“自上而下”方法的光刻技术，构建的固态电子器件。

1.8　未来的技术挑战

纳米技术产品已跨过了四个发展阶段[44, 45]，如图 1.16 所示。每一代产品，以通过对每个现象和加工工艺进行系统控制所制造的第一个商业雏形为标志。

第一代纳米产品（～2001 年）是“被动的纳米结构”，具有代表性的产品是对材料的宏观性能进行剪裁和功能化。主要的目的是制备纳米结构涂层，纳米粒子的分散，纳米结构金属，聚合物及陶瓷。

第二代纳米产品（～2005 年）是可用于机械的、电子的、磁的、光的、生物的以及其它效应的“主动的纳米结构”。并将其集成为微器件和系统，如：新的晶体管，金属氧化物半导体纳电子元件，放大器，靶向药物，化学品，驱动器，人造“肌肉”，以及自组装材料等。

第三代纳米产品（～2010 年）是指可利用各种综合性及组装技术，如：生物组装和机器人技术进行组装 3D 纳米系统。一个关键性的挑战是在纳米尺度上，对元件进行网络化以及分层构建。研究的焦点将偏重于异质纳米结构和超分子系统工程领域，这就包括了直接的多尺度的自组装、人工组织、传感系统、纳米体系内的量子相干、光子或电子自旋的信息处理、纳米机电系统组装等。

第四代（2015～2020 年）将会产生“不同种类分子的纳米系统”，在该系统中，元件的尺寸将减小至分子及大分子量级。在未来，设计新的原子和分子组装是十分重要的，其中包括“设计”大分子、纳米机器、直接的和多尺度的自组装，开发量子控制，用于卫生保健的纳米生物系统，在组织和神经系统水平上的人-机界面。

未来的研究将包括以下主题：用于分子和超分子系统设计的原子操纵，与能量转换相关的光与物质之间的可控作用，开发量子控制的机械-化学分子加工。

1.9　纳米材料的应用

纳米技术囊括了工具、技术和应用的较宽范围，该技术是在纳米尺度上对材料进行操纵，以获得块状材料不具有的新的性能。这些新的性能可为传统技术难以解决的问题，提出新的或改进的解决方案。对于发展中国家而言，这些解决方案可能包括更有效和更廉价的水净化装置、能源、医疗诊断检测以及药物传输系统、耐久性的建筑材料和其它产品。很显然，这将会对社会、环境以及世界带来不可估量的潜在益处。以下对其中一些技术及产品进行简单阐述。

1.9.1 水的净化

用于水净化的纳米技术已经是一个优势领域，因为掺有纳米材料的水处理设备已经得到应用，加之人类的发展对洁净水的需求正面临着压力。通过纳米技术制造的一些水处理设备已投放市场，而另一些正处于研发阶段。

纳滤膜技术已被广泛用于从盐水或有咸味的水中除去可溶性的盐分，除去微小的污染物，水的软化以及废水处理。纳滤膜能够有选择性地除去水中的有害污染物，而使营养物仍然保留在水中。纳米技术中的膜技术预期将降低脱盐成本，因为脱盐成本是制约目前脱盐技术更广泛应用的最大障碍。

碳纳米管过滤器将过滤提高到了一个新的水平，该过滤器有各种用途，它可以将 25nm 的脊髓灰质炎病毒、较大的病原体如大肠杆菌和金黄色葡萄球菌从水中除去。纳米管基的水过滤器可以过滤细菌和病毒，与传统的膜过滤器相比，具有更好的弹性和重复使用性。通过加热或清洗纳米过滤管可使过滤管达到重新使用之目的，并且，水分子流过纳米工程膜的速度要比流过传统的过滤膜快得多[46]。

1.9.2 纳米催化剂

纳米催化剂包括能够提高催化性能的纳米尺寸和纳米结构改性的酶、金属以及其它材料。纳米催化剂，如：二氧化钛及铁的纳米粒子可用于降解有机污染物，并从液体中除去盐分和重金属。人们希望纳电催化剂能将重度污染的水和含盐分高的水净化为饮用、卫生及灌溉用水[47]。

1.9.3 纳米传感器

纳米传感器可以检测到单个细胞，甚至单个原子，纳米传感器比体积大的传感器更灵敏。传统的水质研究通常依靠现场和实验室分析相结合的方法，这就需要专业人员取水质样品，然后送往附近的实验室进行化学和生物分析。将微制造技术和纳制造技术相结合的新式传感器技术有望使用这种体积小、携带方便且精度高的传感器来检测水质的化学和生物参数。

1.9.4 能源

实现低成本的太阳能发电一直是热带国家的渴望，然而用于发电的玻璃及硅质光伏板价格十分昂贵且易碎。纳米技术可以生产出更廉价的可在建筑屋顶上随意展开的光伏膜，如今，有一种新的方法可大量生产纳米结构材料，并用于光伏技术。

纳米材料制造商，Altair 纳米技术公司正在自主研发纳米钛酸盐材料的锂离子电池[48]。因为纳米钛酸盐材料在充电和放电过程中，当离子进入或离开系统时，纳米钛酸盐材料不会发生膨胀或收缩现象。而且充电 9000 次以上时，仍可保持 85%的充电容量，而传统的锂离子电池仅有充电 750 次的使用寿命。加之，纳米钛酸盐电池在约 1min 内充电就可达到电池的 80%电容量。

1.9.5 医药中的应用

纳米技术为靶向药物传输系统、延缓疫苗释放、提升诊断和成像技术等卫生保健和医学研究突破提供了一系列的可能性。

在发展中国家，纳孔膜材料将有助于治疗一些疾病，这些纳孔膜可缓慢地释放药物，这对于远离医院的病人来说非常重要。制造的纳孔膜的孔比药物分子稍大一点，不管胶囊中药物量的多少，纳孔膜能够控制药物的扩散速率。美国的生物制药公司，纳米杀毒剂有限公司，已经研发了一种新的病毒治疗方法，该方法使用一种工程化柔性纳米材料，封装活性药物成分专杀禽流感、流行性感冒等病毒，并在细胞没有感染病毒之前，消除这些病毒[49]。

也许在将来的某一天，纳米机器人将能够在人体内穿行，发现并清除疾病，如动脉粥样硬化斑块。

当我们使用具有刚性的菱形状纳米尺寸的零部件，类似于图 1.17 的分子齿轮和轴承时[50]，我们就能设计和建造完整的纳米机器人。也许在 2020 年，功能强大的纳米医药将会诞生。这些纳米机器人具有全套甲胄的自动控制子系统，如便携式传感器、马达、操作器、电源以及分子计算机。但是，要将这些纳米元件按正确的顺序自发地进行自组装是非常困难的，因为机械结构变得越来越复杂。制造出如此复杂的纳米机器人系统，就需要通过所谓的定位组装来搭建一种分子结构的制造技术。这就需要一个一个地抓取和放置分子零部件，并使它们沿着可控的轨道运动，就像制造汽车组装线上的机器人臂一样。然后，整个动作一遍遍地重复，进行组装其它不同的零部件，直至制造出最终产品，例如，医学纳米机器人。

配备有控制仪表，并具有灵活性的未来的纳米机器人，将能够进行精确的和精准的细胞内的外科手术，这些手术超出了人手直接进行操作的能力。我们期望一些生物相容性的外科纳米机器人能够发现和清除孤立的癌细胞，清除微脉管中的障碍物，并且修复血管内壁细胞，实施“非创伤性的”组织和器官移植，对受损伤的细胞外和细胞内结构实施分子修复，甚至在单个活的人体细胞内进行新旧染色体的交换[51]。

大量研究报道表明纳米技术加速了各种再生疗法，比如针对于骨骼、血管、心脏、软骨、膀胱和脑组织的疗法。由于植入物的表面结构会引起组织细胞特殊的、明显的反应[52]。大多数传统的生物材料具有微米尺度或更大尺度的表面特征[53]。业已发现大多数自然组织的内部或表面具有纳米尺度的特征，所以，可以推测，向传统生物材料的表面掺入纳米形貌可以催生多种细胞类型的功能。鉴于此，许多纳米结构材料被称为仿生纳米材料[54]。例如，与传统的钛表面相比，纳米结构钛植入物表面促进了骨细胞反应，加速了钙的沉积，促进了与周围骨质的结合[55]。但这并非是纳米材料在再生医学中的唯一应用。在软骨组织应用方面，纳米结构的乳酸-羟基乙酸共聚物（PLGA）表面已被证明加速了软骨细胞的黏结和增殖，以及细胞外组织的形成[56]。此外，表面具有纳粗糙度系的人工血管与血管内支架（钛）能够改善内皮（血管内细胞）的细胞功能[57]。

目前，有两种策略可以促进先进的生物材料的发展，第一种策略是通过化学变化［例如，将钛应用于整形外科（骨科），或通过控制药物释放形成植入物表面］[58]。第二种策略涉及改变植入物的物理性质，例如形成纳米表面形貌，以此改变表面粗糙度。基于这些原因，通过有选择性的改变化学及物理因素，对生物材料进行剪裁，以便最大限度地有利于细胞的相互作用（例如：增强形成组织细胞的功能，降低免疫细胞和细菌的功能）。

人们已经开展了各种纳米结构的聚合物和金属（合金）与生物（和细胞）兼容性的研究。与传统的骨科植入材料相比，各种纳米材料具有更大的骨形成细胞（成骨细胞）的功能，如：纳米羟基磷灰石[59]，电纺丝[60]，阳极钛[61]以及纳米结构钛表面[62]。下一步研究的重点在于研究这些材料能够增强骨细胞功能的机理，这些研究提供的确凿证据表明：吸附蛋白质（如玻连蛋白，纤连蛋白和胶原蛋白）的量以及生物活性的改变与纳米结构表面的表面能的增加密切相关[63]。

复 习 题

1. 什么是纳米？
2. 什么是纳米科学？
3. 什么是纳米技术？
4. 纳米科学与纳米科技的研究目标是什么？
5. 什么是纳米材料？
6. 纳米材料与纳米结构材料有什么区别？
7. 零维、一维、二维和块体材料有什么区别？
8. 量子阱、量子线和量子点的含义是什么？
9. 在你周围存在纳米材料吗？请列举一些纳米材料的应用实例。
10. 什么是量子限域？
11. 当前科技的基本局限有哪些？
12. 为什么科学家提出分子电子学？

2. Nanometer Effects of Nanoscale Materials

It is well known that classical mechanics provides an accurate description for the behavior of macroscopic objects. On the other hand, the physics of microscopic systems involving electrons, atoms, and molecules is exclusively quantum mechanics. A fundamental aspect of quantum mechanics is the particle-wave duality, introduced by De Broglie. The quantum mechanics is not required to describe the movement of objects in the macroscopic world because the wavelength associated with a macroscopic object is in fact much smaller than the object's size, and therefore the trajectory of such an object can be excellently derived using the principles of classical mechanics.

With the advances of fabrication techniques, the crossover region between these two fundamental regimes has been the subjects of intense experimental and theoretical investigations. The system that falls into this crossover region is called mesoscopic region and this system is small enough in physical size that its behavior reveals quantum-mechanical effects, although it contains a macroscopic number of particles. In other words, whenever the size of a physical system becomes comparable to the wavelength of the particles that interact with such a system, the behavior of the particles is best described by the rules of quantum mechanics. All the information we need about the particle is obtained by solving its Schrödinger equation. The solutions of this equation represent the possible physical states in which the system can be found.

Some anomalous physical and chemical properties of nanoscale materials, such as mechanical, thermal, optical, and electronic properties are remarkably different from that of bulk materials and from their atomic counterparts. Several new effects have been observed in mesoscopic nanoscaled systems.

2.1 Small size effect

When the size of nanoparticle is comparable to or even less than the optical wavelength, the de Broglie wavelength, the free path length of electrons, or the coherence wavelength of superconducting state, periodic boundary conditions and the translation symmetry of crystal nanoparticles would be broken down, and atomic density near surface of noncrystal nanoparticles decreases under the conditions. Subsequently, the acoustic, magnetic, electronic, thermal, and mechanical characteristics are simultaneously changed so as to cause appearance of some novel phenomena referred to as small size effect.

The most obvious effect of reducing the particle size is the increase in surface and interface areas. The atoms lying at the surfaces or interfaces have different environments both in terms of coordination and chemical properties, since the structure of grain boundaries is significantly different from that of bulk materials.

As far as nanoparticles are concerned, the reduction in melting temperature has been observed experimentally in several cases and has been predicted in many theoretical studies. The physical reason for this reduction can be explained qualitatively by the Lindemann criterion which stipulates that a bulk material will have melted when fluctuations (due to the temperature) in interatomic distances reach a certain value, about 10% of the lattice parameter. If this criterion is assumed to remain valid for nanosystems, it is clear that the many surface atoms will fluctuate more easily due to less restricted in their thermal motions. Thereby the nanoparticles exhibit lower melting temperature. For example, the change in the melting temperature of gold with the particle size is illustrated in Fig.2.1.

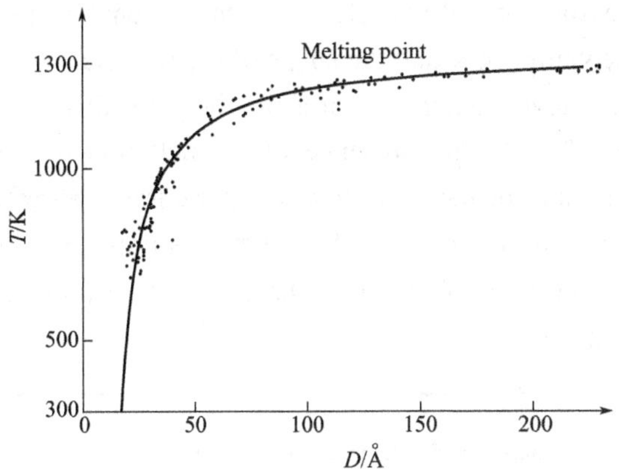

Fig.2.1 Size dependence of the melting temperature of gold

2.2 Quantum size effect

The electron transport through a quite small structure such as quantum dots is governed by the quantum effects.

As the particle size is reduced to, or approach the exciton Bohr radius, the electron energy states near Fermi energy level are drastic changes from quasi-continuous states to discrete states in a metal nanoparticle, or the energy gap drastically widen between discontinuous the highest occupied molecular orbital (HOMO) and the lowest unoccupied molecular orbital (LUMO) in a semiconductor nanoparticle. These phenomena are referred to as quantum size effect.

2.2.1 Relationship between energy gap and particle size

The prediction of discontinuous energy levels was a breakthrough in the area of limited systems. Considering the effects of the discreteness of energy levels on the physical properties of metal particles, Kubo and co-workers[1~5] proposed a famous relation:

$$\delta = 4E_F/(3N) \propto V^{-1} \tag{2.1}$$

Where δ is the energy gap between the nearest neighboring energy levels; E_F is Fermi energy; N is the number of the conduction electrons in the particles; and V is the volume of the particles. If the particles were in the shape of balls, from Eq.(2.1) it would have a relation of $\delta \propto d^{-3}$.

Eq.(2.1) can be divided into three parts to be discussed as follows.

① For the macroscopic particles containing countless atoms: $N \propto \infty$, $\delta \propto 0$;

It can be said that there is a continuous energy level in macroscopic particles.

② For the nanoparticles encompassing numbered atoms: N is finite, the energy gap δ increases rapidly with decreasing particle size. Therefore, the energy gap for the nanoparticles is discrete at low temperatures.

The average energy gap δ close to the Fermi energy could be much larger than the thermal energy k_BT at low temperatures. The discreteness of energy levels would affect obviously the thermodynamic properties of materials, and thus the specific heat; the susceptibility of particles would differ evidently from those of bulk materials.

The bands which overlap in bulk materials may be separated by a gap in small clusters near-spherical particles which are smaller than 10nm are typically called clusters[6]. This leads to the consequence that nanoscopic metal particle may exhibit behavior as a semiconductor or as an insulator, depending on the values of N and the shape of the particle.

A schematic electron level diagram corresponding to this size-induced metal-insulator transition is given in Fig.2.2[7]. In the presence of a small non-zero Kubo gap (δ), there is a continuous transition from an insulator at low temperatures ($\delta \gg kT$) to a semiconductor at intermediate ($\delta = kT$) and to a conductor at high temperatures ($\delta \ll kT$). Only at absolute zero temperature is there a discontinuous first-order phase transition (called the Mott transition) as a function of interatomic distance.

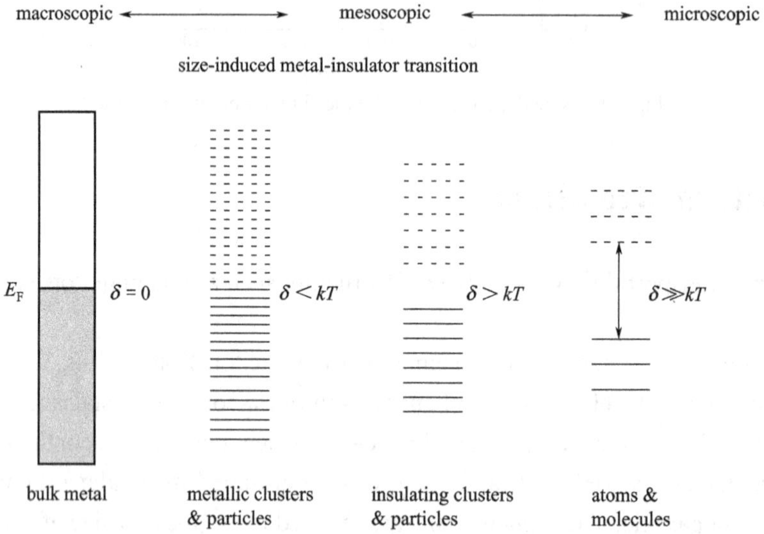

Fig.2.2 Evolution of the band gap and the density of states in different systems [7]

When the energy gap is larger than the thermal, magnetostatic, electrostatic and photonic energies, it was considered that a metallic particle about 10nm occurs the quantum size effect at the temperature about 2K.

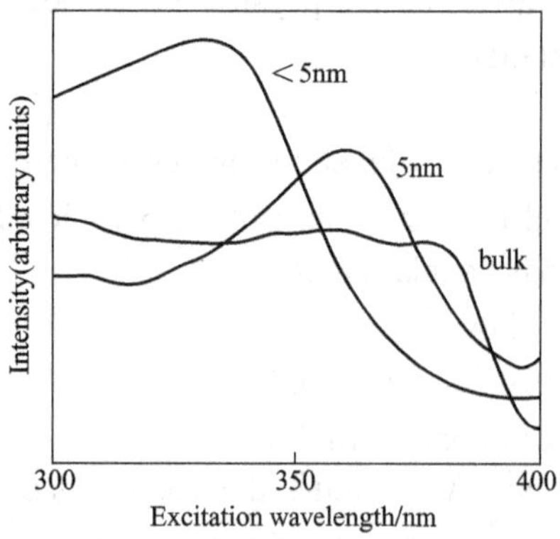

Fig.2.3 Photoluminescence spectra of nanocrystalline ZnO with different crystal sizes in comparison with the bulk material (detection wavelength 550nm)[8]

2.2.2 Application

The quantum size effect will appear when crystallite radius is comparable to or below the exciton radius (generally, the exciton radius of GaAs is about 10nm). Many terms have been used to describe these ultrasmall particles, such as quantum dots, nanocystals, Q particles, clusters and other terms.

Fig.2.3 displays the blue shift in the luminescence spectra as a function of the crystal size of nanocrystalline ZnO (consolidated ZnO crystals separated by grain boundaries)[8]. The blue shift is caused by the quantum size effect. If the crystallite size becomes comparable or smaller than the de Broglie wavelength of the charge carriers generated by the absorbed light, the confinement increases the energy required for absorption. This energy increases the shifts of the absorption/luminescence spectra towards shorter wavelengths (blue).

2.3 Surface effect

The ratio of surface atoms to the total number of atoms will remarkably increase with the decrease of particle size, and the surface bonding energy also increases so as to cause the characteristic changes of nanoparticles called as surface effect.

As the particle size changes, the ratio of surface atoms to the total number of atoms will vary. Considering a homogeneous solid material of compact shape with macroscopic sizes (let us say spherical), most of its properties will be related to its chemical composition and crystal structure. The number of surface atom comprises a negligible proportion in total number of atoms and will therefore play a negligible role in the bulk properties. However, when the size of the object is reduced to the nanometric range, i.e., < 10nm, the proportion of surface atoms is no longer negligible. Hence, at 5nm (around 8000 atoms), this proportion is about 20%, while at 2nm (around 500 atoms), it stands at 50% as shown in Fig.2.4[9]. An empirical law shows as below:

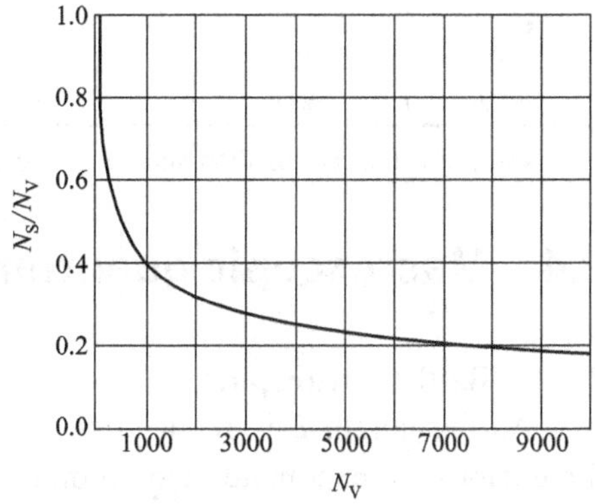

Fig.2.4 Proportion of surface atoms for a spherical particle comprising N_V atoms with N_S at the surface [9]

$$N_S/N_V \approx 1/(2R) \tag{2.2}$$

Where R is the radius in nm. This empirical law gives a proportion of surface atoms of 100% for a size of 1nm. Of course, Eq.(2.2) is no longer valid for smaller dimensions. We shall see that the fact that a large fraction of the atoms are located at the surface of the object will modify the material properties.

The coordination can take a range of values at the surface, 9 for close-packed (111) face. The variation in the number of atoms with different coordinations with value 12 for an atom in the bulk, 8 for (100) face, 7 for atoms on edges and 6 for atoms at corners in fcc metals as a function of the total number of atoms is given in Table 2.1 for a cubo-octahedron of an fcc crystal[10].

The larger number of surface atoms, the higher surface energy, and the lack of atomic coordination number leads to the high chemical activity of the surface atoms. Because the surface atoms possess less stabilized chemical bonds with neighboring atoms, they are very unstable and thus can combine easily with other atoms. It is the reason why the metal nanoparticles can spontaneously burn in air. It is also a fact that the atoms at the surface of the inorganic particles in air could easily adsorb the atoms/molecules, like gases, water and etc.

The surfaces/interfaces play important roles in the physical properties of the nanocapsules, because the environments of the atoms at the surface/interface are totally different from those of the atoms inside the grains [11~14]. The physical behaviors of the atoms located inside the grains could be very close to those of atoms in bulk materials. The properties of the atoms at the surfaces/interfaces could differ from those of bulk materials due to the lack of structural symmetry.

Table 2.1 The number of atoms in different positions in a cubo-octahedral particle and considering successive filled shells

Total number of atoms	Number of surface atoms	Number of corner atoms Z=6	Number of edge atoms Z=7	Number of atoms of type (100) Z=8	Number of atoms of type (111) Z=9
38	32	24	0	0	8
201	122	24	36	6	56
586	272	24	72	24	152
1289	482	24	108	54	296
2406	752	24	144	96	488
4033	1082	24	180	150	728
6266	1472	24	216	216	1016
9201	1922	24	252	294	1352
12234	2432	24	288	384	1736
27534	3632	24	360	600	2648
46929	5882	24	468	1014	4376

Note: Z is the coordination number of the atoms under consideration [10].

2.4 Macroscopic quantum tunnel effect

2.4.1 Ballistic transport

As the length L is reduced from macroscopic dimensions (~millimeters) to nanometric dimensions (~nanometers) or atomic dimensions, the nature of electron transport changes significantly as shown in Fig.2.5[15]. At one end, it is described by a diffusion equation in which electrons are viewed as particles that are repeatedly scattered by various obstacles causing them to perform a "random walk" from the source to the drain in the system of macroscopic dimension. At the other ends, when the dimension of an electronic device is smaller than the electron scattering length (mean-free path), electrons may travel from one electrode to another without encountering any scattering event. Such, an electron movement is called ballistic transport [16]. The ballistic transport of electrons through short and narrow channels results in quantized conductance.

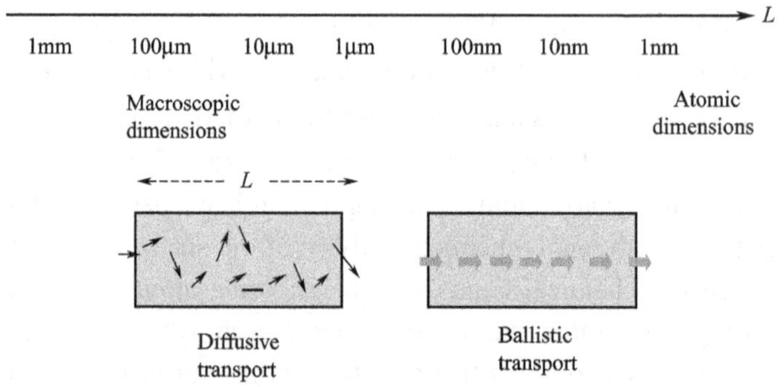

Fig.2.5 Evolution of devices from the regime of diffusive transport to ballistic transport as the channel length L is scaled down from millimeters to nanometers [15]

2.4.2 Tunneling

A classical particle will never be able to progress past a point at which the potential energy exceeds its total energy, and it will be turned back. However, the quantum mechanical principle is described as follows: If the region in which the potential energy exceeds the particle's energy is narrow enough on a quantum scale, the particle can go right through it. This is called "tunneling" as shown in Fig.2.6. Another example, Fig.2.7 shows the wave packet of a particle passing right through a region where the peak potential exceeds the particle's expectation energy [17]. The particle has a chance of passing through because its motion is governed by the Schrodinger equation, instead of the equations of classical physics.

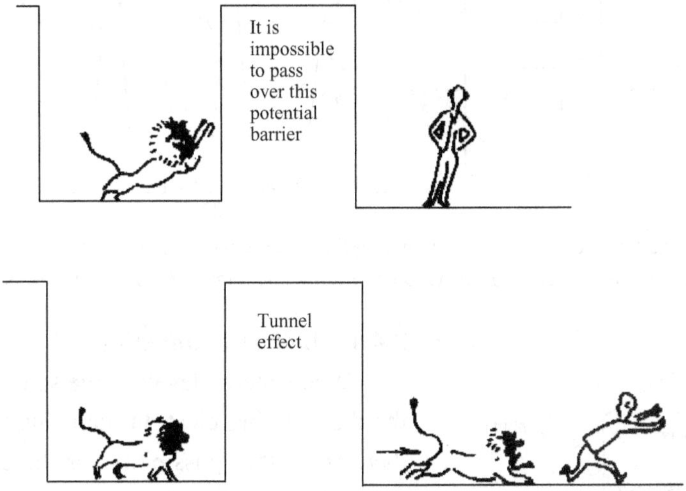

Fig.2.6 Differences between classical theory and quantum theory

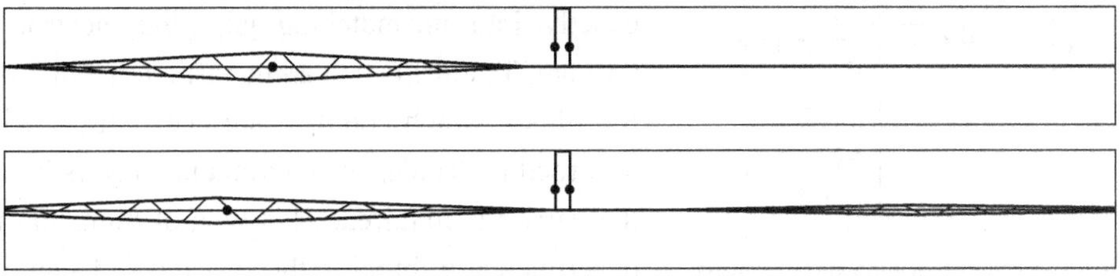

Fig.2.7 Tunneling of wave packet of a particle [17]

2.4.3 Resonance tunneling

Resonance tunneling devices are based upon the phenomenon of an electron tunneling through a complex barrier having intermediate electron states. For instance, consider the electron transfer through the structure consisting of a conducting (or semiconducting) quantum dot by thin insulating barriers separated from two metal electrodes. Fig.2.8 illustrates the schematic energy diagrams of this structure [18]. The total separation (w) between metal electrodes is large enough to reduce the probability of direct tunneling as shown in Fig.2.8(a). However, if the Fermi level (E_F) in the source contact matches one of the energy levels (E_i) in the quantum dot, the tunneling will take place through a much thinner barrier w^*, via this intermediate level E_i,. Such resonance conditions can be achieved by the application of external dc bias, as shown in Fig.2.8(b). The resulting I-V characteristics in Fig.2.8(b) show peaks, corresponding to the energy levels E_1, E_2, E_3, and so on.

The realization of this idea was achieved in GaAs/ AlGaAs layered structures produced by molecular beam epitaxy[18~20].

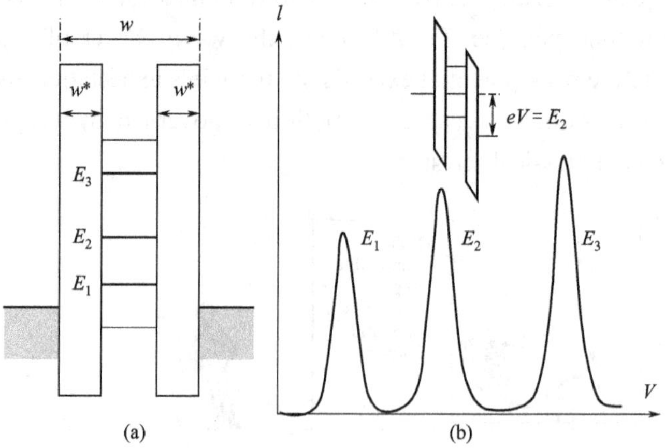

Fig.2.8 The scheme of resonance tunneling through the barrier having intermediate electron states (a) band diagram of the resonance tunneling barrier; (b) *I-V* characteristic [18]

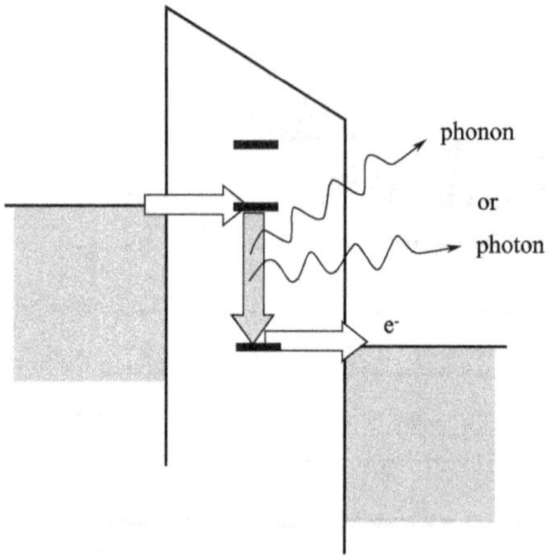

Fig.2.9 The scheme of inelastic electron tunneling, which involves the excitation of phonons (or photons)[18]

2.4.4 Inelastic tunneling

When energy levels in the source are higher than in the drain during electron tunneling through the barrier, there is another possibility for the electron transfer, as shown in Fig.2.9[18]. The excess energy can be released by an exciting phonon, and if the energy of phonon fits exactly into mismatching gap, the electron will go through. To fulfill such an inelastic tunneling process, a third body which could be surface states of impurities, adsorbed molecules, or intermediate layers integrated in the barrier is required. This type of tunneling has been experimentally found in the structures of semiconductor (or metal)/ oxide/ metal[21~24]. The electron energy can be partially consumed to excite the vibrations (phonons) of the impurity molecules on the metal/oxide interfaces.

2.4.5 Tunnel effect

The tunnel effect means that microscopic particle has an ability to traverse potential barrier.

2.4.6 Macroscopic quantum tunnel effect

It is found that some macroscopic physical quantities, for instance, magnetization intensities of nanoparticles and the magnetic flux in quantum interference device, also have the tunnel effect that is referred to as macroscopic quantum tunnel effect.

When the size of the nanoparticles is reduced to be less than certain critical value, the system could be treated as zero (or quasi-zero) dimension. In this case, the quantum effects become dominant, which originate not only from the discreteness of the energy levels but also from the quantum interference as well as the quantum tunneling in a system. It is said that the threshold for the semiconductor integrate circuit is about 50nm due to macroscopic quantum tunnel effect.

References

[1] Z.D. Zhang, Encyclopedia of Nanoscience and Nanotechnology, 6 (2004) American Scientific Publishers, ISBN: 1-58883-062-4, P77-160.

[2] R. Kubo, J. Phys. Soc. Jpn., 17(1962) 975.

[3] A. Kawabata, R. Kubo, J. Phys. Soc. Jpn., 21(1966) 1765.

[4] A. Kawabata, J. Phys. (Paris) Colloq., 38 (1977) 2.

[5] R. Kubo, A. Kawabata, S. Kobayashi, Annu. Rev. Mater. Sci., 14 (1984) 49.

[6] E. Roduner, Nanoscopic Materials Size-dependent Phenomena, (2006) The Royal Society of Chemistry, ISBN-10: 0-85404-857-X (Hardback), ISBN-13: 978-0-85404-857-1 (Electronic).

[7] P.P. Edwards, R.L. Johnston, C.N.R. Rao, On the size-induced metalinsulator transition in clusters and small particles, in Metal Clusters in Chemistry, Wiley,Weinheim, 3 (1999).

[8] H. Gleiter, Nanostruct. Mater., 6 (1995) 3.

[9] C. Brechignac, P. Houdy, M. Lahmani, Nanomaterials and Nanochemistry, (2007) Springer-Verlag Berlin Heidelberg, ISBN 978-3-540-72992-1, P4.

[10] C. Brechignac, P. Houdy, M. Lahmani, Nanomaterials and Nanochemistry, (2007) Springer-Verlag Berlin Heidelberg, ISBN: 978-3-540-72992-1, P284.

[11] R.K. Kawakami, E.J. Escorcia-Aparicio, Z.Q. Qiu, Phys. Rev. Lett., 77(1996) 2570.

[12] H.J. Choi, Z.Q. Qiu, J. Pearson, S.J. Jiang, D. Li, S.D. Bader, Phys. Rev. B, 57 (1998) R12713.

[13] R.K. Kawakami, M.O. Bowen, H.J. Choi, E.J. Escorcia-Aparicio, Z.Q. Qiu, Phys. Rev. B, 58 (1998) R5924.

[14] E. J. Escorcia-Aparicio, H.J. Choi, W.L. Ling, R. K. Kawakami, Z. Q. Qiu, Phys. Rev. Lett., 81 (1998) 2144.

[15] Z.K. Tang, P. Sheng, Nanoscale Phenomena Basic Science to Device Applications, (2008) Springer Science and Business Media, LLC ISBN-13: 978-0-387-73047-9 (Hardback), ISBN-13: 978-0-387-73048-6 (Electronic).

[16] H.S. Nalwa, Encyclopedia of Nanoscience and Nanotechnology, 9 (1) (2004) American Scientific Publishers, ISBN: 1-58883-057-8, P387.

[17] L. van Dommelen, Fundamental Quantum Mechanics for Engineers, (2004).

[18] A. Nabok, Organic and inorganic nanostructures. (2005) Artech House Inc. ISBN: 1-58053-818-5.

[19] K.B. Nichols, et al., "Fabrication and Performance of In0.53Ga0.47As/AlAs Resonant Tunneling Diodes Overgrown on GaAs/AlGaAs Heterojunction Bipolar Transistors," (1994) Compound Semiconductors.

[20] H. Goronkin, U. Mishra, (eds.), Institute of Physics Conference Series, 141 (1995) 737-742.

[21] W.K. Lye, et al., Quantitative Inelastic Tunneling Spectroscopy in the Silicon Metal-Oxide-Semiconductor System, Appl. Phys. Lett., 71(17) (1997) 2523-2525.

[22] L.S. Braginskii, E.M. Baskin, Inelastic Resonant Tunneling, Phys. Solid State, 40(6) (1998) 1051-1055.

[23] S. Okur, J.F. Zasadzinski, Modification of Al-Oxide Tunnel Barriers with Organic Self-Assembled Monolayers, J. Appl. Phys., 85(10) (1999) 7256-7262.

[24] F. Jimenez-Molinos, et al., Physical Model for Trap-Assisted Inelastic Tunneling in Metal-Oxide-Semiconductor Structures, J. Appl. Phys., 90(7) (2001) 3396-3404.

Review questions

1. What is the small size effect?

2. What is the quantum size effect?

3. What is the surface effect?

4. What are tunnel effect and macroscopic quantum tunnel effect?

5. How do you think the band gap and the density of states in different systems such as bulk metal, nanoparticles and single atom?

6. What does the blue shift mean?
7. What is ballistic transport?
8. Please describe the basic principle of resonance tunneling.
9. What is inelastic tunneling?
10. What is tunneling?

Vocabulary

acoustic [ə'kuːstɪk] adj. 声学的；音响的；听觉的；n. 原声乐器；不用电传音的乐器
anomalous [ə'nɒm(ə)ləs] adj. 异常的；不规则的；不恰当的
ballistic [bə'lɪstɪk] adj. 弹道的；射击的
barrier ['bærɪə] n. 障碍物，屏障；界线；vt. 把…关入栅栏
condense [kən'dens] vi. 浓缩；凝结；vt. 使浓缩；使压缩
cyclic ['saɪklɪk; 'sɪk-] adj. 环的；循环的；周期的
coordination [koˌɔrdɪ'neʃən] n. 协调，调和；对等，同等
configuration [kənˌfɪgə'reɪʃ(ə)n; -gjʊ-] n. 配置；结构；外形
capsule ['kæpsjuːl; -sjʊl] n. [医] 胶囊；[航] 太空舱；adj. 压缩的；概要的；vt.压缩；简述
criterion [kraɪ'tɪərɪən] n. 标准；准则；准据；规范
edge [edʒ] n. 边缘；优势；刀刃；锋利；vt. 使锐利；vi. 缓缓移动；侧着移动
electrostatic [ɪˌlektrə(ʊ)'stætɪk] adj. 静电的；静电学的
empiric [em'pɪrɪk; ɪm-] n. 经验主义者；[古] 江湖医生；adj. 经验主义的
exciton ['eksɪtɒn; ɪk'saɪ-; ek-] n. 激子；激发性电子
fluctuation [ˌflʌktʃʊ'eɪʃ(ə)n; -tjʊ-] n. 起伏，波动
fraction ['frækʃ(ə)n] n. [数] 分数；部分；小部分；稍微
homogeneous [ˌhɒmə(ʊ)'dʒiːnɪəs; -'dʒen-] adj. 均匀的；齐次的；同种的
inelastic [ˌɪnɪ'læstɪk] adj. 无弹性的；无适应性的；不能适应的
inorganic [ɪnɔː'gænɪk] adj. 无机的；无生物的
luminescence [ˌluːmɪ'nes(ə)ns] n. [物] 发冷光，发光，冷光，荧光
macroscopic [ˌmækrə(ʊ)'skɒpɪk] adj. 宏观的；肉眼可见的
magnetization [ˌmægnətɪ'zeʃən] n. 磁化
magnetostatic [mægniːtəʊ'stætɪk] adj. 静磁的
negligible ['neglɪdʒɪb(ə)l] adj. 微不足道的，可以忽略的
octahedron [ˌɒktə'hiːdrən; -'hed-] n. 八面体
qualitative ['kwɒlɪtətɪv] adj. 定性的；质的，性质上的
resonance ['rez(ə)nəns] n. 共振；共鸣；反响
spectra ['spektrə] n. 光谱；范围
spontaneous [spən'teiniəs] adj. 自发的；自然的；无意识的
susceptibility [səˌseptɪ'bɪlɪtɪ] n. 敏感性；感情；磁化系数
symmetry ['sɪmɪtrɪ] n. 对称（性）；整齐，匀称
thermodynamic [ˌθɜːməʊdaɪ'næmɪk] a. 热力学的，使用热动力的
threshold ['θreʃəʊld; 'θreʃˌhəʊld] n. 入口；门槛；开始；极限；临界值
tunnel ['tʌnl] n. 隧道；坑道；洞穴通道；vt. 挖；在…挖掘隧道；vi. 挖掘隧道；打开通道

2. 纳米材料的纳米效应

众所周知，经典力学对宏观物体的行为给出了准确的描述。另一方面，包括电子、原子和分子在内的微观体系的物理学涉及专门的量子力学。量子力学基本原理是德布罗意提出的波-粒二象性，宏观物体的运动无法用量子力学来描述；因为宏观物体的波长，事实上要比物体本身的尺寸要小得多，因此，宏观物体的运动轨迹可以用经典力学进行完美的描述。

随着制备技术的发展，处于宏观与微观两个领域之间的交叠区已经成为实验和理论深入

研究的学科领域，该交叉领域的体系称为介观领域，虽然该体系包含了大量的粒子，但实际上该体系的尺寸相当小，以至于该体系的材料显示出量子力学效应。换言之，当一个物理系统的尺寸与粒子的波长相当时，粒子与系统会发生相互作用，则粒子的行为可通过量子力学定律来描述。我们所需要的粒子的所有信息，可通过对薛定谔方程进行求解而获得，方程的解代表着系统内的物理量。

纳米材料具有一些异常的物理和化学性能，如：力学、热学、光学和电学性能，截然不同于块体材料以及由原子组成的微观材料的性能。在介观的纳米体系中，人们已经观察到了数个新效应。

2.1 小尺寸效应

当纳米粒子的尺寸与光波长、德布罗意波长、电子的自由程长度或者超导态的相干波长相当或更小时，晶体纳米粒子的周期性边界条件和平移对称性遭到破坏，非晶体纳米粒子表面附近的原子密度减小，材料的声、磁、电、热和力学特性同时发生了改变，从而引起一些新的现象的出现，称为小尺寸效应。

当粒子的尺寸减小时，最显著的特征是表面积和界面积增大。因为纳米晶的晶界结构完全不同于块体材料，位于表面或界面的原子具有不同的配位环境和化学环境。

就纳米粒子而言，通过许多实验的观察及多种理论的论证，纳米粒子的熔点有所降低。对于熔点的降低可用 Lindemann 准则给予定性的解释，该准则认为，在一定温度下当原子间距波动达到一个临界值时，该临界值约为晶粒参数的 10%时，块状材料就会熔化。如果假设这个准则在纳米体系中仍然有效，显而易见，由于纳米材料表面的原子受到的约束更少，因为热运动，表面原子的波动也就更加容易，因而纳米粒子显示出更低的熔点。例如，图 2.1 给出了金的熔点随着粒子尺寸的改变曲线。

2.2 量子尺寸效应

电子传输可通过一个微小的结构，如量子点，这种传输受到量子效应的支配。

当粒子尺寸减小到或接近于激子玻尔半径时，金属纳米粒子费米能级附近的电子能级由准连续态变化为离散能级，而半导体纳米粒子中最高占据分子轨道（HOMO）与最低未占分子轨道（LUMO）之间的能隙急剧变宽，这些现象称为量子尺寸效应。

2.2.1 能隙与粒子尺寸的关系

在限域体系的领域，不连续能级的预言是一个突破。考虑到金属纳米粒子物理性能中能级的离散性，Kubo 和他的同事们提出了一个著名的公式[1~5]：

$$\delta = 4E_F/(3N) \propto V^{-1} \tag{2.1}$$

式中，δ是相邻能级间的能隙；E_F是费米能级；N是粒子中传导电子的数量；V指颗粒的体积。若微粒呈球形，从公式(2.1)可推导出这样的关系$\delta \propto d^{-3}$。

公式(2.1)可分为以下三个部分进行讨论。

① 对于由无数个原子组成的宏观粒子：$N \propto \infty$，$\delta \propto 0$。

可以认为，在宏观粒子中存在着连续的能级。

② 对于有限个原子组成的纳米粒子：N 是有限的，能隙 δ 随粒子尺寸的减小而急剧增大。因此，在低温下，纳米粒子的能隙是离散的。

在低温下，当材料的平均能隙δ接近于费米能级时，其能隙就会远远大于材料本身的热能k_BT，能级的离散化将会严重影响材料的热力学性质，如：比热容以及磁化率明显不同于块体材料。

在块体材料中相互重叠的带，在一个小的簇（粒径小于 10nm 的近乎球形的粒子被称为簇）材料中，或许被带隙所隔开[6]。依据纳米粒子的数量（N）和粒子的形状，金属纳米粒子可能变成半导体或者绝缘体。

尺寸诱导金属向绝缘体转变，如图 2.2 所示[7]。当 Kubo 带隙（δ）是一个很小的非零值时，材料存在一个连续的转变过程，低温下（$\delta \gg kT$）的绝缘体可转变为中温下的半导体（$\delta = kT$）；中温下的半导体（$\delta = kT$）还可继续转变为高温下的导体（$\delta \ll kT$）。只有在绝对零度时，存在不连续的一级相变（称为 Mott 转变）与原子间距的函数关系。

当能隙大于热能、静磁能、静电能、光子的能量时，在温度约为 2K 时，尺寸约为 10nm 的金属粒子就会出现量子尺寸效应。

2.2.2 应用

当晶体的半径接近或小于激子半径（通常 GaAs 的激子半径约为 10nm）时，就会出现量子尺寸效应。有许多术语用于描述这些超微粒子，如量子点、纳米晶、Q 粒子、簇等。

发光谱图的蓝移与纳米 ZnO 晶体尺寸的函数关系（固化的 ZnO 晶粒被晶界所隔开）如图 2.3 所示[8]。蓝移是由量子尺寸效应所引起的。当纳米粒子的粒径尺寸接近或小于由吸收光所产生的载流子的德布罗意波长时，限域作用导致需要吸收更多的能量，这就使得吸收光谱及发光光谱向更短的波长方向移动（蓝移）。

2.3 表面效应

纳米粒子的表面原子数与总原子数之比随粒径的减小而显著增大，表面键合能亦随之增大，从而引起纳米粒子特性的变化称为表面效应。

当微粒的尺寸变化时，表面原子数与总原子数之比也随着发生变化。考虑具有宏观尺寸（以球形为例）紧密堆积的均质固体材料，大多数性能将与它的化学组成和晶体结构相关。在块体材料中，表面原子数与总原子数相比，可以忽略不计，因而，表面原子对材料的块体性质没有什么影响。然而，当材料的尺寸减小至纳米范围时，当小于 10nm 时，表面原子所占的比例将不能忽视，当尺寸约为 5nm 时（表面原子约 8000 个），表面原子比例约为 20%，当尺寸小至 2nm 时（表面原子约 500 个），所占比例达到 50%，如图 2.4 所示[9]。以下为一个经验公式：

$$N_S / N_V \approx \frac{1}{2R} \tag{2.2}$$

式中，R 是颗粒的半径，nm。从该经验公式可知，当粒径等于 1nm 时，则表面原子的比例将达到 100%。当然，对于小于 1nm 尺度的粒子，该公式将不再有效。我们应该认识到，当表面原子的比例在总原子数中占到很大的份额时，材料的性能将会发生改变。

表面原子的配位数有一定的范围，对紧密堆积的（111）晶面，表面原子的配位数为 9。对于面心立方（fcc）的金属晶体而言，原子的配位数随所在位置的不同而变化，在块体中，原子的配位数为 12，在（100）晶面上原子的配位数为 8，晶边处原子的配位数则为 7，晶格顶角处的原子配位数为 6。表 2.1 给出了面心立方晶体中正八面体的原子配位数[10]。

粒子表面的原子数越多，则表面能就越大，由于表面原子缺少配位数，导致表面原子具有高的化学活性。由于表面原子与相邻的原子结合的化学键不太牢固，从而造成这些原子很不稳定，容易与其它原子结合，这就是金属纳米粒子暴露在空气中能够自发燃烧的原因所在。无机粒子暴露在空气中很容易吸附其它原子或分子，如吸附空气中的气体分子和水分子等。

表面/界面对纳米胶囊的物理性质起着十分重要的作用，这是由于表面/界面原子所处的环境与晶粒内部中的原子所处的环境是完全不同的[11~14]。处于晶粒内部的原子的物理性能与块体材料中的原子的物理性能非常相近。由于缺乏结构对称性，表面/界面原子的性能与块体材料中原子的性能截然不同。

2.4 宏观量子隧道效应

2.4.1 弹道传输

当粒子的长度（L）从宏观尺度（mm）减小至纳米尺度（nm）或原子尺度时，电子传输的属性将会发生重大的变化，如图 2.5 所示[15]。在长的一端（宏观尺度一端），电子的传输可通过扩散方程来描述，电子被认为是一个粒子，在传输中会遇到各种障碍而反复地散射，从而引起电子在一个宏观的系统中传输时，电子从进到出表现为“随机游动”的特点。在另一端（纳米尺度一端），当电子器件的尺寸小于电子散射长度（平均自由程）时，电子从一个电极传输到另一个电极时，不会遇到任何散射问题，这样的电子运动称为弹道传输[16]。电子的弹道传输通过短而窄的隧道就会导致量子化的电导。

2.4.2 隧穿

一个经典粒子将决不能越过比它自身总能量大得多的势垒，（当遇到势垒）粒子就会被挡住。然而，量子力学原理认为：当大于粒子能量的势垒区域窄到足以达到量子尺度，粒子就可以直接穿过该势垒，这就称为“隧穿”，如图 2.6 所示。另外一个例子，图 2.7 是一个粒子波包直接穿过比粒子本身能量要大得多的势垒区[17]。该粒子波包有一种隧穿的机会，是因为它的运动受薛定谔方程的支配，而不受经典力学方程的支配。

2.4.3 共振隧穿

共振隧穿器件是基于电子隧穿，通过具有一个中间电子态的复杂势垒。例如，考虑电子传输通过一个涂有薄的绝缘层（势垒）的导体量子点（或半导体量子点）的结构，该量子点被两个电极所分开。这个结构的能量示意图如图 2.8 所示[18]。两个电极之间距离（w）宽到足以避免它们之间发生隧穿的可能性，如图 2.8(a) 所示。然而，若源极中的费米能级（E_F）与量子点上的某个能级（E_i）相匹配时，隧穿就会发生，即：电子就会从源极端，经由量子点的中间能级（E_i），穿过相当薄的双势垒（w^*），到达漏极。通过施加一个外部的直流偏压，可实现这样的共振隧穿，如图 2.8(b) 所示，加偏压后得到的曲线特征峰对应于 E_1、E_2、E_3 等能级。

利用分子束外延法制备的层状结构 GaAs/AlGaAs，可实现这个构想[18~20]。

2.4.4 非弹性隧穿

当源极的能级比漏极的能级高得多（不匹配）时，若电子要隧穿通过势垒时，电子的传输存在另外一种可能性，如图 2.9 所示[18]。通过激发声子可释放出多余的能量，若声子的能量恰好等于不匹配的带隙能时，电子将发生隧穿。为了实现这个非弹性隧穿过程，就需要一种对表面态进行掺杂、吸附分子、势垒中组合的中间层的第三种材料。实验中发现，由半导体（或金属）/氧化物/金属组成的结构材料可发生这种隧穿[21~24]。用于激发位于金属/氧化物界面处的杂质分子的振动（声子）需要消耗掉一部分电子的能量。

2.4.5 隧道效应

微观粒子具有穿过势垒的能力称为隧道效应。

2.4.6 宏观量子隧道效应

研究发现一些宏观物理量，如：纳米粒子的磁化强度和量子相干器件中的磁通量，亦具有隧道效应，称为宏观量子隧穿效应。

当纳米粒子的尺寸减小至某个临界值以下时，该系统可认为是零维（或准零维）。在这种情况下，量子效应起主导作用，这不仅源于能级的离散化，而且来自于体系中的量子相干以及量子隧道效应。据报道，由于宏观量子隧道效应，半导体集成电路的阈值约为50nm。

复习题

1. 什么是小尺寸效应？
2. 什么是量子尺寸效应？
3. 什么是表面效应？
4. 什么是隧道效应和宏观量子隧道效应？
5. 你是如何理解不同系统，如：块体金属、纳米粒子和单个原子中的能隙和态密度的？
6. 蓝移是指什么？
7. 什么是弹道传输？
8. 请描述共振隧穿的基本原理。
9. 什么是非弹性隧穿？
10. 什么是隧穿？

3. Properties of Nanoscale Materials

Nanoscale materials are produced in various forms such as nanopowders, nanostructured particulates, fibers, tubes and thin films. Nanoscale materials have many interesting properties including mechanical, thermal, magnetic, electronic and optical properties.

3.1 Mechanical properties

It has well known that grain refinement often leads to an improvement in the properties of metals and alloys. For example, reducing grain size lowers the ductile-to-brittle transition temperature in steel. Of interest is the influence of grain size reduction on the mechanical behavior of materials. The empirical Hall-Petch equation[1, 2] relates yield stress σ_y to average grain size d.

$$\sigma_y = \sigma_0 + kd^{-1/2} \tag{3.1}$$

Where σ_y is the yield stress; σ_0 is a friction stress; k is a constant; and d is the grain diameter. A similar relationship also exists between hardness and grain size.

$$H = H_0 + kd^{-1/2} \tag{3.2}$$

In order to explain these empirical observations, several models have been proposed, which involve either dislocation pileups at grain boundaries or grain boundary dislocation networks acting as dislocation sources. In all cases the Hall-Petch effect is related to the phenomena of dislocation generation and dislocation motion in materials. Extrapolating this law to grains about ten nanometers, very high values of the hardness are predicted. For instance, nanophase Cu (d = 6nm) shows a hardness five times greater than that of bulk Cu (d = 50 μm). The yield stress of nanophase Pd (d = 7nm) is five times greater than that of bulk metal (d = 100 μm).

We can say that the hardness increases as the grain-size drops. However, stress-strain curves on small tensile samples show little change of the yield stress. Several recent reviews have summarized the mechanical behavior of nanostructured materials in terms of hardness and yield stress[3~5]. It is the evident that as the grain size is reduced down to the nanoscale regime (<100nm), hardness typically increases with decreasing grain size.

3.1.1 Positive Hall-Petch slopes

The hardness of nanomaterials increases with decreasing particle size as shown in Fig.3.1. That is, HV is proportional to $d^{-1/2}$.

3.1.2 Negative Hall-Petch slopes

For the Ni-P curve in Fig.3.2, the hardness of nanomaterial decreases with decreasing particle size. The dislocations in nanostructures might not be the thermodynamically stable defects; higher internal compressive strains and reduced vacancy concentration are the characteristics of the nanostructures. These factors were reported to contribute towards the deviations in the mechanical properties of the nanomaterials and an inverse Hall-Petch relation[7].

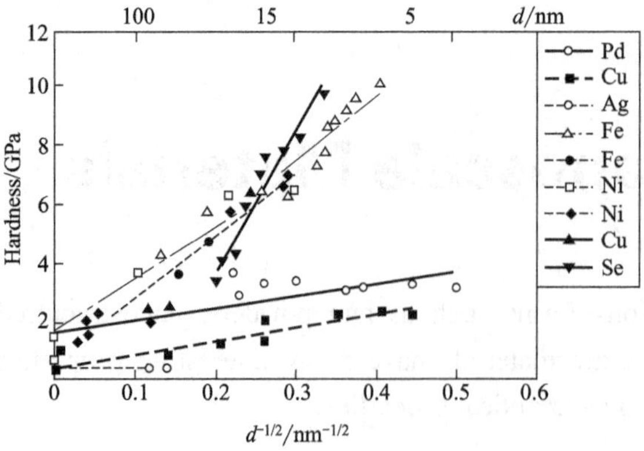

Fig.3.1 Room-temperature hardness versus grain size for nanophase metals compared with the hardness of their coarse-grained counterparts (values near origin)[6]

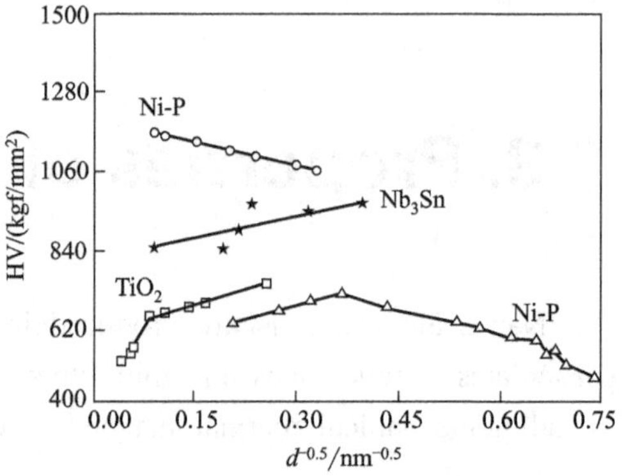

Fig.3.2 Hardness versus grain size for nanophase metals

3.1.3 Positive and negative Hall-Petch slopes

Much of the early work was concerned with room temperature micro-hardness measurements. Fig.3.3 shows the results obtained Vickers hardness of Ni-P electrodeposits at room temperature [8], and also shows the results on nanocrystalline Pd and Cu produced by the inert gas condensation technique [9]. It can be seen from Fig.3.3 [10] that the hardness of some samples exhibit initial increases and followed by significant decreases with decreases of particle sizes. The observed decreases in hardness are contrary to the Hall-Petch behavior[8, 11]. However, Erb et al. [12, 13] in an extensive study of electrodeposited Ni found a strong deviation from linearity below about 25nm grain sizes and a negative Hall-Petch slope between 11nm and 6nm as depicted in Fig.3.4.

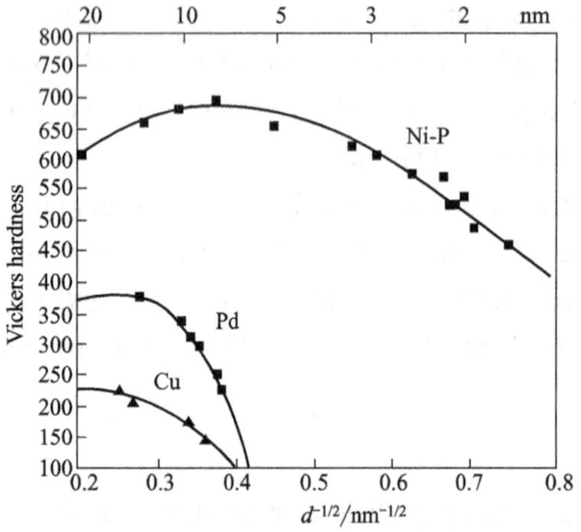

Fig.3.3 Vickers hardness measurements for nanocrystalline Ni-P, Pd and Cu[10]

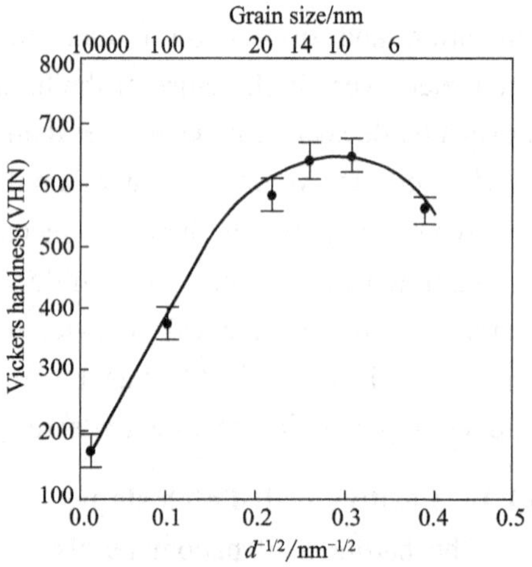

Fig.3.4 Hardness of nanocrystalline Ni as a function of grain size [11, 12]

The properties of bulk materials often change dramatically with addition of nano ingredients. Hardness of composites made from particles of nano-size ceramics or metals (<100nm) can suddenly become much stronger than predicted by existing materials-science models. For example, metals with a so-called grain size of around 10 nanometers are seven times harder and tougher than their ordinary counterparts with grain sizes in the hundreds of nanometers.

3.2 Thermal properties

Kubo[14] predicted in his theory that the thermal behavior of metal particles at low temperatures could be evidently different from that of bulk materials, because the average energy gap δ close to the Fermi energy could be much larger than the thermal energy $k_B T$ (where k_B is the Boltzmann constant and T is temperature) due to the discontinuity of the energy levels in the limited systems.

Some experimental results show that a decrease in size below about 5nm results in a dramatic decrease in the melting point of Au nanoparticles. The result indicates that the melting point is lowered by nearly 900K since the regular order of the lattice is rearranged and destroyed.

Nanocrystals are also known as quantum dots. Specific heats of nanocrystalline materials have been studied by many researchers. In most instances, experimental measurements indicate that nanoparticles exhibit enhanced specific heat as compared to the bulk material as shown in Fig.3.5 [15, 16].

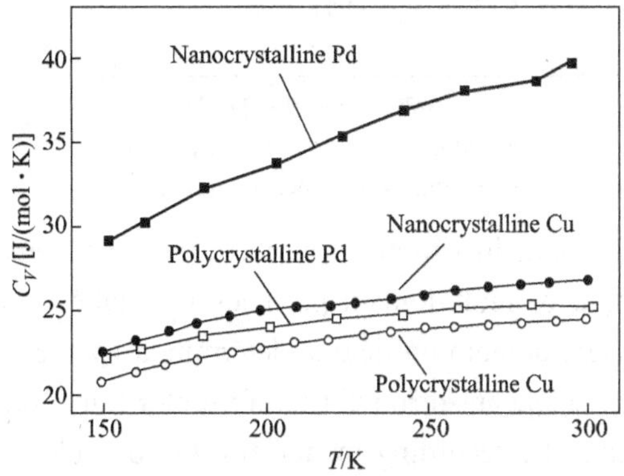

Fig.3.5 Specific heat of polycrystalline and nano crystalline palladium and copper at high temperatures[15, 16]

3.3 Magnetic properties

Strong magnetic materials such as ferro- and ferromagnetic alloys are by far the most useful materials and are commonly categorized by the coercive force H_c. Generally, materials with $H_c <$ 1000A/m are considered magnetically soft, materials with H_c between 1kA/m and 30kA/m are semihard, and materials with $H_c >$ 30kA/m are considered hard magnetic materials.

Magnetic materials based on iron, cobalt, or nickel, have the behaviour of soft magnetic materials owing to their weak anisotropy. The coercivity is low and the variation of the magnetization with the applied field is dominated by the wall displacement process. In rare earth-transition metal compounds, there are strong 3d-4f interactions. They result from hybridization between 5d electrons in the rare earth atoms and 3d electrons in the transition atoms. By this mechanism, the magnetism of the rare earths is preserved at high temperature. In systems where the magnetocrystalline anisotropy is uniaxial, there may be a strong coercivity. This kind of hard magnetic material is used in high-performance permanent magnets. The best known is the compounds $SmCo_5$, with hexagonal symmetry, and $Nd_2Fe_{14}B$, with quadratic symmetry.

When a magnetic material is placed in a magnetic field, it is magnetized with the change in the applied magnetic field strength as shown in Fig.3.6[17]. Magnetic solid of bulk or powders materials are not magnetized originally unless they are exposed to a magnetic field, since the original directions of the magnetic dipoles in magnetic domains of bulk material or in powder are random. Thus the magnetization of magnetic solid is brought about by the orientation of dipoles. As the magnetic field increases, the degree of orientation of the dipole increases, and the saturation

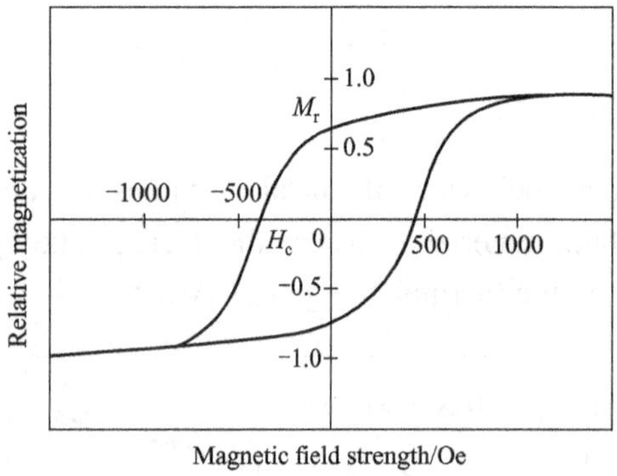

Fig.3.6 Magnetization curve: M_r, residual magnetization; H_c, coercive force[17]

magnetization can be attained at the perfect orientation of the dipoles. Even though the magnetic field is decreased to zero, the material will have a certain magnetization. This is so-called residual or remnant magnetization. A magnetic field of opposite direction is required to bring the magnetization to zero. This magnetic field is called the coercive force (H_c). Coercive force is essential for magnetic recording media.

As the size of the particle decreases, the number of the magnetic domains in the particle decreases. The domain structures in particles are schematically shown in Fig.3.7[18]. Generally, the coercive force of multidomain particles is smaller than single domain particles. The coercive force of a single-domain particle is determined by the so-called magnetocrystalline anisotropy together with the shape anisotropy. Therefore elongated single domain particles are preferentially employed for magnetic recording media. So, the particles available for magnetic media must be single-domain particles having high saturation magnetization and proper coercive force. But, if the particle becomes sufficiently small, the magnetic moment shows no preferential orientation due to the thermal agitation exhibiting superparamagnetic property as shown in Fig.3.7(a).

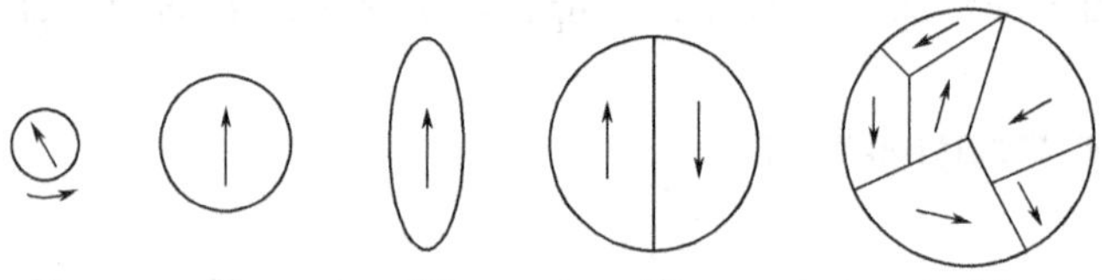

Fig.3.7 Schematic picture of magnetic domain structures of magnetic particles, (a) superparamagnetic; (b), (c) single-domain; (d) two-domain; and (e) multidomain particle [18]

3.4 Electronic properties

The characteristic electronic properties of the nanostructures are a result of tunneling currents, purely ballistic transport and the coulomb blockade effects [19].

Those nanoparticles called "quantum dots" or "artificial atoms" should have novel properties in many respects. The most spectacular one is to be expected in relation to the so-called single electron tunneling (SET). This means that electron transitions between particles can principally be performed one by one as Licharev predicted[20].

A collective motion of electrons (current I) in a bulk metal is induced by applying a voltage U. Ohm's law describes the linear relation between U and I by $U = IR$ (R is resistance of the material).

All relations described above depend on the existence of a band structure, that is, the presence of freely mobile electrons. The band structure begins to change if the dimension of a metal particle becomes small enough. Discrete energy levels finally play a dominant rule. As a consequence, Ohm's law is no longer valid. How does the current-voltage behavior change if we adopt a zero-dimensional metal particle?

To answer this, single nanoparticle has to be investigated with respect to their *I-U* characteristic. This is a big challenge! In principle, individual particle is positioned between two electrodes to which an increasing/decreasing voltage can be applied. However, to observe electronic details, two electrodes directly contacted with one metal particle have to be avoided. Instead, two capacitances (non-conducting material usually is a coating covered on particle) between two electrodes and a particle are used. Fig.3.8(a) illustrates the experimental conditions.

If an electron is transferred to the particle, its Coulomb energy E_C increases by $E_C = e^2/(2C)$, which C is the capacitance of the particles. Thermal motion of the atoms in the particle can initiate the changes of the charge and the Coulomb energy, so that further electrons may tunnel uncontrolled. To produce single electron tunneling (SET) processes, the thermal energy of the particle must be much smaller than Coulomb energy, in order to keep the energy on the particle $kT \ll e^2/(2C)$, With an additional charge, a voltage $U = e/C$ is produced, leading to a tunneling current $I = U/R_t = e/(R_tC)$ (where R_t = tunnel resistance)[21].

The current-voltage characteristic of an ideal quantum dot shows no current up to $U_{Coulomb} = \pm e/(2C)$ (Coulomb blockade). If this value is reached, an electron can be transferred. Following an electron tunneling process occurs if the Coulomb energy of the quantum dot is compensated by an external voltage of $U_{Coulomb} = \pm ne/(2C)$. An idealized "staircase", resulting from the repeated tunneling of single electrons, is produced as shown in Fig.3.8(b). The step height ΔI corresponds to $e/(RC)$ and the width ΔU corresponds to $e/(2C)$. It becomes possible to charge and to discharge in a quantum dot via a quantized manner, which is a condition for use of such particles in future to create new generation computer.

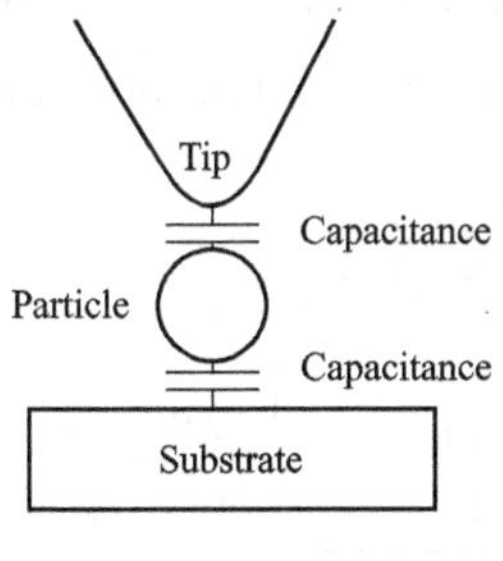

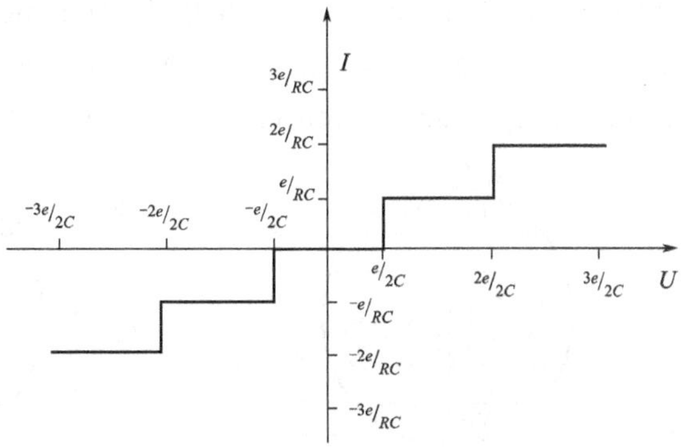

Fig.3.8 (a) Schematic diagram of the tip-nanoparticle arrangement. An STM tip and metal substrate contact a ligand monolayer surrounding; (b) Staircase shaped *I-U* curve obtained from a nanoparticle double tunnel junction

The quantization of the energy levels for Al nanoparticles (<10nm) at temperatures (<1K) is observed in a simple tunneling device [22, 23].

Electronic measurements for single metal nanoparticles are now often done by coating the particle with molecular ligands known as monolayer protected clusters (MPCs). The MPCs are located on metallic substrates and contacted with a scanning tunneling microscope (STM) tip[24]. The STM tip and metal substrate serve as the electrical contacts, while the ligand monolayer surrounding the nanoparticle provides the capacitance layer. This creates a double tunnel junction, where the staircase shaped *I-U* curve is observed.

3.5 Optical properties

Three-dimensional quantum confinement effects occur in semiconductor nanocrystallites when the particle size approaches the bulk exciton Bohr radius. This confinement gives rise to interesting new effects, leading to novel optical properties. These properties of semiconductor microcrystallites make them potentially attractive for applications in optoelectronic devices, such as optical data storage and high-speed optical communication.

3.5.1 Photochemical and photophysical processes of nanomaterials

The photophysical and photochemical processes in semiconductor nanoparticles are quite analogous to those of molecules that are well known. It has become customary to describe these processes by drawing the valence and conduction band levels inside circles, representing the spherical clusters, as shown in Fig.3.9[25].

After a semiconductor nanoparticle absorbs the energy of a photon, an electron (e) in the valence band is excited to the conduction band and the hole (h) is left behind in the valence band. Both the electron (e) and the hole (h) relax to trapped states e_t and h_t, normally at the surface [Fig.3.9(a)]. Alternatively, if the particle is in contact with electron acceptor molecules A, the electron is transferred to A to form A^-, and the particle contacts with suitable hole acceptor (electron donor) molecules D to form D^+ [Fig.3.9(b)].

The interest is concentric core-shell heterostructures consisting of two or more semiconductors with different energy levels. An example is given in Fig.3.9(c) where the shell material has a larger band gap. In this case, when electron in shell is excited, both electron and

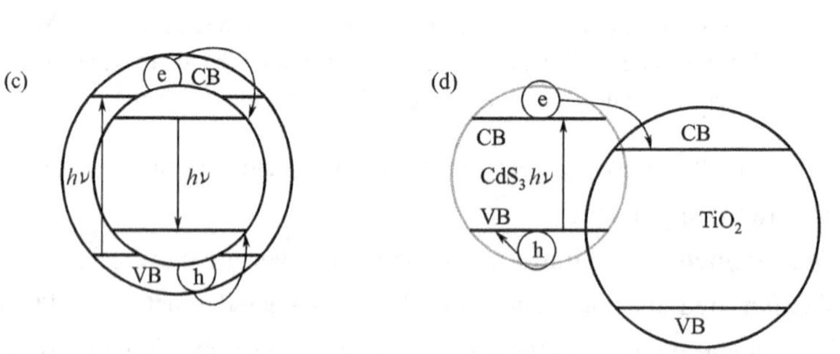

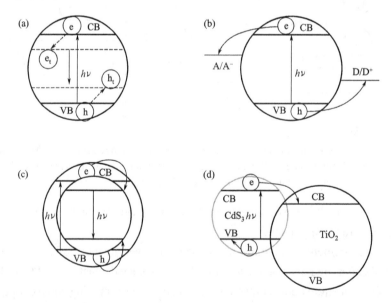

Fig.3.9 Schematic representation of typical photophysical and photochemical processes in semiconductor nanoparticles. The circles represent the nanoparticles, and the horizontal lines are the edge energies of the valence (VB) and conduction (CB) bands[25]

hole relax to the core particle where they are protected from reaction with quenchers in the outer environment. The core-shell heterostructures in Fig.3.9(d) are opposite to Fig.3.9(c) and show an excellent optical property.

3.5.2 Absorption and luminescence spectra

In the absorption process, the electrons are excited across the band gap, from the valence band to the conduction band, by providing an appropriate energy greater than the band gap of the material. These excited electrons must decay back to the valence band by radiative or non-radiative thermal processes.

Non-radiative processes are the predominant routes of de-excitation when the electron-phonon coupling is strong. Phonon coupling provides quasi-continuous states and the electrons can occupy successively to de-excite to the ground state. In the absence of strong phonon coupling, the electrons have to make a radiative transition that often corresponds to energies close to the visible region of the electromagnetic spectrum. This phenomenon, known as fluorescence, can occur in nanocrystals, for example, the electron-hole recombination can take place by involving the electronic states at the band edge, surface, defect and other trap states, or states on doped impurity atoms, giving rise to fluorescence at different energies for the same material.

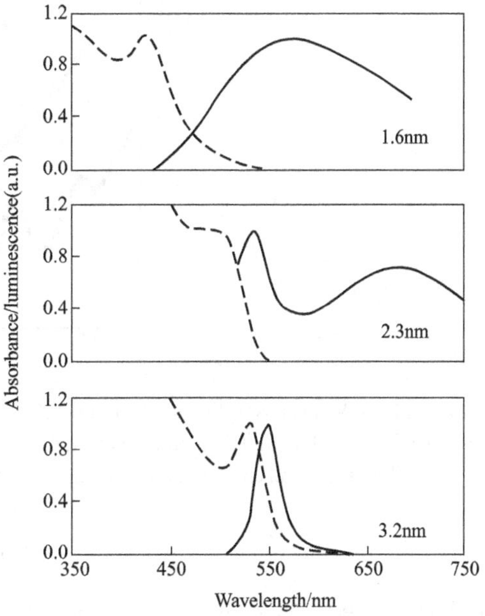

Fig.3.10 Absorption (dotted lines) and luminescence spectra (full lines) of three monodisperse CdSe nano particle dispersions in toluene with diameters [26]

Absorption and luminescence spectra of three monodisperse CdSe nanoparticles are shown in Fig.3.10[26]. It is obvious that the absorption edge and maximum peaks shift to higher energy as the particle size decreases. The luminescence of the 3.2nm particles is dominated by narrow, near-band-edge or shallow trap luminescence, while deep-trap emission dominates in the 1.6nm particles. As the particle size decreases, surface interactions dominate the luminescence properties.

3.5.3 Ultraviolet-visible absorption spectroscopy

Band gaps of all semiconducting and insulating materials increase monotonically with decreasing size of the nanocrystallites.

The increase of band gap is most easily observed in the UV-Vis absorption spectra, that is, the absorption edge shifts systematically to lower wavelengths. The absorption spectra for different sizes of (a) ZnS and (b) CdS nanocrystals were shown in Fig.3.11[27~29]. It is observed an increase of the band gap with decreasing size. One can also clearly find a sharply excitonic peak near the absorption edge, particularly in the spectra from the smaller sized nanocrystallites.

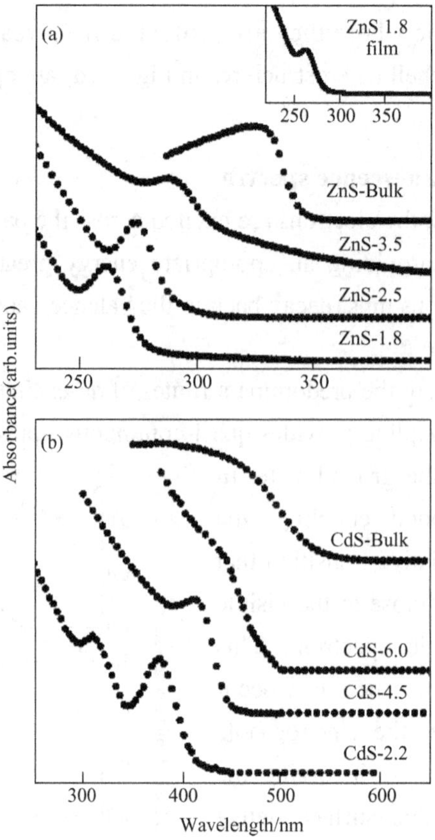

Fig.3.11 UV-Vis absorption spectra of (a) ZnS nanocrystallites of different sizes: 3.5nm, 2.5nm and 1.8nm diameters in dimethyl formamide solution (inset shows the absorption spectrum of 1.8nm ZnS nanocrystalline film) and (b) CdS nanocrystallites of various sizes: 6.0nm, 4.5nm and 2.2nm diameters[27~29]

References

[1] E.O. Hall, Proc. Phys. Soc. London, B 64 (1951) 747.
[2] N.J. Petch, J. Iron Steel Inst., 17 (1953) 25.
[3] R.W. Siegel, Analusis Mag., 24 (1996) M10.
[4] R. W. Siegel, Mater.Sci.Forum, 851 (1997) 235-238.
[5] E.Y. Gutmanas, L.I. Trusov, I. Gotman, Consolidation, Microstructure and Mechanical Properties of Nanocrystalline Metal Powders, Nanostr. Mater., 8 (1994) 893-901.
[6] R.W. Siegel, G.E. Fougere, Nanostruct.Mater., 6 (1995) 205.
[7] H.W. Song, S.R. Guo, Z.Q. Hu, Nanostruct Mater., 11 (1999) 203.
[8] E.O. Hall, Proc. Roy. Soc. London, B64 (1951) 474.
[9] A.H. Chokshi, A.H. Rosen, J. Karch, H. Gleiter, Scripta Met. Mater., 23 (1989) 1679.
[10] G. Palumbo, U. Erb, K.T. Aust, Scripta Met. Mater., 24 (1990) 2347.
[11] N.J. Petch, J. Iron Steel Inst., 174 (1953) 25.
[12] A.M. El-Sherik, U. Erb, G. Palumbo, K.T. Aust, Scripta Met. Mater., 27 (1992) 1185.
[13] U. Erb, Nanostr. Mater., 6 (1995) 533.
[14] R. Kubo, J. Phys. Soc. Jpn., 17 (1962) 975.
[15] J. Rupp, R. Birringer, Physical Review B, 36 (15) (1987) 7888.
[16] K.J. Klabunde, Nanoscale Materials in Chemistry, (2001) John Wiley & Sons, Inc. ISBNs: 0-471-38395-3 (Hardback), 0-471-22062-0 (Electronic), P268.

[17] B.L.S. Komarneni, Chemical processing of ceramics, (2005) Taylor & Francis Group, LLC, ISBN-10: 1-57444-648-7 (Hardback), ISBN-13: 978-1-57444-648-7 (Electronic), P195.
[18] B.L.S. Komarneni, Chemical processing of ceramics, (2005) Taylor & Francis Group, LLC, ISBN-10: 1-57444-648-7 (Hardback), ISBN-13: 978-1-57444-648-7 (Electronic), P196.
[19] V.N.T. Satyanarayana, A.S. Kuchibhatla, Karakoti, D. Bera, S. Seal, One dimensional nanostructured materials, Prog. Mater Sci., 52 (2007) 699-913.
[20] K.K. Licharev, T. Claeson, Sci. Am., 266 (1992) 50.
[21] K.J. Klabunde, Nanoscale Materials in Chemistry, (2001) John Wiley & Sons, Inc. ISBNs: 0-471-38395-3 (Hardback), 0-471-22062-0 (Electronic).
[22] C.B. Murray, C.R. Kagan, M.G. Bawendi, Annu. Rev. Mater. Sci., 30 (2000) 545.
[23] R.P. Andres, T. Bein, M. Dorogi, S. Feng, J.I. Henderson, C.P. Kubiak, W. Mahoney, R.G. Osifchin, R. Reifenberger, Science, 272 (1996) 1323.
[24] S. Ogawa, F.-R. F. Fan, A.J. Bard, J. Phys. Chem., 99 (1995) 11182.
[25] E. Roduner, Nanoscopic Materials Size-dependent Phenomena, (2006) The Royal Society of Chemistry, ISBN-10: 0-85404-857-X (Hardback), ISBN-13: 978-0-85404-857-1 (Electronic), P44.
[26] E. Roduner, Nanoscopic Materials Size-dependent Phenomena, (2006) The Royal Society of Chemistry, ISBN-10: 0-85404-857-X (Hardback), ISBN-13: 978-0-85404-857-1 (Electronic), P67.
[27] J. Nanda, B.A. Kuruvilla, D.D.Sarma, Phys. Rev. B, 59 (1999) 7473.
[28] J. Nanda, D.D. Sarma, J. Appl. Phys., 90 (2001) 2504.
[29] C.N.R. Rao, A. Muller, A.K. Cheetham, The Chemistry of Nanomaterials: Synthesis, Properties and Applications. (2004) Wiley-VCH Verlag GmbH & Co. KGaA, ISBN: 3-527-30686-2, P395.

Review questions

1. What does the positive Hall-Petch slopes mean?
2. What does the negative Hall-Petch slopes mean?
3. Please give a sample to illustrate the positive and negative Hall-Petch slopes.
4. Why is the melting point of Au nanoparticles lower than that of bulk Au?
5. What is the definition of coercive force?
6. What is residual or remnant magnetization?
7. What is Coulomb energy?
8. Please indicate the physical meaning of "staircase" in Fig.3.7. What do the height and the width represent in Fig.3.7, respectively?
9. What are the semiconductors of core-shell heterostructures?
10. The UV-Vis absorption spectra for CdS nanocrystallites of various sizes: 6.0nm, 4.5nm and 2.2nm, which direction do they shift (red shift or blue shift) with decreases of particle size?

Vocabulary

agitation [ædʒɪ'teɪʃ(ə)n] n. 激动；搅动；煽动；烦乱
analogous [ə'næləgəs] adj. 类似的；[生] 同功的；可比拟的
anisotropy [ˌænaɪ'sɒtrəpɪ] n. 各向异性
blockade [blɒ'keɪd] vt. 封锁；n. 阻塞
brittle ['brɪt(ə)l] adj. 易碎的，脆弱的；易生气的
capacitance [kə'pæsɪt(ə)ns] n. 电容；电流容量
colloidal [kə'lɔɪdl] adj. 胶体的；胶质的；胶状的
concentric [kən'sentrɪk] a. 同中心的
cobalt ['kəʊbɔːlt; -ɒlt] n. 钴；钴类颜料；由钴制的深蓝色
coercive [kəʊ'ɜːsɪv] adj. 强制的；胁迫的；高压的
coercive force 矫顽力；抗磁力
Coulomb ['kuːlɒm] n. 库仑（电量单位）
customary ['kʌstəm(ə)rɪ] adj. 习惯的；通常的；n. 习惯法汇编

decay [dɪ'keɪ] n. 衰退，衰减；腐烂，腐朽；vt. 使腐烂，使腐败；使衰退，使衰变
deformation [ˌdiːfɔː'meɪʃ(ə)n] n. 变形
densification [ˌdensifi'keiʃən] n. 密实化；封严；稠化
deviation [diːvɪ'eɪʃ(ə)n] n. 偏差；误差；背离
dipole ['daɪpəʊl] n. 偶极；双极子
dislocation [ˌdɪslə(ʊ)'keɪʃ(ə)n] n. 转位；混乱；脱臼；位错
dynamical [daɪ'næmɪkl] adj. 动力学的
ductile ['dʌktaɪl] adj. 柔软的；易教导的；易延展的
electrodeposit [ɪˌlektrəʊdɪ'pɒzɪt] vt. 电沉积；n. 电沉淀物
elongate ['iːlɒŋgeɪt] vt. 拉长；使延长；使伸长；vi. 拉长；伸长；adj. 伸长的
emission [ɪ'mɪʃ(ə)n] n. （光、热等的）发射，散发；喷射；发行
extrapolate [ɪk'stræpəleɪt; ek-] vt. 外推;推断；vi. 外推；进行推断
fluorescence [flʊə'res(ə)ns; flɔː-] n. 荧光；荧光性
fracture ['fræktʃə] n. 破裂，断裂；骨折；vi. 破裂；折断；vt. 使破裂
heterostructure [ˌhetərəʊ'strʌktʃə] n. [电子] 异质结构
hexagonal [hek'sægənəl] adj. 六边的，六角形的
illustrative ['ɪləstrətɪv; ɪ'lʌst-] adj. 说明的；作例证的；解说的
ingredient [ɪn'griːdɪənt] n. 原料；要素；组成部分；adj. 构成组成部分的
inverse ['ɪnvɜːs; ɪn'vɜːs] n. 相反；倒转；adj. 相反的；倒转的；vt. 使倒转；使颠倒
ligand ['lɪg(ə)nd] n. [化] 配合基
linearity [ˌlɪnɪ'ærətɪ] n. 线性；线性度；直线性
luminescence [ˌluːmɪ'nes(ə)ns] n. [物] 发冷光，荧光
magnetic [mæg'netɪk] adj. 有吸引力的；有磁性的；地磁的
magnetization [ˌmægnətɪ'zeʃən] n. 磁化
micrometric [ˌmaikrəu'metrik] adj. 测微的；测微术的
monodisperse [ˌmɔnəudis'pəːs] adj. [化] 单分散的
nickel ['nɪk(ə)l] n. 镍；镍币；五分镍币；vt. 镀镍于
nucleate ['njuːklɪət] adj. 有核的；vt. 使成核；vi. 成核
optical ['ɒptɪk(ə)l] adj. 光学的；眼睛的，视觉的
optoelectronic [ˌɒptəʊɪˌlek'trɒnɪk] adj. 光电子的
paramagnetic [pærəmæg'netɪk] adj. 顺磁性的；常磁性的；n. 顺磁物质；顺磁体
phonon ['fəʊnɒn] n. [物] 声子
pileup ['paɪlˌʌp] n. 连环相撞；堆积
predominant [prɪ'dɒmɪnənt] adj. 主要的；卓越的；支配的；有力的；有影响的
quadratic [kwɒ'drætɪk] adj. 二次的
quench [kwen(t)ʃ] vt. 熄灭，淬火；解渴；结束；冷浸；vi. 熄灭；平息
refinement [rɪ'faɪnm(ə)nt] n. 精制；文雅；提纯；强化
remnant ['remnənt] n. 剩余；adj. 剩余的
rigorous ['rɪg(ə)rəs] adj. 严格的，严厉的；严密的；严酷的
saturation [sætʃə'reɪʃ(ə)n] n. 饱和；色饱和度；浸透；磁化饱和
segregation [ˌsegrɪ'geɪʃ(ə)n] n. 隔离，分离；种族隔离
slope [sləʊp] n. 斜坡；倾斜；斜率；vi. 倾斜；逃走；vt. 倾斜；使倾斜
spectacular [spek'tækjʊlə] adj. 壮观的，惊人的；公开展示的
staircase ['steəkeɪs] n. 楼梯，阶梯
symmetry ['sɪmɪtrɪ] n. 对称（性）；整齐，匀称
tensile ['tensaɪl] adj. [物] 拉力的；可伸长的；可拉长的
uniaxial [ˌjuːnɪ'æksɪəl] adj. [物] 单轴的
vacancy ['veɪk(ə)nsɪ] n. 空缺；空位；空白；空虚
voltage ['vəʊltɪdʒ; 'vɒltɪdʒ] n. [电] 电压

3. 纳米材料的性能

人工制备的纳米材料有多种存在形式，如纳米粉体、纳米结构颗粒、纳米纤维、纳米管以及薄膜。纳米材料具有许多奇特的性能，包括力学、热学、磁学、电学和光学性能。

3.1 力学性能

众所周知，晶粒细化常常会导致金属或合金性质的变化。例如，在钢中，减小晶粒尺寸

可降低钢材的延展性-脆性的转变温度。引人关注的是晶粒尺寸减小，对材料的力学行为有所影响。经验公式 Hall-Petch 方程[1, 2]，表达了屈服应力σ_y与平均晶粒尺寸 d 之间的关系。

$$\sigma_y = \sigma_0 + kd^{-1/2} \tag{3.1}$$

式中，σ_y 是屈服应力；σ_0 是摩擦应力；k 是一个常数；d 是晶粒直径。而且物质的硬度和晶粒尺寸之间存在相似的关系式。

$$H = H_0 + kd^{-1/2} \tag{3.2}$$

为了解释这些经验观察结果，人们提出了若干个模型，如作为位错源的晶界位错塞积模型及晶界位错网状模型。在所有的例子中，Hall-Petch 效应与材料中位错的产生和移动现象有关。将该规律外推至粒径约 10nm 的晶粒，预计该尺寸的晶粒将具有很高的硬度。例如，纳米相 Cu（$d = 6$nm）的硬度比块体 Cu（$d = 50\mu$m）高五倍多。纳米相 Pd（$d = 7$nm）的屈服应力比相应的块体（$d = 100\mu$m）金属 Pd 高五倍多。

随着晶粒尺寸的减小，材料的硬度增大。然而，具有小的拉伸应力样品的应力-应变曲线表明，其屈服应力几乎没什么变化。最近，一些综述对纳米结构材料中的硬度和屈服应力的力学行为给予了总结[3~5]，证实了当晶粒尺寸减小至纳米范围（<100nm），材料的硬度随粒径的减小而增大。

3.1.1 正的 Hall-Petch 斜率关系

纳米材料的硬度随粒径尺寸的减小而增大，如图 3.1 所示，也就是说，HV 与 $d^{-1/2}$ 成正比关系。

3.1.2 负的 Hall-Petch 斜率关系

如图 3.2 所示的 Ni-P 的曲线，纳米材料的硬度随粒径尺寸的减小而减小。纳米结构中的位错可能不是造成材料热力学稳定性的缺陷。较高的内部压应变和更少的空位浓度是纳米结构的特点。据报道，这些因素导致纳米材料的力学性能发生偏离，形成负的 Hall-Petch 斜率关系[7]。

3.1.3 正-负 Hall-Petch 斜率关系

许多早期的工作涉及在室温下的微硬度测量。图 3.3 给出了在室温下，电沉积 Ni-P 的维氏硬度[8]，以及利用惰性气体冷凝技术制备的纳米 Pd 和 Cu 的维氏硬度[9]。从图 3.3 可知[10]，有些样品的硬度随着粒子尺寸的减小呈先增大后急剧减小的趋势，硬度减小是负的 Hall-Petch 行为[8, 11]。然而，Erb 等人[12, 13]通过大量的电沉积 Ni 的研究发现，当 Ni 的晶粒小于 25nm 时，硬度与晶粒尺寸的线性正相关偏离很大，且当晶粒尺寸在 11～6nm 时，呈负的 Hall-Petch 斜率关系，如图 3.4 所示。

块体材料的性能通常随着纳米成分的加入而发生急剧的变化。由纳米尺寸的陶瓷或金属颗粒（<100nm）制成的复合材料的硬度发生突变，比现有的材料科学模型预测的硬度要大得多。例如，一个粒径约为 10nm 的金属，其硬度和韧性比粒径为几百纳米的颗粒要高 7 倍多。

3.2 热学性能

Kubo[14]在他的理论中预测，低温下金属粒子的热学行为与块体材料有明显的区别，因为金属粒子的平均能隙 δ 接近于费米能级，该能隙 δ 远远大于材料的热能 k_BT（其中 k_B 是玻尔兹曼常数；T 为热力学温度），是由于在限域的体系中，形成了不连续的能级。

一些实验结果表明，粒径减小至 5nm 以下，就会导致 Au 纳米粒子的熔点急剧下降。结果指出金属的熔点降低到约 900K 的原因是晶格点阵的重排及破坏所致。

纳米晶也被称为量子点。许多研究者们研究了纳米晶的比热容。大多数的实验测试表明：与块体材料相比，纳米粒子具有更大的比热容，如图 3.5 所示[15, 16]。

3.3 磁学性能

强磁材料如铁和铁磁合金，是到目前为止最有用的材料，通常以矫顽力 H_c 对磁性材料进行分类。通常，矫顽力 H_c<1000A/m 的材料被认为是软磁性材料，矫顽力 H_c 在 1～30kA/m 之间的材料被认为是半硬性磁性材料，而矫顽力 H_c >30kA/m 的材料被认为是硬磁性材料。

由于它们弱的各向异性，Fe、Co、Ni 基的磁性材料表现出软磁材料的行为。小的矫顽力和磁性随着施加的磁场而变化的性质由畴壁的移动过程所控制。在稀土-过渡金属化合物中，存在 3d-4f 轨道的强相互作用。这就导致了稀土原子中的 5d 电子与过渡金属中 3d 电子发生杂化。通过这种机理，在高温下稀土元素仍具有磁性。在磁晶各向异性的单轴系统中，将会有很强的矫顽力。这种硬磁材料可用作高性能的永磁性材料。最熟知的是六方对称性的 $SmCo_5$ 化合物和二重对称性的 $Nd_2Fe_{14}B$。

当把磁性材料放置于磁场中时，其被磁化的程度会随施加的磁场强度而变化，如图 3.6 所示[17]。当将块体或粉体的磁性固体置于磁场中，它们就会被磁化，因为块体或粉末的磁畴中的磁偶极子的初始方向是随机分布的，而磁性固体的磁化过程就是使磁性偶极子发生定向运动。当磁场强度增加时，偶极子的定向程度越高，在磁畴的方向完全一致时，就达到了饱和磁化。即使当磁场强度降低到零时，材料仍具有一定的磁化强度。这就是所谓的剩余或残余磁化强度。当施加一个反向的磁场，可使剩余磁化强度消失为零。这个反向磁场称为矫顽力，常以 H_c 表示。矫顽力对磁记录媒介至关重要。

随着粒子尺寸的减小，粒子中的磁畴数就会减少。图 3.7 是粒子的磁畴结构示意图[18]。一般地，多畴粒子的矫顽力小于单畴粒子的矫顽力。单畴粒子的矫顽力是由所谓的磁晶各向异性与形状各向异性共同决定的。因此，细长的单畴粒子优先应用于磁记录材料。具有很高的饱和磁化强度、适中矫顽力的单畴粒子适合作为磁记录材料。但是，当粒子变得足够小，由于热扰动，磁矩不可能有择优取向，而展示出超顺磁性，如图 3.7（a）所示。

3.4 电学性能

纳米结构独特的电学性能是隧穿电流、完全的弹道传输和库仑阻塞效应的结果[19]。

那些被称为“量子点”或“人造原子”的纳米粒子具有在很多方面新奇的性能。其中最惊人的是与所谓的单电子隧道（SET）相关的性质。按照 Licharev 的预测，这意味着电子在粒子之间的传输方式是一个接着一个地进行[20]。

块体金属中，电子的聚集定向运动（电流 I）是由施加的电压（U）所产生的，欧姆定律 $U=IR$ 描述了电压 U 和电流 I 之间的线性关系（R 是材料的电阻）。

以上描述的所有关系均因能带结构的存在，也就是说，可自由移动的电子的存在。如果一个金属粒子的尺寸变得足够小时，能带结构开始变化，最终离散能级占主导地位。其结果，欧姆定律不再有效。若以零维的金属粒子为例，那么它的电流-电压会如何变化呢？

为了回答这个问题，必须研究单个纳米粒子的 I-U 特征，这是一个大的挑战！原理上，将单个粒子放置在两个可调变电压（可增大或减小的电压）的电极之间。为了观测到电子传输的细节，应该避免直接将两个电极与一个金属粒子直接相连。取而代之的是在两电极之间使用两个电容器（绝缘材料涂在粒子表面）和一个粒子，图 3.8（a）给出了实验条件。

当一个电子传输到粒子上时，粒子的库仑能 E_C 增大到 $E_C = e^2/(2C)$，其中 C 指粒子的电容。粒子中原子的热运动可促使电荷和库仑能的变化，以至于电子将进一步可自由地隧穿。为了产生单电子隧穿过程（SET），粒子的热能必须远远小于库仑能，为了保证粒子的能量 $kT \ll e^2/(2C)$，当增加一个电荷时，产生电压 $U = e/C$，导致隧穿电流 $I = U/R_t = e/(R_tC)$(R_t 为隧道电阻)[21]。

一个理想量子点的电流-电压特征是，当 $U_{Coulomb} = \pm e/(2C)$ (库仑阻塞) 时才有电流产生。若达到该电压值，电子就能从电极传输到粒子。如果量子点的库仑阻塞能通过一个外加电压 $U_{Coulomb} = \pm ne/(2C)$ 补偿时，电子隧穿过程就会发生。由于单个电子的多次隧穿，一个理想的“阶梯”就会产生，如图 3.8（b）所示。一个阶梯的高度 ΔI 相当于 $e/(RC)$，宽度 ΔU 相当于 $e/(2C)$。通过一个量子化方式，一个量子点的充电或放电成为可能，这种粒子的使用是未来新一代计算机的基础。

在单电子隧穿器件中，观察到了小于 10nm 的 Al 纳米粒子在低温（<1K）下的能级量子化[22, 23]现象。

通过把被称为单层保护分子簇（MPCs）的配合物涂在粒子表面，进行单个金属纳米粒子的电子测量。该单层保护分子簇位于金属基质上并与一个扫描隧道显微镜（STM）的针尖相连接[24]。STM 针尖和金属基质分别作为电触头（两电极），而包裹着纳米粒子的配体单分子层作为电容层。这就创造出一个双隧道结，可观测到阶梯状的伏安（I-U）曲线。

3.5 光学性能

当粒子尺寸接近块体材料的激子玻尔半径时，半导体纳米晶就会出现三维量子限域效应。这个限域引起了一些令人关注的新效应，从而产生新的光学性能。半导体微晶的这些性能在光电子器件中应用潜力巨大，如：光学数据存储和高速光学通信。

3.5.1 纳米材料的光化学和光物理过程

半导体纳米粒子的光化学和光物理过程与那些众所周知的分子非常类似。习惯上通过是在一个代表球形粒子的圆圈里画一个价带和导带，如图 3.9 所示[25]。

当一个半导体纳米粒子吸收一个光子的能量后，价带上的一个电子被激发跃迁到导带，价带上留下一个空穴，电子（e）和空穴（h）弛豫到表面上的诱捕态 e_t 和 h_t [图 3.9（a）]。另外，如果粒子是与电子接受体分子 A 相接触，电子传输给 A, A 形成 A^-，当颗粒与空穴（h）的接受体分子 D（电子施主）接触时，空穴（h）从颗粒传输给 D 而形成 D^+ [图 3.9（b）]。

由两种或多种不同能级的半导体组成的同中心核-壳结构的异质结引起了人们的兴趣。在图 3.9（c）给出的例子中，壳材料具有较大的能隙，在这种情况下，当壳层电子被激发，电子和空穴弛豫到核粒子上，核粒子上的电子和空穴被壳保护，防止与外界环境中的猝灭剂发生反应。图 3.9（d）中的核-壳异质结和图 3.9（c）中相反，显示出优异的光学性能。

3.5.2 吸收光谱和发光光谱

在吸光过程中，当吸收的能量大于材料带隙能量时，电子被激发而穿越带隙，从价带跃迁至导带。这些激发的电子通过辐射或非辐射的放热过程，衰减返回到价带。

当电子-声子耦合很强时，非辐射过程是主要的衰减途径。与声子结合时可释放出准连续态的能量，电子就可以连续地去除激发回到基态。当与能量高的声子耦合时，电子不得不发生辐射传输，该传输的能量接近电磁光谱可见光区的能量，纳米晶出现荧光现象。例如，电子-空穴的复合可发生在带边、表面、缺陷和其它诱捕的电子态中，或掺杂在其它原子的能态中，对于同种材料可产生不同能量的荧光。

图 3.10[26]所示的是三种单分散 CdSe 纳米粒子的吸收光谱和发射光谱图。从图中可见，随着粒子的尺寸减小，其吸收边和最大吸收峰均移向更高的能量。尺寸为 3.2nm 的粒子的发射光谱以窄的、近带边或浅阱的发光光谱为主，而尺寸为 1.6nm 的粒子则以深阱发光光谱为主。当粒子尺寸减小时，表面的相互作用主导着发光光谱的性质。

3.5.3 紫外-可见吸收光谱

所有的半导体和绝缘体材料的带隙随纳米晶尺寸的减小单调性地增加。

然而，带隙增大在紫外-可见吸收光谱上最容易被观察到，那就是吸收边系统性地向较短波长方向位移。图 3.11 的两个例子分别为不同尺寸的 ZnS（a）和 CdS（b）纳米晶的吸收光谱[27~29]。它们的带隙均随粒子尺寸的减小而增大。在吸收边附近可以清楚地发现一个尖锐的激子峰，尤其是更小尺寸的纳米晶。

复 习 题

1. 什么是正的 Hall-Petch 斜率关系？
2. 什么是负的 Hall-Petch 斜率关系？
3. 请举例说明正的和负的 Hall-Petch 斜率关系。
4. 为什么纳米 Au 粒子的熔点小于块体 Au 的熔点？
5. 什么是矫顽力？
6. 什么是剩余的或残余的磁化强度？
7. 什么是库仑能？
8. 请说明图 3.7 中“阶梯”的物理含义，图 3.7 中“阶梯”的阶跃高度和宽度分别指什么？
9. 什么是半导体的核-壳异质结？
10. 观察纳米晶 CdS 的不同尺寸（6.0nm、4.5nm 和 2.2nm）的紫外-可见漫反射光谱，随着粒径的减小，它们的光谱移向何方（红移或蓝移）？

4. Synthesis of Nanoscale Materials

During the past decades, scientists have developed techniques for synthesizing and characterizing of many new materials with at least one dimension on the nanoscale, including nanoparticles, nanolayers, and nanotubes. The design and fabrication of nanoscale materials with controlled properties are still a significant and ongoing challenge within nanoscience and nanotechnology. The most applicable fabrication procedures include physical techniques such as arc discharge, heat evaporation, laser/electron beam heating, sputtering, and chemical techniques such as chemical vapor deposition, solid-state reactions, sol-gel, precipitation, hydrothermal and etc.

Feymann[1] said in one of his famous lectures "There is plenty of room at the bottom" in which he referred to the possibility that computing advances may not have to stop with the gradually diminishing stature of Moore's Law for silicon technologies. The international technology roadmap has projected that silicon technology could easily continue scaling down at least until the middle of the next decade. To continue beyond this point, we have to think of alternative technologies. Already we have been seeing some alternatives in the form of carbon nanotube devices and molecular switches. These provide novel challenges to the technologists who physically try to make them. These new technologies not only bring forth new physical challenges, but also imply system level challenges.

In a way, silicon technologies are already at the threshold of nano-technological wonders. In recent years, most semiconductor companies have manufactured devices at the 90nm scale and some going to 65 nanometers. Some companies have even figured out how to scale below these numbers. As we scale down in size, we also scale up in number. There are very wide ways to fabricate nanomaterials. It is important to understand some features of the processes used in the synthesis of nanomaterials because the processing route often dominates the behavior of any given material.

4.1 "Top-down" and "bottom-up" approaches

Productions of new nanomaterials are achieved by either "top-down" or "bottom-up" approaches.

The "top-down" approach involves that the bulk materials are fabricated to nanometer size range by employing processes such as grinding, milling, etching and lithography. Therefore, the "top-down" approach involves size reduction by the application of a force. The degree of control and refinement in size reduction processes directly influences the properties of the materials produced[2]. The commercial scale productions of nanomaterials today primarily involve the "top-down" approaches.

In the "bottom-up" approaches, atoms, molecules and even nanoparticles themselves can be used as the building blocks for the creation of complex nanostructures by altering the size of the building blocks, controlling their surface and internal structure, and then controlling their organization and assembly, whereas these processes are essentially controlled in complex chemical synthesis.

In the overall "bottom-up" R&D roadmap, two general pathways can be defined. The first is a hard-tech approach[3] which pertains to the massive and complex equipments to induce manipulations of atom-by-atom or molecule-by-molecule, and then organize relatively simple building blocks into applicable nanostructural devices. The second is the soft-tech approach, which includes the design of complex building blocks via self-assembly by performing manipulations on a macroscopic scale[4]. This approach uses relatively inexpensive and commonly accessible equipment that might be contrary to the hard-tech approach[5]. As shown in Fig.4.1 in terms of scale, the "top-down" method is opposite to "bottom-up" method, and biological processes are essentially intermediate between "top-down" and "bottom-up" processes.

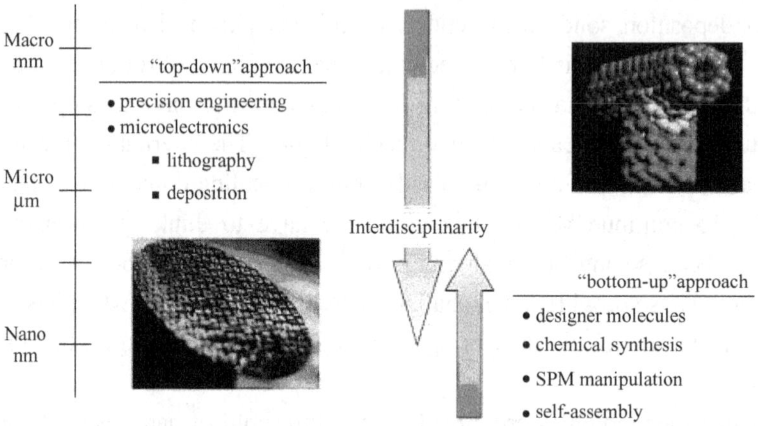

Fig.4.1 Schematic representation of the "top-down" and "bottom-up" approaches

In this chapter, the synthesis methods mainly include the solid phase method, physical vapor deposition method (PVD), chemical vapor deposition (CVD), liquid phase synthesis method, synthesis of nanostructural materials by consolidation of nanopowder, template-assisted self-assembly nanostructured materials and toward greener nano synthesis.

4.2 Solid phase method

4.2.1 Mechanically milling

Mechanically milling method actually belongs to solid-state processing of materials and is an example of "top-down" approach. One nanofabrication process of major industrial importance is high energy ball milling, also known as mechanical attrition or mechanical alloy. Several different mills are schematically shown in Fig.4.2[6].

Attrition mill is also known as the attritor or stirred ball mill. The balls are set in motion by rotation of the central shaft on which secondary arms are fixed, and the cylinder itself is fixed. The coarse grained materials in the form of powder are crushed mechanically in rotating drums by hard steel or tungsten carbide balls, usually under controlled atmospheric conditions to prevent unwanted reactions such as oxidation. This repeated deformation could largely cause reductions in grain size via the formation and organization of grain boundaries within the powder particles.

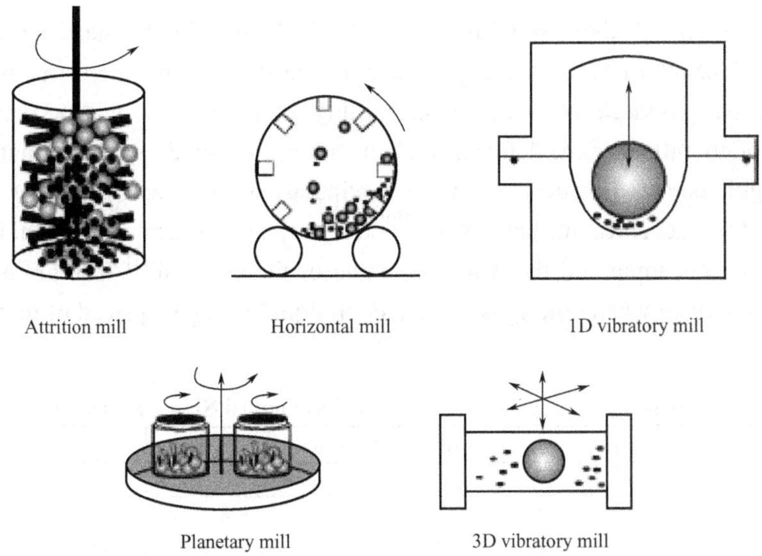

Fig. 4.2 Different types of ball mill: attrition mill, horizontal mill, planetary mill, the 1D and 3D vibratory mills [6]

Horizontal mill means the cylinder rotates around its horizontal axis. The combined effects of the centrifugal force induced by this rotation and gravity cause the balls to rise and fall onto the powder particles.

1D vibratory mill is also known as a shaker mill. The vessel is set in vertical oscillatory motion. Under this action, the 1kg ball rises then falls back onto the powder particles.

Planetary mill is that the containers are fixed on a table that rotates. The containers themselves also rotate in modern mills, this rotation being either coupled or uncoupled with respect to the rotation of the table.

3D vibratory mill is also known as a three-axis shaker. These operate according to the same principle of 1D vibratory mill, but this time in a more complex way due to the 3 vibrational degrees of freedom. The balls collide with the sidewalls of the container (friction and impacts), but also with its floor and ceiling.

High-energy milling forces can be obtained by using high frequencies and small amplitudes of vibration. High energy ball mills, in which the balls are in permanent relative motion, differ from the mills traditionally used in industry to grind materials in either dry or wet conditions. Low energy mills can be used to fabricate certain materials by mechanosynthesis, but the milling time then becomes excessively long, from a few hundred to a thousand hours, which significantly reduces their industrial potential, compared with process times of a few hours or a few tens of hours for high energy mills. However, the production capacity of high energy mills needs to be adapted to industrial requirements.

It is noteworthy that contamination from the milling tools (Fe) and atmosphere (trace elements of O_2, N_2 in rare gas) have become major problems causing surface and interface contamination fabricated materials during mechanical attrition. Mechanical attrition and mechanical alloying have been used to commercial manufactures on large scale[7, 8].

Mechanical alloying is a high-energy ball-milling process in which elemental blended powders are continuously welded and fractured to achieve alloying at the atomic level[9]. The influencing factors of the mechanical alloying/milling processes include milling time, charge ratio, milling environment, and the internal mechanics specific. A recent study of the microstructural evolution of nanocrystalline Ni produced by mechanical milling indicates that milling environment and temperature strongly influence the deformation behavior of the powders. The formation of irregular flake-shaped agglomerates is attributed to the continuous welding and fracturing of the powder particles during the mechanical milling process[10]. The rate of structural refinement is dependent on the mechanical energy input and the work hardening of the material[11]. The grain size of Ni has been sufficiently refined when milling is conducted in liquid nitrogen instead of methanol, as listed in Table 4.1[12].

Table 4.1 Characteristic powders properties of mechanical milled Ni in different milling conditions[12]

Milling environment	Milling time /h	Agglomerate size/μm			Grain size/nm	Aspect ratio
		d^*_{10}	d^*_{50}	d^*_{90}		
Methanol	5	27	80	144	90 ± 65	1.67
Methanol	10	15	35	77	82 ± 25	2.77
Liquid nitrogen	10	20	43	84	19 ± 11	1.47

Note：d^*_{10}, d^*_{50}, d^*_{90} represent the particle size corresponding to 10wt%, 50wt%, and 90wt% of the cumulative particle size distribution.

4.2.2 Solid-state reaction[13]

Solid-state reactions offer the possibility of generating nanoparticles by controlled phase transformations or reactions of solid materials. The advantage of this technology is its simple production route. The solid-state reaction technique is a convenient, inexpensive, and effective preparation method for monodisperse oxide nanoparticles in high yield.

This is the most traditional approach for manufacturing many ceramic powders such as $BaTiO_3$. It involves milling $BaCO_3$ with TiO_2 to achieve comminution and mixing. This mixture is then calcined at 900～1200℃ to generate $BaTiO_3$. The reaction mechanism has been described as following[14~16].

① Interface reaction by diffusion of Ba into TiO_2 to form a $BaTiO_3$ shell (layer) around TiO_2; the shell also limits Ba diffusion into TiO_2 further.

$$BaCO_3 + TiO_2 \longrightarrow BaTiO_3 + CO_2$$

② Formation of the orthotitanate phase.

$$BaCO_3 + BaTiO_3 \longrightarrow Ba_2TiO_4 + CO_2$$

$$2BaCO_3 + TiO_2 \longrightarrow Ba_2TiO_4 + 2CO_2 \text{ (less likely)}$$

③ Final formation of $BaTiO_3$.

$$Ba_2TiO_4 + TiO_2 \longrightarrow 2BaTiO_3$$

The high calcination temperature is needed to promote the interdiffusion of barium and titanium, which results in the solid-state reaction to form barium titanate, as shown in Fig.4.3[17].

Sintering can be defined as the consolidation of a dispersed material under the action of heat. If part of a material reaches its melting point during this densification, it is called liquid-phase sintering; otherwise, it is called solid-state sintering. Moreover, if the material is chemically synthesized during sintering, this is called reactive sintering.

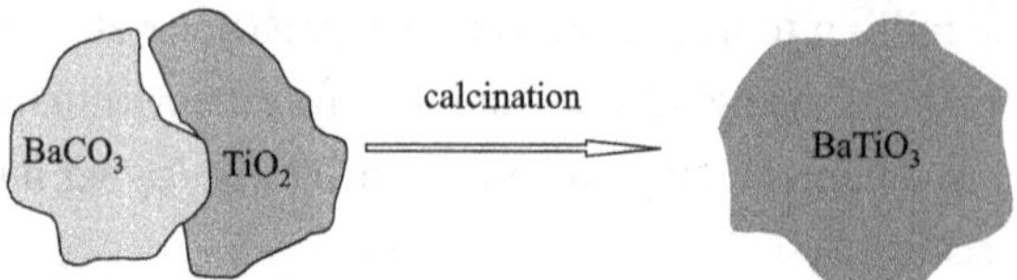

Fig.4.3 Solid-state reaction through interdiffusion to form $BaTiO_3$ during calcinations[17]

4.3 Physical vapor deposition (PVD) method

Physical vapor deposition(PVD) method involves the conversion of solid material into a gaseous phase by physical processes, and then the material is cooled and redeposited on a substrate with perhaps some modification, such as reaction with a gas. Examples of PVD conversion processes include thermal evaporation, laser ablation or pulsed laser deposition (where a short nanosecond pulse from a laser is focused onto the surface of a bulk target) and sputtering (the removal of a target material by bombardment with atoms or ions). PVD can be used to fabrication thin films, multilayers, nanotubes, nanofilaments or nanometer-size particles.

4.3.1 Thermal evaporation PVD method

(1) Vapor phase expansion from an oven

One example of a PVD technique is vapour phase expansion, which relies on the expansion of a high pressure vapor phase through a jet into a low pressure ambient background to produce supersaturation of the vapor. Flow rates can approach supersonic speeds and the process can lead to the nucleation of extremely small clusters, ranging from a few atoms upwards, which can be analyzed using mass spectrometry. Although the quantities of such clusters are low, these can be of great use for studying the physics of small metallic clusters, which often are composed of magic numbers of atoms due to their very stable geometric and electronic configurations. A schematic diagram of such a system is shown in Fig.4.4[18].

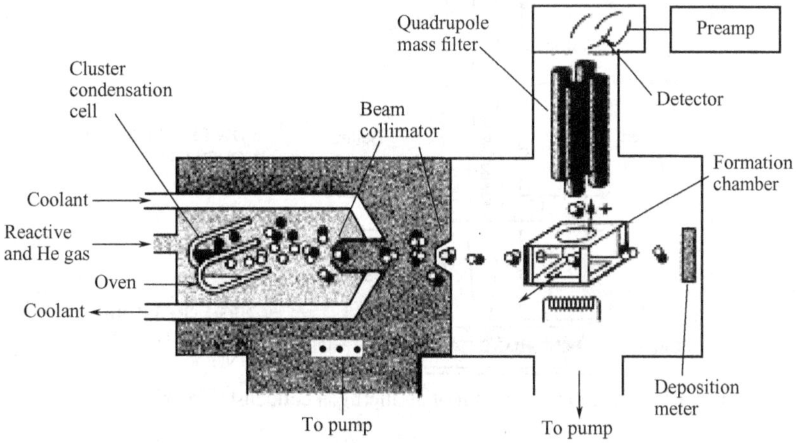

Fig.4.4 Schematic diagram of cluster formation via vapour phase expansion from an oven source[18]

(2) Inert gas evaporation-condensation technique

Another PVD process is the inert gas evaporation-condensation technique, in which nanoparticles are formed via the evaporation of a metallic source in an inert gas and then condensation. Although the technique is old, the application to the production of truly nanoscaled powders is relatively recent. Here a material, often a metal material, is evaporated from a temperature-controlled crucible into inert gas environment of a low pressure [ultra-high vacuum (UHV) chamber]. The metal vapour cools through collisions with the inert gas species becomes supersaturated and then nucleates homogenously. The particle size is usually in the range 1～100nm and can be controlled by varying the inert gas pressure. Particles can be collected on a cold finger cooled by liquid nitrogen, scraped off and compacted to produce a dense nanomaterial. Fig.4.5 shows a typical experimental apparatus for the production of nanopowders by the inert gas evaporation-condensation technique. The method consists of evaporating a metallic source by resistive heating (radio frequency heating, an electron or laser beam as the heating source) inside a chamber which has been previously evacuated to about 10^{-7}Torr and backfilled with inert gas to a low pressure. The vapor metal atoms migrate from the hot source to the cooler, and collide with inert gas in the chamber losing their kinetic energy. Finally, particles deposit on the surface of a cooler. Most applications of the inert gas condensation technique utilize the liquid nitrogen as cooling source to enhance the deposition efficiency.

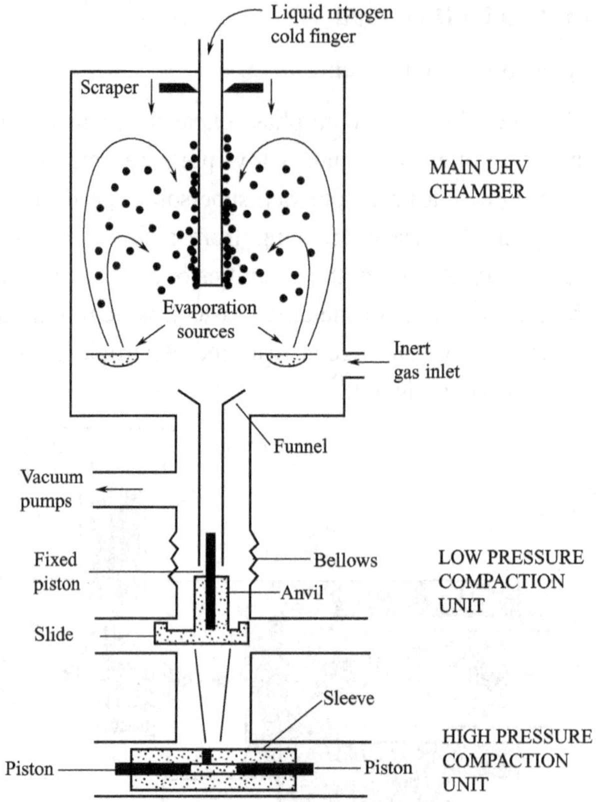

Fig.4.5 Schematic diagram of an inert gas condensation apparatus[18]

(3) Thermal evaporation of solid powders

Belt-like quasi-one-dimensional nanostructures (called nanobelts) of semiconducting oxides have been synthesized by simply evaporating the desired commercial metal oxide powders at high temperatures. The as-synthesized oxide nanobelts are pure, structurally uniform, single-crystalline, and mostly free from dislocations.

Several nanostructures of ZnO are synthesized by a solid state thermal sublimation process under certain experimental conditions, such as by controlling of the growth kinetics, local growth temperature, and the chemical composition of the source materials. Fig.4.6 shows the nanostructures of ZnO such as nanohelixes, nanobows, nanopropellers, nanowires, and nanocages[19].

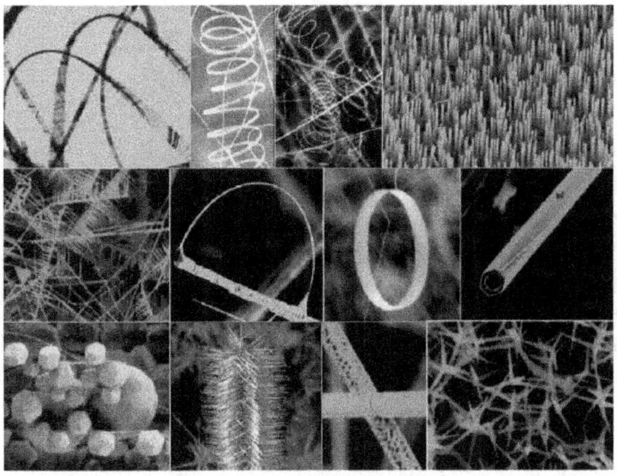

Fig.4.6 Nanostructures of ZnO synthesized under controlled conditions by thermal evaporation of solid powders[19]

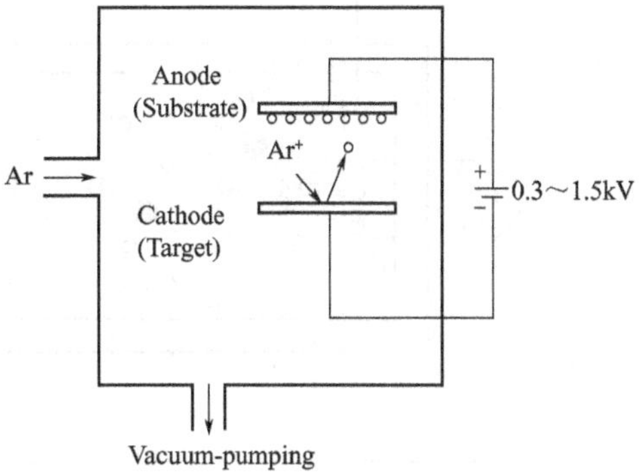

Fig. 4.7 DC cathode sputtering. The glow discharge between the anode and the cathode ionized the argon atoms to bombard the target and eject surface atoms

4.3.2 Plasma-assisted PVD method

There are several different types of plasma-assisted PVD methods described as follows.

(1) Sputtering technique of direct current (DC) glow discharge

The simplest sputtering process is cathode sputtering as shown schematically in Fig. 4.7. A plasma is produced in a chamber, in which an inert gas (usually argon Ar) at low pressure (1.5～15Pa) is introduced by applied a direct current (DC) of voltages (0.3～1.5kV) between the cathode and the anode.

The glow discharge between the cathode and the anode produces a plasma which is positive Ar^+ ions with high energy (> 30eV). This plasma (Ar^+) is accelerated by the electric field and crosses the chamber to bombard target material, so that the atoms of the target materials eject from the surface to sputter on the surface of substrate (anode). The cooling and the condensation of these atoms in inert gases result in the formation of nanoparticles, or thin films on the substrate. However, only a very small fraction of the gas atoms are ionized (～0.01%). The ion flux bombarding the target is therefore rather small with deposition rates at the substrate likewise (～0.1μm/h). The advantages of sputtering as follows: no crucible is needed; particles of metals with a high melting point can be prepared; particles of alloys can be synthesized using active gases; granular thin films can be prepared; the narrow size distribution of the particles can be controlled well; nanocomposite materials can be produced if several different materials are used as targets and etc. The voltage, current, gas pressure, and target are the most important factors which affect the formation of nanoparticles.

Sputtering utilizes a mechanical activation instead of a thermal process to dissociate particles from a source (the target)[20].

DC glow discharge also involves the ionization of gas atoms by electrons emitted from a heated filament as shown in Fig.4.8. The gas ions in the plasma are then accelerated to produce a directed ion beam. If the gas is a reactive precursor gas, this ion beam is used to deposit directly onto a substrate. Alternatively an inert gas may be used and the ion beam allowed striking a target material that sputters neutral atoms onto a neighbor substrate as shown in Fig.4.8[18].

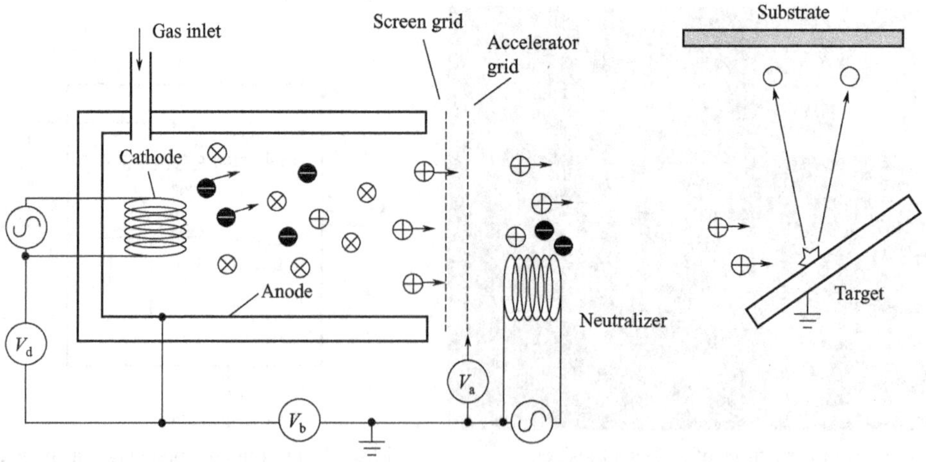

Fig.4.8 Schematic diagram of a DC glow discharge apparatus in which gas atoms are ionized by an electron filament and either deposit on a substrate or cause sputtering of a target[18]

(2) Magnetron sputtering

Magnetron sputtering involves the creation of plasma by the application of a large DC potential between two parallel plates in Fig.4.9[18]. A static magnetic field is applied near a sputtering target for ionization inert gas to produce positive plasma. This plasma is accelerated by the magnetic field to bombard target material so that the neutral atoms are sputtered from the target to deposit on the surface of a substrate. A further benefit of the magnetic field is that it prevents secondary electrons produced by the target from impinging on the substrate and causing heating or damage. The deposition rates produced by magnetron are high enough (about 1 μm/min) to be industrially viable. Multiple targets can be rotated so as to produce a multilayered coating on the substrate.

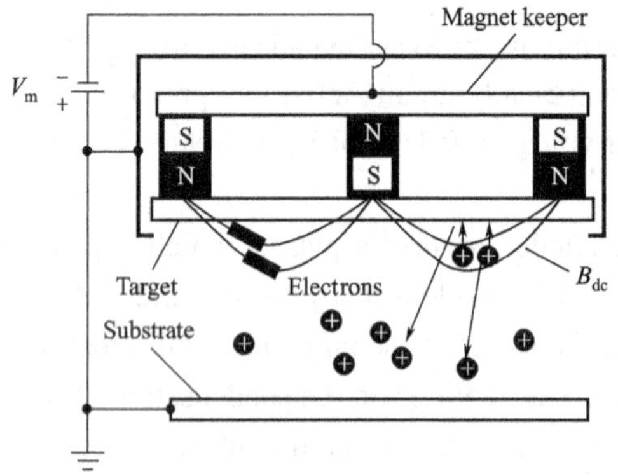

Fig.4.9 Schematic diagram of a magnetron sputtering apparatus that uses a magnetic field to contain ionized gas molecules close to a sputtering target, giving large deposition rates[18]

Recently, Anpo et al. [21] have succeeded in the preparation of visible light-responsive TiO_2 thin films for the separate evolution H_2 and O_2 from water under visible light irradiation by applying a RF (radio frequency)-magnetron sputtering method as shown in the schematic diagram in Fig.4.10. The system is equipped with a substrate (quartz or Ti foil) connecting with heater. The target-to-substrate distance was fixed at the value in the region of 70～90mm. The TiO_2 target is bombard by Ar^+ plasma generated by a RF power of 300W in Ar atmosphere and the neutral atoms are sputtered from the target to deposit on the surface of a substrate. The substrate temperature was held at 200～600℃. Pt deposition was also performed by the RF-magnetron sputtering method under a substrate temperature of 25℃.

(3) Vacuum arc deposition

Vacuum arc deposition involves a cathode constructed from the target material and an anode attached to an igniter. An arc which is self-sustaining with low-voltage and high-current was generated

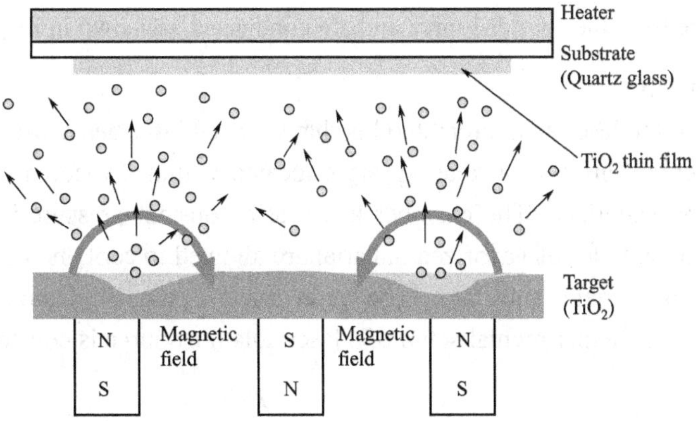

Fig.4.10 Schematic diagram of RF-magnetron sputtering method[21]

by contacting a cathode with igniter.

The arc ejects predominantly ions and micrometer-sized droplets from a small area on the cathode. The ions in the arc are accelerated towards a substrate and deflected using a magnetic field if desired. Any large particles are filtered out before deposition as shown in Fig.4.11[18]. A vacuum cathodic arc can operate without a background gas under high-vacuum conditions. However, the deposition can be achieved by introducing a background gas such as nitrogen. The high ion energy produces dense films even at low substrate temperatures and consequently arc technology is commonly used for the deposition of hard coatings. As well as DC sources, plasmas can also be produced using radio frequency and microwave power. These methods have the advantage of being able to provide higher current densities and higher deposition rates.

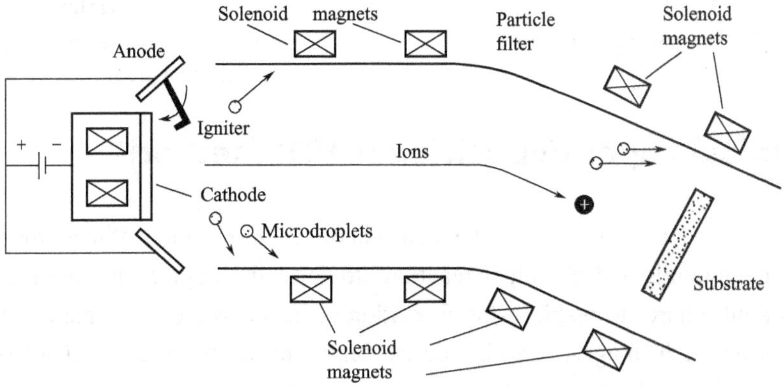

Fig.4.11 Schematic diagram of a vacuum arc deposition apparatus[18]

(4) Plasma sputtering by using plasma gun

In the plasma sputtering process a metal target material is bombarded with high-energy ions of an inert gas such as argon (Ar) or krypton (Kr) ionized. The source of high-energy ions is plasma, which can be generated by direct current (DC) or radio frequency power sources (plasma gun). The high-energy ions created in the plasma are accelerated to the target and bombard it causing the ejection of predominantly neutral atoms and a small fraction of ions from the target surface[22].

Plasma is ionized gases consisting of positive ions, negative ions, electrons and neutral species produced by direct current discharges of plasma gun.

The ejected materials are also of high energy and can be collected on a cooled substrate where their energy is not sufficient to form a thin film. The ejected particles can also be swept away by the inert gas atmosphere from the sputter source and are condensed as shown in Fig.4.12.

4.3.3 Laser ablation

The principle of the laser ablation method is that a pulsed laser beam with very high energy is focused via an external lens onto a rotating target located within a vacuum chamber causing the target material to be vaporized. The chamber keeps at a constant pressure in the atmosphere of helium gas. The nanoparticles of vaporized material are allowed to cool and condense before being collected on an appropriate substrate. The laser-driven process demonstrates to be zero contamination. The basic experimental setup of a laser ablation source is constructed by Yoshida et al. as shown in Fig.4.13[23].

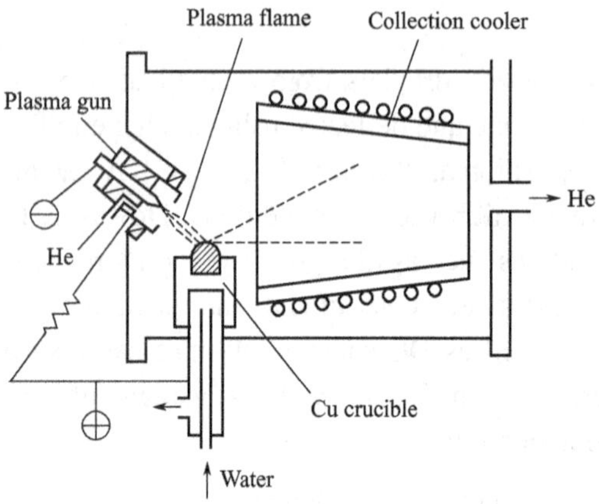

Fig.4.12 A typical plasma sputtering process

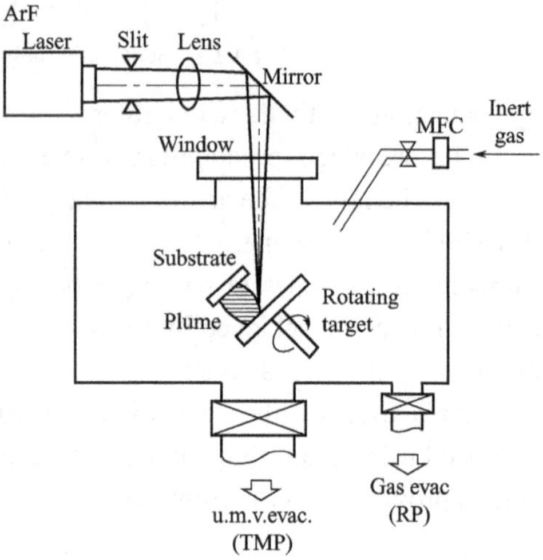

Fig.4.13 General setup for laser ablation and vaporization production of nanoparticles[23]

4.4 Chemical vapor deposition (CVD) method

The CVD method is a process of chemical vapor reaction which utilizes the evaporation of volatile metal compounds and chemical reactions to form the compounds desired and the rapid cooling in gas atmospheres to result in the formation of nanoparticles. The nanoparticles produced by CVD have several advantages, like high purity, high uniformity, small particle size, narrow size distribution, good dispersal, and high chemical reaction activity.

This mature method can deposit single layer, multilayer, composite, nanostructured, functional coating materials, polycrystalline and amorphous materials at a relatively low processing temperature and general vacuum. The CVD process involves various chemical reactions of gaseous compounds in an activated environment (with heat, light, or plasma), followed by the deposition of a stable solid product. Because of the versatile nature of CVD, various types of chemical reactions are involved. Gas phase (homogeneous) reactions and surface (heterogeneous) reactions are intricately mixed. Gas phase reactions become progressively important with increasing temperature and partial pressure of the reactants. An extreme high concentration of reactants will make gas phase reactions predominantly, leading to homogeneous nucleation. The wide variety of chemical reactions can be grouped into the following types: pyrolysis, reduction, oxidation, compound formation, disproportionation, and reversible transfer, depending on the precursors used and the deposition conditions applied. Examples of the above chemical reactions are given below.

The vapor decomposition method is a so-called single compound heat decomposition method. The compound is heated to evaporate and decompose, as in the following reaction:

$$A(g) \longrightarrow B(s) + C(g)$$

① Pyrolysis or thermal decomposition:

$$SiH_4\ (g) \longrightarrow Si\ (s) + 2H_2\ (g) \quad at\ 650℃$$

$$Ni(CO)_4\ (g) \longrightarrow Ni\ (s) + 4CO\ (g)\ at\ 180℃$$

Here shows a sample of spray pyrolysis.

A spray pyrolysis process is also called aerosol pyrolysis[24] or vapor pyrolysis. It consists of three parts: aerosol generation, particle formation inside a furnace, and particle collection, as shown in Fig.4.14. A homogeneous liquid solution (organic, aqueous, or a mixture) containing the reactants is converted into an aerosol using air, nitrogen gas, or a mixture of gases. The sizes of the droplets formed in the aerosol depend on the types of aerosol generators used, the carrying gas volume, and the nature of the liquid[25, 26]. An ultrasonic generator usually gives very fine droplets that may be suitable for making nanoparticles[24]. The aerosol is then carried by the gas into a preheated furnace, where the liquid droplets go through solvent evaporation, solute precipitation, solute decomposition, and oxidation sintering to final particles. Eventually, the particles are carried out by gases cooling as shown in Fig.4.14[24, 27].

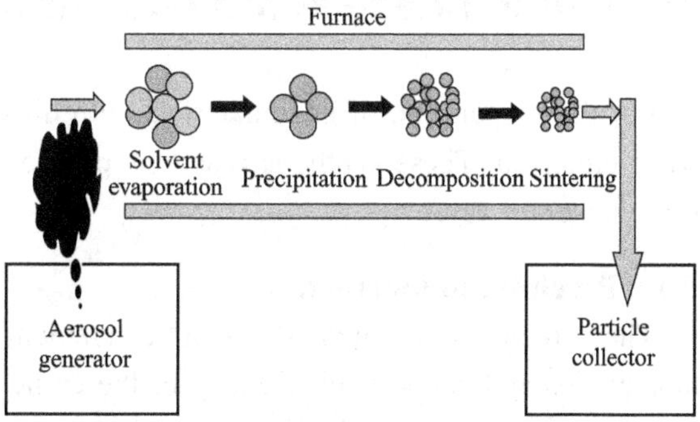

Fig.4.14 Schematic representation of an aerosol pyrolysis process[24, 27]

② Reduction:

$$SiCl_4\ (g) + 2H_2\ (g) \longrightarrow Si\ (s) + 4HCl\ (g)\ at\ 1200℃$$

$$WF_6\ (g) + 3H_2\ (g) \longrightarrow W\ (s) + 6HF\ (g)\ at\ 300℃$$

③ Oxidation:

$$SiH_4\ (g) + O_2\ (g) \longrightarrow SiO_2\ (s) + 2H_2\ (g)\ at\ 450℃$$

$$4PH_3\ (g) + 5O_2\ (g) \longrightarrow 2P_2O_5\ (s) + 6H_2\ (g)\ at\ 450℃$$

④ Compound formation:

$$SiCl_4\ (g) + CH_4\ (g) \longrightarrow SiC\ (s) + 4HCl\ (g)\ at\ 1400℃$$

$$TiCl_4\ (g) + CH_4\ (g) \longrightarrow TiC\ (s) + 4HCl\ (g)\ at\ 1000℃$$

⑤ Disproportionation:

$$2GeI_2\ (g) \longrightarrow Ge\ (s) + GeI_4\ (g)\ at\ 300℃$$

⑥ Reversible transfer:

$$As_4\ (g) + As_2\ (g) + 6GaCl\ (g) + 3H_2\ (g) \longrightarrow 6GaAs\ (s) + 6HCl\ (g)\ at\ 750℃$$

The versatile chemical nature of the CVD process is further demonstrated by the fact of film deposition. Many different reactants or precursors can be used as different chemical reactions. For example, silica film is attainable through any of the following chemical reactions using various reactants:

$$SiH_4\ (g) + O_2\ (g) \longrightarrow SiO_2\ (s) + 2H_2\ (g)$$

$$SiH_4\ (g) + 2N_2O\ (g) \longrightarrow SiO_2\ (s) + 2H_2\ (g) + 2N_2\ (g)$$

$$SiH_2Cl_2(g) + 2N_2O(g) \longrightarrow SiO_2(s) + 2HCl(g) + 2N_2(g)$$
$$Si_2Cl_6(g) + 4N_2O(g) \longrightarrow 2SiO_2(s) + 3Cl_2(g) + 4N_2(g)$$
$$Si(OC_2H_5)_4(g) \longrightarrow SiO_2(s) + 4C_2H_4(g) + 2H_2O(g)$$

Different films can be deposited when the ratio of reactants and the deposition conditions are varied. For example, both silica and silicon nitride films can be deposited from a mixture of Si_2Cl_6 and N_2O, depending on the ratio of reactants and deposition conditions.

Although CVD is a nonequilibrium process controlled by chemical kinetics and transport phenomena, equilibrium analysis is still useful in understanding the CVD process.

4.5 Liquid phase synthesis method

For the preparation of nanomaterials with different morphologies, liquid phase syntheses are always preferred. These synthetic reactions proceed at relatively lower temperatures and therefore require lower energies.

4.5.1 Precipitation method

The precipitation method is a convenient and powerful technique for synthesizing nanoparticles and nanocapsules. Many of the earliest syntheses of nanoparticles were achieved by the coprecipitation of soluble products from aqueous solutions and followed by thermal decomposition to produce nanooxides. Coprecipitation reactions involve the simultaneous occurrence of nucleation, growth, coarsening, and/or agglomeration processes. Due to the difficulties in isolating each process for independent study, the fundamental mechanisms of coprecipitation are still not thoroughly understood[28].

(1) Synthesis of metals from aqueous solutions

The precipitation of metals from aqueous or nonaqueous solutions typically requires the chemical reduction of a metal cation. Reducing agents take many forms, such as H_2, solvated ABH_4 (A is alkali metal), hydrazine hydrate ($N_2H_4 \cdot H_2O$), and hydrazine dihydrochloride ($N_2H_4 \cdot 2HCl$).

As a matter of convention, the favorability of oxidation-reduction processes is reflected in the standard electrode potential, $E^\ominus$. It is well known that $E^\ominus$ is the corresponding electrochemical half-reaction.

Since the $E^\ominus$ values of all reactions are closely related to the $E^\ominus$ values of H_2, the half-reaction of H_2 and $E^\ominus$ are defined at standard temperature and pressure (STP).

$$2H^+ + 2e^- \longrightarrow H_2, \quad E^\ominus = 0.00V$$

Numerous metal ions can be reduced from aqueous solution to the metallic state in the presence of gaseous H_2 with proper adjustment of pH.

The electrochemical half-reaction and $E^\ominus$ for borohydride ion are given by the chemical reaction:

$$B(OH)_3 + 7H^+ + 8e^- \longrightarrow BH_4^- + 3H_2O, \quad E^\ominus = -0.481V$$

Borohydride ions, however, should be employed judiciously as a reducing agent, particularly in aqueous systems[29~32]. The use of borohydride specifically for the precipitation of metal nanoparticles has recently been reviewed[33].

Hydrazine hydrate is freely soluble in water, but since N_2H_4 is basic, the chemically active free-ion is normally represented as $N_2H_5^+$.

$$N_2H_4 \cdot H_2O \longrightarrow N_2H_5^+ + OH^-$$

Or, in the case of hydrazine dihydrochloride.

$$N_2H_4 \cdot 2HCl \longrightarrow N_2H_5^+ + H^+ + 2Cl^-$$

The standard reduction potential for the hydrazinium ion, $N_2H_5^+$, is

$$N_2 + 5H^+ + 4e^- \longrightarrow N_2H_5^+, \quad E^{\ominus} = -0.23V$$

In theory, the reduction of any metal with an $E^{\ominus}$ more positive than −0.481V or −0.23V, respectively, should be possible at room temperature, given a sufficient excess of reducing agent and proper control of pH. With respect to precipitating metals from solution, this would obviously include many first-row transition metal ions, such as Fe^{2+}, Fe^{3+}, Co^{2+}, Ni^{2+}, and Cu^{2+}, but also many second- and third-row transition metals, as well as most post-transition elements and a few nonmetals. The reduction of gold cations to gold metal is thoroughly studied the metal precipitation reaction.

Gold cations, usually in the form of $AuCl_4^-$, are easily reduced by gaseous H_2 although $AuCl_4^-$ is so strongly oxidizing ($E^{\ominus}$ = +1.002V) that weaker reducing agents such as carboxylates or alcohols are usually sufficient.

$$2AuCl_4^-(aq) + 3H_2(g) \longrightarrow 2Au(s) + 6H^+(aq) + 8Cl^-(aq)$$

Tan et al. [34] have recently reported the synthesis of Au, Pt, Pd, and Ag nanoparticles by reduction with potassium bitartrate; all of the products form stable colloids with the addition of a suitable stabilizing agent.

(2) Synthesis of metal nanomaterials from nonaqueous solvents

In a similar reaction, silver nanoparticles have been prepared by reduction of $AgNO_3$ or $AgClO_4$ by *N,N*-dimethylformamide (DMF), where 3-(aminopropyl) trimethoxysilane served as the stabilizing agent[35]. The reaction probably involves the oxidation of DMF to a carboxylic acid:

$$HCONMe_2 + 2Ag^+ + H_2O \longrightarrow 2Ag + Me_2NCOOH + 2H^+$$

In this case, the size of the silver nanoparticles could be varied from 6nm to 20nm by adjustment of the temperature as well as the [DMF]/[Ag] molar ratio. The silane-based stabilizing agent is susceptible to hydrolysis and condensation, particularly at elevated temperatures and under acidic conditions, and some of the Ag particles exhibited thin SiO_2 coatings, although this apparently did not lead to agglomeration Fig.4.15.

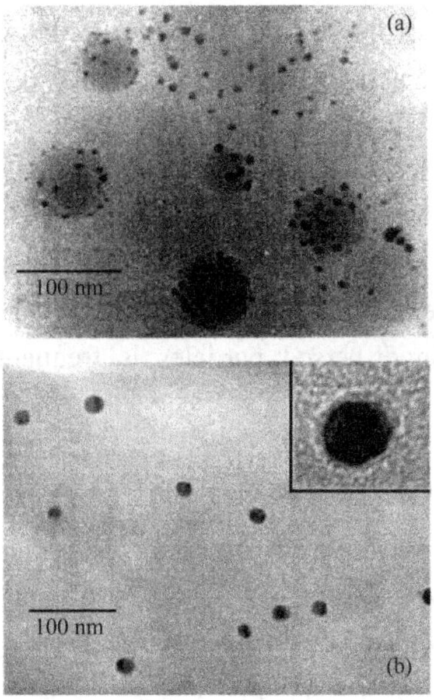

Fig.4.15 TEM image of Ag nanoparticles prepared in DMF (a) at room temperature and (b) under reflux conditions. Both samples are capped with 3-(aminopropyl) trimethoxysilane that sometimes forms a thin silica shell, as demonstrated by the inset in image (b)[35]

Citrate reduction of gold ions leads to the formation of spherical particles, but the analogous reaction with silver ions shown in Fig.4.16 can yield large silver particles (60~200nm) having a wide range of morphologies, depending upon the reaction conditions due to citrate's additional role as a complexing agent[35]. The formation of citrate-silver complexes influences crystal growth and even facilitates photochemical reactions that convert spherical silver nanocrystals to triangular nanoprisms. The research results demonstrated that excess citrate ions dramatically slow the growth of silver nanoparticles as the molar amount of silver ion is held constant. A series of pulse experiments indicated the citrate ions to form the complex with Ag_2^+ dimer in the early stages of the reaction[36].

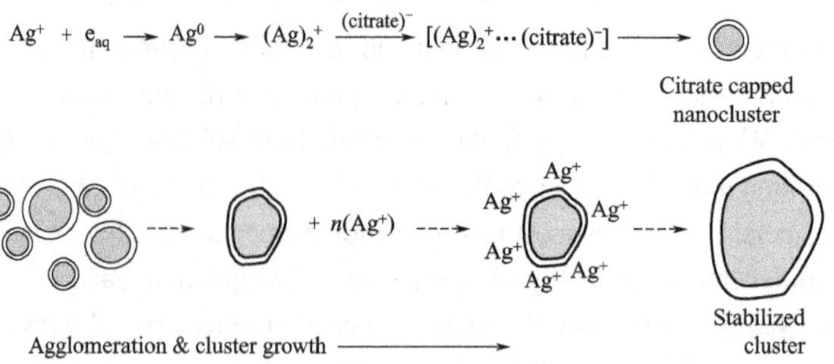

Fig.4.16 Large silver nanoparticles prepared by the citrate reduction route. Excess silver ions permit fusion of smaller particles into a large and stable cluster[36]

4.5.2 Solvethermal method

The term "solvethermal" means reactions in liquid or supercritical media at temperatures higher than the boiling point of the medium. Hydrothermal reaction is one of solvethermal reactions. To carry out reactions at temperature higher than the boiling point of the reaction medium, pressure vessels (autoclaves) are usually required.

It must be noted that the liquid structure of the solvent is essentially unchanged at above or below the boiling point because the compressibility of the liquid is quite small, but as it is near the critical point, the structure of the solvent is drastically altered by changes in the solvent density.

Various nanomaterials have been prepared by solvethermal reactions: metals, metal oxides, chalcogenides, nitrides, phosphides, oxometalate clusters, organic-inorganic hybrid materials, and carbon nanotubes.

Most of the solvethermal products are nano- or microparticles with well-defined morphologies. The distribution of the particle size of the product is usually quite narrow, and formation of monodispersed particles is frequently reported. When the solvent molecules or additives are preferentially adsorbed on (or have a specific interaction with) a certain surface of the products, growth of the surface is prohibited and therefore products with unique morphologies may be formed in solvethermal reactions. Thus nanorods, nanowires, nanotubes, and sheets of various types of products have been obtained solvethermally.

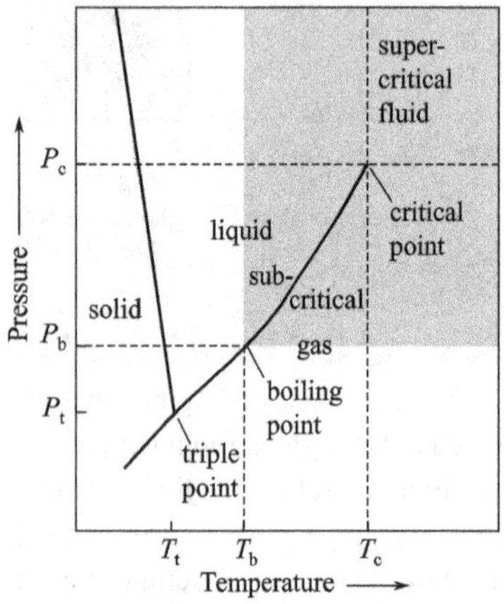

Fig.4.17 Schematic phase diagram of water

(1) Reaction solvents

① Water Water (boiling point bp 100℃; critical temperature T_c, 374℃; critical pressure P_c, 221bar) is the most widely applied reaction medium for solvethermal reactions.

If one relies on the phase diagram of water, only the

branch between liquid and gaseous water is essential for further considerations (see Fig.4.17). Below 100℃ the equilibrium vapor pressure of liquid water is below 1bar. Above the boiling point the important hydrothermal pressure range is available. The pressure of the closed system is given by temperature, corresponding to the equilibrium line in Fig.4.17. About 374.1℃, corresponding to a pressure of 221bar, gas and liquid phase can not be distinguished any more (critical point). Near the critical point, the properties of fluids change radically. The supercritical fluid becomes much more compressible than the gas and denser than the liquid. The properties of these fluids can be adjusted via the pressure and temperature to favour either the gaseous or the liquid characteristics. Thus, at constant temperature, an increase in pressure leads to an increase in density ρ and hence a densification of the fluid. Therefore, temperature, water pressure and the time of reaction are the three principal physical parameters in hydrothermal processing.

② Liquid ammonia Liquid ammonia (bp, 78℃; T_c, 132.4℃; P_c, 112.9bar) is also used for the solvethermal synthesis of nitrides. Metastable or otherwise unobtainable nitride materials were reported to be formed by this method.

③ Organic medium Various organic solvents have been applied for the synthesis of inorganic materials. From a thermodynamic point of view, all organic compounds have an inherent tendency to decompose into carbon, hydrogen (or water, if the organic compound has oxygen atoms), nitrogen and so on at high temperatures.

Solvent thermal reactions in alcohols are sometimes called “alcohol thermal” reactions; this word is derived from alcoholysis based on the similarity between “hydrothermal” and “hydrolysis”.

The solvethermal reaction in glycols is called a “glycothermal” reaction. Ethylene glycol (1,2-ethanediol) is the simplest glycol. This compound is stable at high temperatures. Table 4.2 shows the critical temperatures and pressures for several commonly encountered compounds[37].

Table 4.2 Critical coordinates (T_c, P_c) of a few fluids

Compound	T_c/℃	p_c/bar	ρ_c/(kg/m^3)	Compound	T_c/℃	p_c/bar	ρ_c/(kg/m^3)
CO_2	31.2	73.8	468	C_6H_{12}	279.9	40.3	270
NO_2	36.4	72.4	457	C_6H_6	289.5	49.2	304
NH_3	132.4	112.9	235	C_7H_8	320.8	40.5	290
H_2O	374.1	221	317	CH_4O	240	79.5	275
C_2H_4	9.5	50.6	220	C_2H_6O	243.1	63.9	280
C_2H_6	32.5	49.1	212	C_3H_8O	235.6	53.7	274
C_3H_8	96.8	42.6	225	C_3H_6O	235	47.6	273
C_5H_{12}	196.6	33.7	232				

(2) Solvethermal synthesis

Metal oxides can be synthesized by various solvethermal techniques such as solvethermal dehydration of metal hydroxides, solvethermal decomposition of metal alkoxides, solvethermal synthesis of mixed oxides, solvethermal crystallization of amorphous oxides, solvethermal ion exchange or solvethermal intercalation, and solvethermal oxidation of metals. Solvethermal techniques for synthesizing crystalline particles of metal oxide are summarized in Fig.4.18[38].

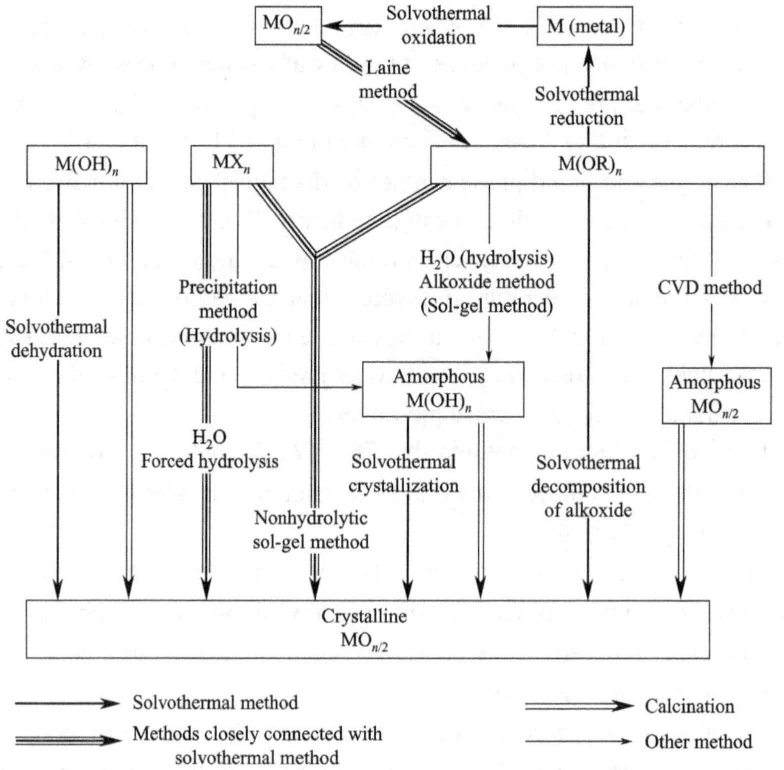

Fig.4.18 Solvethermal routed for synthesis of crystalline particles of metal oxide[38]

(3) Hydrothermal synthesis

The term "hydrothermal" comes from the earth sciences, where it implies a regime of high temperatures and water pressures. In typical hydrothermal research one needs a high-temperature, high-pressure apparatus called an autoclave or bomb. Hydrothermal synthesis involves water both as a catalyst and occasionally as a component of solid phases in synthesis at elevated temperatures (greater than 100℃) and pressures (more than a few atmospheres).

A popular autoclave is composed of stainless steel shell with a teflon-lined used in the laboratory to perform subcritical solvethermal synthesis as shown in Fig.4.19[39]. At present, one can get many kinds of autoclaves to cover different pressure-temperature ranges and volumes. The first and foremost parameters to be considered in selecting a suitable reactor are the experimental temperature and pressure conditions.

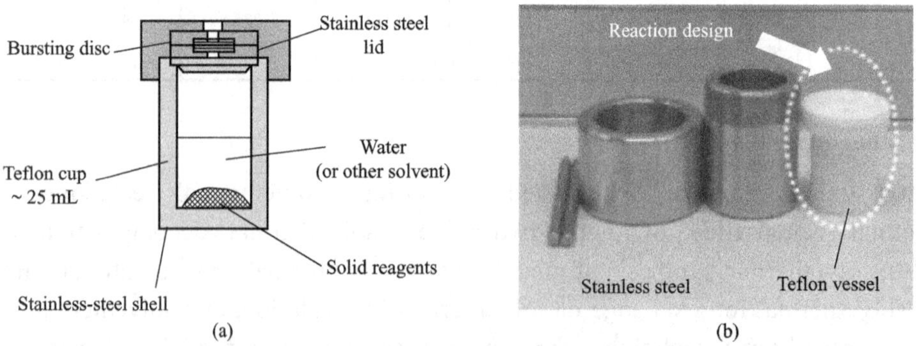

Fig.4.19 A schematic of a teflon-lined, stainless steel autoclave typically used in the laboratory to perform subcritical solvethermal synthesis. (a) A profile; (b) A set of autoclaves[39]

Ilmenite ($FeTiO_3$) is a very stable mineral. Extraction of titanium dioxide (TiO_2) from such ores has potential. Using 10mol/L KOH or 10mol/L NaOH mixed with ilmenite in a ratio of 5:3 (ilmenite:water) under reaction conditions of 500℃ and 300kgf/cm^2 (pressure), ilmenite was decomposed completely after 63h. If the ratio of ilmenite to water is 5:4, under the same conditions, 39h is needed to decompose the ilmenite. Reaction equations show as follows in the case of KOH solution.

$$3FeTiO_3 + H_2O \longrightarrow Fe_3O_4 + 3TiO_2 + H_2$$

$$nTiO_2 + 2KOH \longrightarrow K_2O(TiO_2)_n + H_2O,\ n = 4 \text{ or } 6$$

(4) Hydrothermal synthesis in supercritical water

Hydrothermal synthesis (HTS) is used to produce synthetic materials imitating natural geothermal processes [40,41]. The reaction equilibrium of metal salt aqueous solutions changes with temperature, and results in the formation of metal hydroxides or metal oxides. Supercritical water (SCW) provides an excellent reaction medium for hydrothermal synthesis, since it allows to vary the reaction rate and equilibrium by shifting the dielectric constant and solvent density with pressure and temperature. One of the expected benefits is higher reaction rates and smaller particles. The reaction products have to be not soluble in SCW. The HTS-SCW process is usually operated as follows: a metal salt aqueous solution is prepared, pressurized and heated. The pressurized metal salt solution and a supercritical water stream are combined in a mixing point, which leads to rapid heating and subsequent reaction. After the solution leaves the reactor, it is rapidly quenched and in-line filters remove larger particles. Cooling water is directly fed to the reactor to quench the reaction as shown in Fig.4.20. Two different process modes have been proposed: the first uses a batch reactor and is characterized by a long reaction time; the second uses a flow reactor that assures continuous operation.

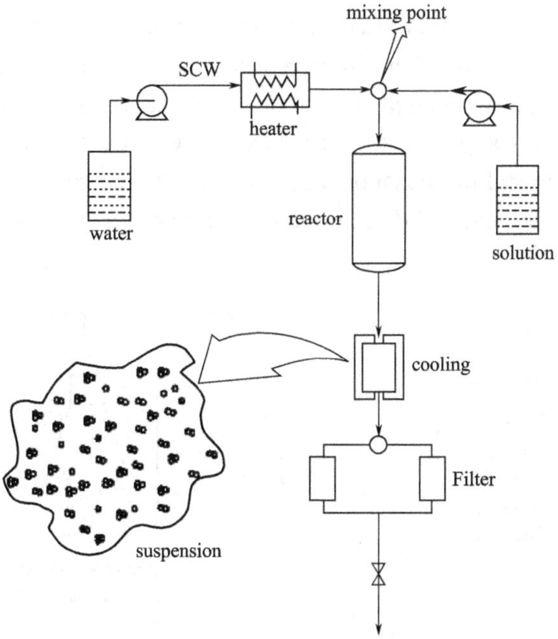

Fig.4.20 A schematic representation of the HTS-SCW (hydrothermal synthesis in supercritical water) process [44]

Cabanas et al. [42,43] used a small continuous reactor to produce $Ce_{1-x}Zr_xO_2$ by HTS-SCW starting from mixtures of cerium ammonium nitrate and zirconium acetate. According to these authors, the continuous reactor allows a better control of the experimental conditions when compared to the batch process.

HTS-SCW can produce very small nanometric particles; however, this process requires challenging operative conditions since supercritical point for water is located at 374℃ and 220 bar. To operate in SCW, stainless steel with special characteristics is required due to the simultaneous application of high pressures and high temperatures. Moreover, SCW is a strong oxidizing agent not only for processed materials, but also for the elements of the plant which it is in contact with. HTS-SCW can be used only for compounds that are stable at high temperatures.

4.5.3 **Freeze-drying method** (Cryochemical synthesis method)

Various kinds of low-temperature processing have been used in materials science and technology. Among these, the cryochemical method is believed to be the first from the work of Landsberg and Campbell who dealt with the synthesis of nanocrystalline powders of W and W-Re alloys[45].

The freeze-drying method involves following steps: ①preparation of a solution; ②rapid freezing of the solution; ③freeze-drying to remove water and ④ thermal decomposition and heat treatment. Tretyakov and Shlyakhtin[46] have provided an excellent review of recent progress in the cryochemical synthesis of oxide materials.

(1) Freezing process

Freezing of aqueous solutions in cryochemical synthesis is performed via fast cooling. This process is usually realized by spraying the solution into liquid nitrogen via a nozzle or ultrasonic nebulizer under intense mechanical stirring. The product of freezing consists of soft agglomerated spherical granules 0.01mm to 0.5mm in size, and it is separated from the refrigerant by evaporation.

The easy evaporation of liquid nitrogen reduces the cooling rate of freezing droplets and therefore several authors have proposed the use of cold hexane instead of nitrogen in Fig.4.21. In the freezing process, the cryocrystallization product of the salt solution may consist of ① a fine mechanical mixture of crystalline hexagonal ice and crystalline salt or its higher hydrate; ② a mixture of crystalline ice and solid amorphous solution; or ③ an amorphous, chemically homogeneous solid.

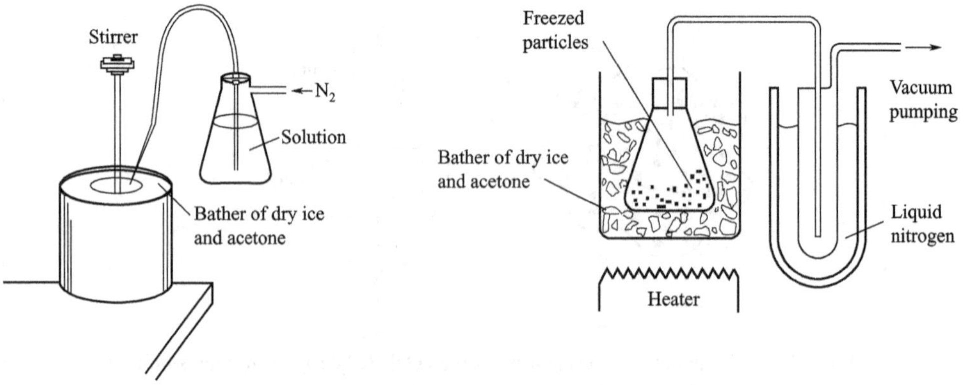

Fig.4.21 Schematic diagram of freeze-drying process

(2) Drying process

The next crucially important stage of synthesis is the freeze-drying processes. Low temperatures and reduced pressures are selected such that the drying time is minimized, while preventing the formation of liquid phase in the macroscale by melting of ice-salt eutectics. A considerable advantage of this process over other processes is the availability of a wide range of industrially produced freeze-drying machines. Hence this process is especially suitable for small laboratories as shown in Fig.4.21.

The main technique of solvent elimination from the frozen products is freeze-drying based on the direct transformation of solid ice into vapor, avoiding liquid phase formation. Formation of liquid usually degrades the morphological homogeneity of the cryocrystallization products achieved during fast freezing.

The sublimation is a considerably endothermic process. Freeze-drying at a reasonable rate needs a permanent external heat supply. Processing conditions of freeze-drying can be evaluated using pressure-temperature phase diagrams. According to Gibbs' rule, the triple point (coexistence of three states: ice, liquid, and vapor) in the phase diagram of water can be achieved at $T = 273.15$K and p = 610.5Pa (in Fig.4.17). So the coexistence of ice and water vapor is necessary in the ice sublimation process. The ice can be evaporated (sublimed) without melting under an external heat supply if the pressure in the system doesn't exceed 610.5Pa (4.58Torr). The usual temperature of freeze-drying is 253K to 273K, corresponding to equilibrium water vapor pressure values of 100Pa to 610Pa[47].

(3) Thermal decomposition

In most cases, freeze-drying is not the final stage of the synthesis process. Its product needs further processing in order to be converted to the final oxide or metal powder. These subsequent thermal decomposition and powder processing techniques are the same or similar to other chemical methods of powder synthesis. The temperatures of decomposition of the individual components of salt mixtures are usually shifted to lower temperatures compared to pure individual salts due to formation of solid solutions. Formation of the complex oxides usually occurs at the last stages in the thermal decomposition of salt precursors, although the range of possible formation temperatures is rather wide from 200 to 900℃. The cryocrystallization product consists of a highly dispersed mechanical mixture of the individual compounds, so that chemically homogeneous complex oxide is formed only during thermal processing at the final stage of the synthesis.

(4) Spray freeze-drying technique

Spray freeze-drying (SFD) technique, which is considered as one of the advanced ways for the fabrication of nanoparticles, has been widely used for the synthesis of nanostructures of biological medicaments and inorganic materials [48~51]. The particles obtained by this method generally possess the intriguing advantages, such as molecular scale homogeneity because of flash freezing in the cryogen, and minimal agglomeration because of the sublimation of the water ice in the low-temperature vacuum drying condition [52,53]. So far, this system has been used to synthesize the nanostructures of energetic materials.

1,1-diamino-2,2-dinitroethylene (FOX-7) 3D grid nanostructures with different unit sizes were also prepared by the freeze-drying technique [54]. The preparation process of FOX-7 3D grid nanostructures is shown in Fig.4.22(a). The typical SEM images in Fig.4.22(b) and (c) indicate that the resulting FOX-7 3D structures are constructed from one-dimensional nanostructures. The size and reticular structures of the resulting nano-FOX-7 could be easily controlled by adjusting the concentration of the aqueous solution of raw materials. The further investigation on a possible formation mechanism of this structure reveals that a high degree of supercooling under the liquid nitrogen conditions leads to a high nucleation rate, and the small particles and ice are rapidly formed due to the rapid heat transfer between the aqueous solution and the cryogen. After the sublimation of ice, the small particles with a relatively high energy start to aggregate by selfassembly, leading to the formation of 3D grid structures. A scheme of this growth process is summarized in Fig.4.22(d). More importantly, this facile synthesis strategy may represent a general approach for the preparation of nanostructures of other water-solubility energetic compounds.

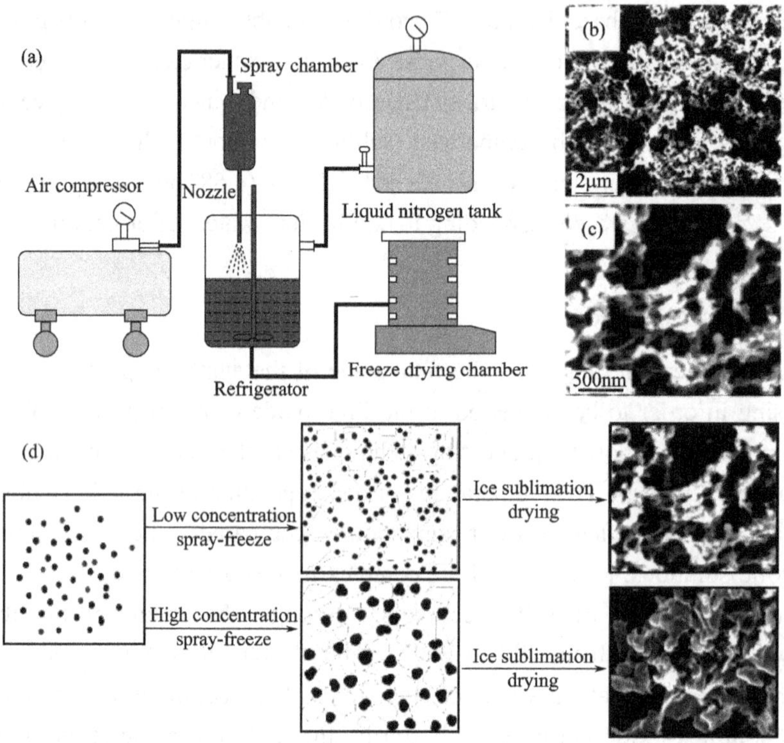

Fig.4.22 (a) Flow chart of FOX-7 quasi-three-dimensional grids fabricated via SFD technique; (b) low magnification; (c) high magnification of SEM images of FOX-7 nanostructures prepared with CFOX-7 = 0.1 g/L; (d) a probable formation process of FOX-7 network structures under the same supercooling rate [54]

4.5.4 Sol-gel method

The sol-gel method is a process consisting of solution, sol, gel, solidification, and heat treatment of metal organic or inorganic compounds, which is applicable to preparation of metal oxides and other compounds.

Traditionally, sol-gel process involves hydrolysis and condensation of metal alkoxides. Metal alkoxides have the general formula $M(OR)_x$ and the alkoxyl group (OR) depends on an alcohol. The general synthesis of metal alkoxides involves the reaction of metal species (a metal, metal hydroxide, metal oxide, or metal halide) with an alcohol. For example, $Si(OR)_4$ is easily prepared from $SiCl_4$ and the appropriate alcohol, as was reported in the original work of Ebelmen[55].

$$SiCl_4 + 4ROH \longrightarrow Si(OR)_4 + 4HCl$$

Where R = CH_3, C_2H_5, or C_3H_7, the resulting products in the absence of water is a stable sol of $Si(OR)_4$.

After introduction of water into the system, the hydrolysis initiates, that is, a hydroxide group from water substitutes for an alkoxide group and a free alcohol is formed. Subsequently, condensation takes place.

$$M(OR)_4 + H_2O \longrightarrow M(OR)_3(OH) + ROH$$

$$2M(OR)_3(OH) \longrightarrow (RO)_3M—O—M(OR)_3 + H_2O$$

These condensation reactions result in the formation of gel. The detail involves two hydrolyzed fragments join together and either an alcohol or water is released.

The degree of hydrolysis of the alkoxide precursor strongly influences the structure of the Si—O—Si network. Because OH^- is a marginally better leaving group than —OR, the condensation process can be tailored to favor the formation of dimers, chains, or 3D agglomerates described in Fig.4.23. As a result, Si—O—Si chains tend to preferentially form in the early stages of the polymerization process, followed by subsequent branching and cross-linking of the chains during aging. These gels are made of small polymeric entities (diameter 0.5nm to 5nm), which are more or less aggregated and densely packed according to the process (hydrolysis-polycondensation rate, initial alkoxide-to-solvent and alkoxide-to-water ratios).

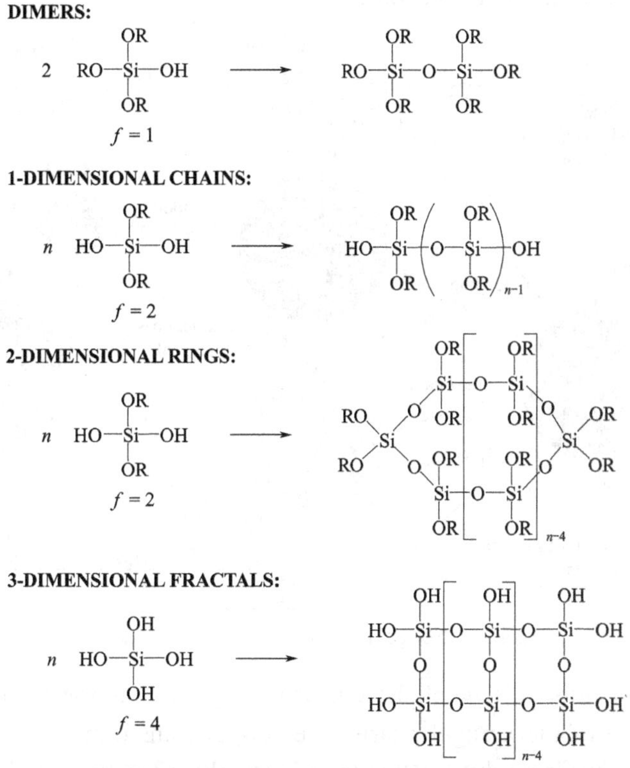

Fig.4.23 Precursors favoring the formation of dimer and 2D and 3D structures in sol-gel processing of silicon alkoxides

Sol-gel routes can be used to prepare pure, stoichiometric, dense, equiaxed, and monodispersed particles. For example, amorphous TiO_2 particles with average particle size of 70nm to 300nm were prepared by controlled hydrolysis of titanium tetraisopropoxide. Monodispersed amorphous powders were synthesized using titanium tetraethoxide, where only homogeneous nucleation occurred without flocculation. Depending on the reaction kinetics and subsequent aging (post-synthesis) of particles, nanoscale oxide particles synthesized can be amorphous or crystalline. These nanostructured oxide particles can be carburized or nitrided to form non-oxide powders at reduced temperatures and shorter reaction time due to the significant amount of highly active surfaces.

Factors that need to be considered in a sol-gel process are solvent, temperature, precursors, catalysts, pH, additives, and mechanical agitation. These factors can influence the kinetics, growth reactions, hydrolysis, and condensation reactions. The solvent influences the kinetics and conformation of the precursors, and the pH affects the hydrolysis and condensation reactions. Acidic conditions favor hydrolysis, which means that fully or nearly fully hydrolyzed species are formed before condensation begins. Under acidic conditions there is a low crosslink density. By varying influence factors of the reaction rates of hydrolysis and condensation, the structure and properties of the gel can be tailored. In the traditional sol-gel methods, the condensation reaction may be reversible and special efforts are required to completely remove water or polar solvents.

The classification of sol-gel technology and products was given by Blinker and Scherrer[56] as shown in Fig.4.24. The main idea of the sol-gel process is the spontaneous formation of gel from a sol by hydrolysis and condensation reactions.

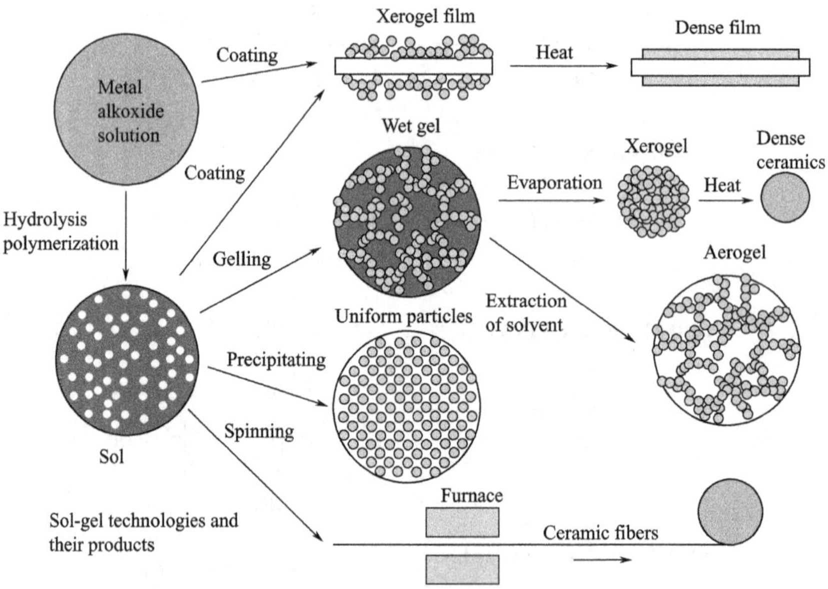

Fig.4.24 Sol-gel technology and their products[56]

Further transformation of the gel phase is driven by the evaporation of the solvent, and the subsequent formation of the xerogel phase. The sol-gel transformation may take place in the bulk of the solution, but it works much more effectively when the solution is spread over the surface of solid substrate. Thin xerogel films (in the range of 100nm) can be formed on the solid substrates by dip coating, spin coating, or spraying of the solution. Heating of the xerogel completely removes solvent molecules leading to further aggregation of inorganic clusters and the formation of solid materials in the form of thin films. Subsequent repeating of the routine allows the formation of thicker multilayered films. A quick, supercritical drying carried out at high temperature leads to the formation of the aerogel, an extremely porous (greater than 75% porosity) material. On the other extreme, a very slow evaporation of the solvent at ambient conditions causes the precipitation of solid phase, and eventually yields fine, uniform particles. The sol-gel was implemented in the industrial processing of ceramic materials in the form of sheets, tubes, and fibers (Fig.4.24)[56~59].

In non-hydrolytic sol-gel methods, the traditional hydrolysis and condensation reactions are replaced by direct condensation reactions or transesterification reactions. The reaction may be direct condensation of metal halide and alkoxides or condensation of a metal alkoxide and a metal carboxylate. The transesterification reaction involves reaction of a trialkylsilyl acetate and a metal alkoxide. Nonhydrolytic reactions do not involve water or polar solvents.

4.5.5 Microemulsions method

(1) Fundamentals of microemulsions

The combination of water, oil, surfactant, and an alcohol or amine-based cosurfactant produces clear, seemingly homogeneous dispersion system which is called as "micro emulsions"[60, 61]. In this case the oil phases are simple long-chain hydrocarbons. The surfactants are long-chain organic molecules with a hydrophilic head (usually an ionic sulfate or quarternary amine) and lipophilic tail. The surfactant that is still used extensively is cetyltrimethyl ammonium bromide (CTAB) as shown in Fig.4.25.

The amphiphilic nature of the surfactant such as CTAB makes it miscible in both hydrocarbons and water. The surfactant, through ion-dipole interactions with the polar cosurfactant, forms spherical aggregates in which the polar (ionic) ends of the surfactant molecules orient toward the center. The cosurfactant acts as an electronegative "spacer" that minimizes repulsions between the positively charged surfactant heads in Fig.4.26[60].

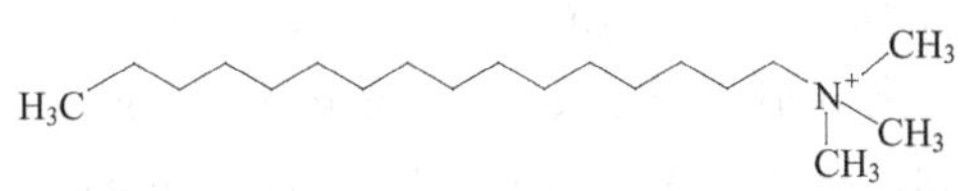

Fig.4.25 Structure of cetyltrimethyl ammonium bromide (CTAB), a common surfactant used in microemulsions

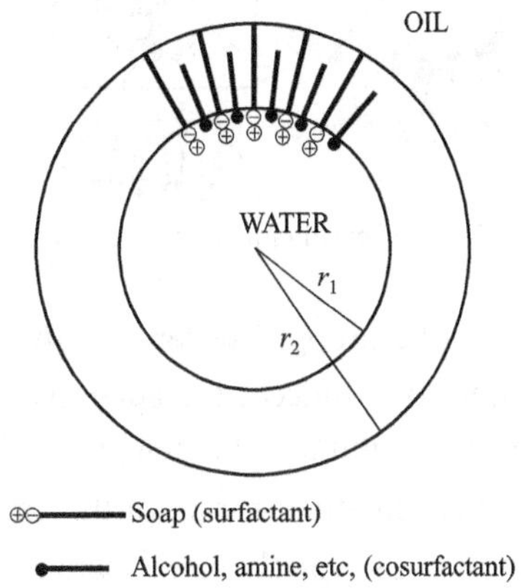

Fig.4.26 Schulman's model of the reverse micelle[60]

Surfactants are molecules with a hydrophilic polar head and a hydrophobic hydrocarbon chain. Such molecules tend to spontaneously position themselves at the interface of two immiscible media, such as water and oil to form different aggregates. The special composition of these molecules induces a reduction in surface tension between oil and water. The shape of the surfactant will play an important role in the form of eventual product assembly. Surfactants can also self-organize into micelles and other structures when a critical concentration (termed as the critical micelle concentration or CMC) is reached.

If the surfactant has a very bulky polar head with a short chain, the surfactant then has a cone shape, and the chains will associate together to form spherical aggregates called micelles (see Fig.4.27[62]). When micelles form at low concentrations, they are spherical and their diameter is determined by the length of the hydrocarbon chain and the polar head. It can be formed the monomers, micelle and reverse micelle in the different conditions.

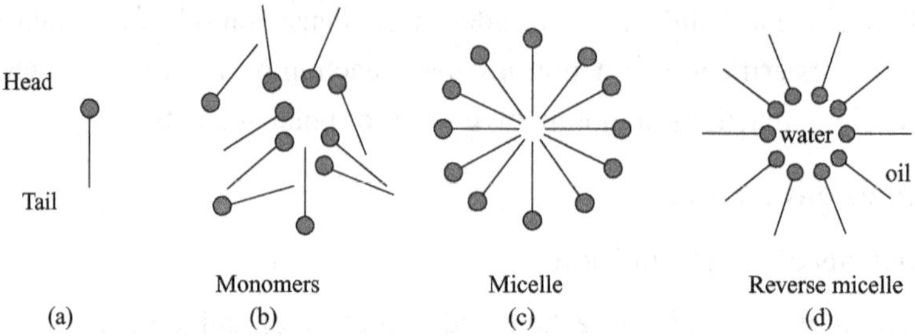

Fig.4.27 Formation of a reverse micelle. (a) Surfactant with small polar head; (b)Monomers; (c) Micelle; (d) A reverse micelle formed[62]

In general, the thermodynamically unstable dispersions of two or more immiscible liquids can be kinetically stabilized by forming an oil-in-water or water-in-oil emulsion at the liquid-liquid interface. Multiple emulsions, water-in-oil or water-in-oil-in-water, may also be formed as shown in Fig.4.28. Nanoemulsion is considered to be the emulsion of a continuous phase with disperse-phased droplets between 1nm and 100nm in diameter.

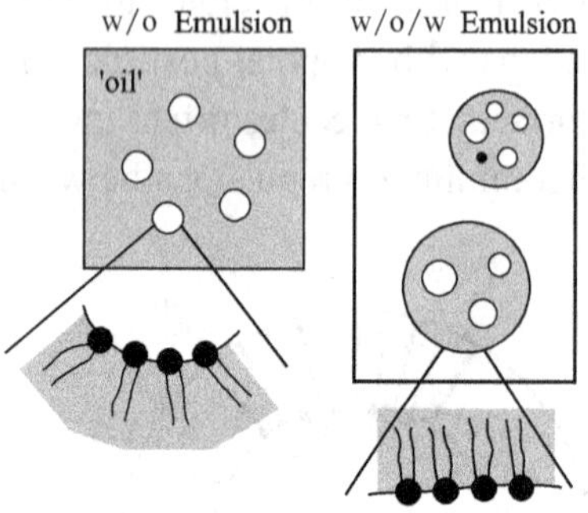

Fig.4.28 Different emulsion systems

(2) Reverse micelles (spherical nanoreactors)

Reverse micelles (sometimes known as water-in-oil microemulsions) are a single layer of surfactant molecules entrapping solubilized water pools in a hydrocarbon solvent as shown in Fig.4.28. The size of the water pool depends on the amount of entrapped water at a given surfactant concentration. These droplets can be considered as nanoreactors. Indeed, the aqueous core of the droplet provides a nanoscale reactive medium with controlled volume in the range 1～500nm^3.

A schematic diagram of this process is represented in Fig.4.29[63]. The reverse micelles can be employed as variable-size nanoreactors because it has two properties, i.e., size control and exchange of aqueous contents. The small size of reverse micelles subjects them to continuous Brownian motion even at room temperature. When two solutions are mixed, aqueous core (in reverse micelles) exchange induced by Brownian motion brings the two reactants of metal salt and reducing agent into contact so that they can react with each other in Fig.4.29[63].

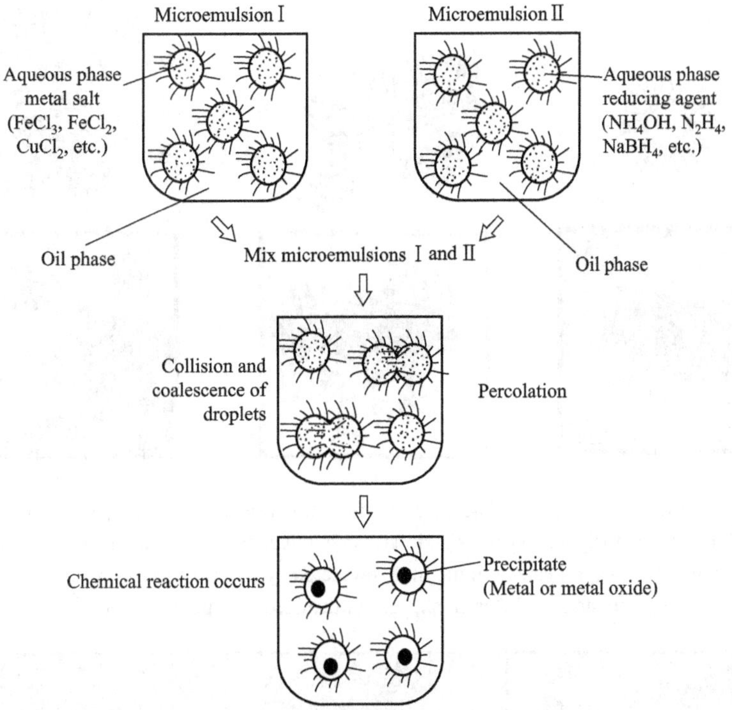

Fig.4.29 Proposed mechanism for the formation of metal particles by the microemulsion approach[63]

Many experimental results revealed that the reverse micelles as nanoreactors can be used to fabricate many nanosized compounds, such as nanosemiconductors, nanometals, nanooxides, and nanoalloys. In fact, for Ⅱ～Ⅵ semiconductors such as CdS, CdTe, CdMnS, or ZnS[64], particle sizes vary from 2nm to 4nm, whereas for metals[53], it lies in the range 2nm to 6nm for silver and 2nm to10nm for copper. The size control of these nanoparticles is obtained for the smallest water droplets in oil (between 0.6nm and 6nm), whereas for the largest templates (6nm to 12nm), no change is observed in the size of the resulting nanoparticles. This can be explained in terms of the structure of the water contained within the micelle[65]. Indeed, for low water contents (w = 10%), the water molecules in the micelle core are strongly associated with counter ions. When the water content is increased (w>10%), the water is no longer primarily involved in a hydration process and it is said to be free.

Controlling the shape of nanocrystals is a real challenge. Some published results indicate that ternary system (water-surfactant-oil) served as templates leads to self-assemblies of surfactants with different shapes. The shape of the colloidal template composed of functionalized surfactants was partially responsible for imposing the shape of the nanocrystals[66, 67]. Fig.4.30 shows that a template of interconnected cylinders produces spherical copper nanocrystals and some cylindrical ones[68].

Let us consider effect of the sodium halide concentration on the shape of copper nanocrystals in the ternary system $Cu(AOT)_2$-water-isooctane (AOT: sodium bis(2-ethylhexyl) sulfosuccinate). In the interconnected cylinder region, small cylindrical nanocrystals are obtained [see Fig.4.31(a)]. Adding less than 2×10^{-3}mol/L of chlorine ions to the colloidal solution leads to the production of long copper rods as shown in Fig.4.31(b) and (c)[69~71]. These long rods have the same crystal structure as the small cylinders.

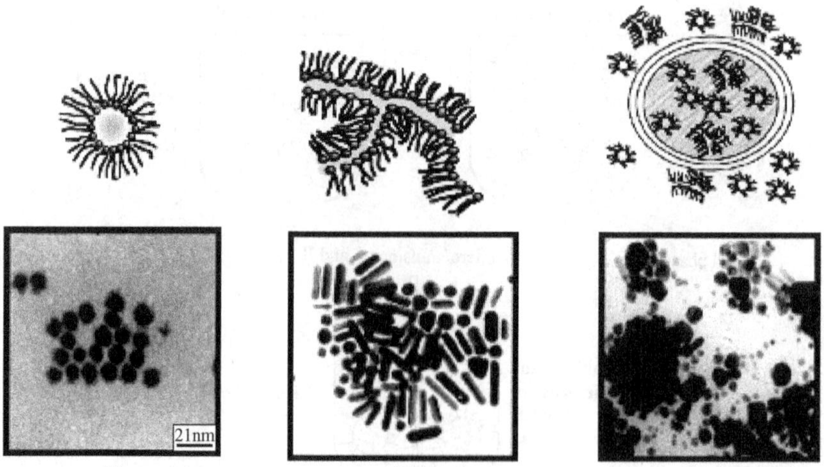

(a) (b) (c)

Fig.4.30 Copper nanocrystals synthesized in different phases of the water/surfactant/oil system. (a) Spherical nanocrystals synthesized in reverse micelles; (b) Spherical and cylindrical nanocrystals synthesized in the interconnected cylinder phase; (c) Mixture of nanocrystals of different shapes synthesized in superaggregates[68]

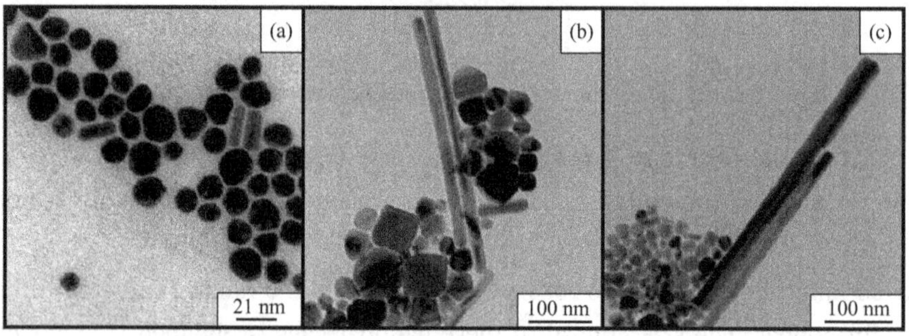

Fig.4.31 Effect of NaCl concentration on the aspect ratio of copper nanocylinders. (a) [NaCl] = 0mol/L; (b) [NaCl] = 5 $\times 10^{-4}$mol/L; (c) [NaCl] = 1.1 $\times 10^{-3}$mol/L[71]

If NaBr is added instead of NaCl, nanocubes are obtained (see Fig.4.32[72]). However, their size does not vary with the bromine ion concentration. With the addition of NO_3^-, a whole range of nanocrystal shapes and sizes is obtained, while with F^-, small cubes are synthesized[73].

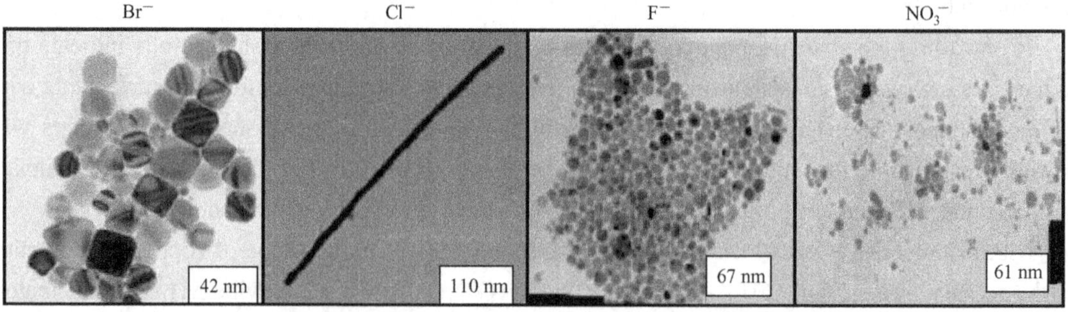

Fig.4.32 Effect of different anions on the shape of copper nanocrystals[72]

4.5.6 Microwave-assisted synthesis

Microwaves lie in the electromagnetic spectrum between infrared waves and radio waves. They have wavelengths between 1mm and 1m with frequency range from 300MHz to 30GHz. The typical bands for industrial applications are (915±15)MHz and (2450±50) MHz[74, 75].

Almost all the microwave synthesis reported are currently conducted at 2450MHz (the corresponding wavelength is 12.24cm). One reason is that near to this frequency, the microwave energy absorption by liquid water is maximal. Another probable reason is that 2450MHz magnetrons are mostly often used in the available commercial microwave chemistry equipments. Generally, materials can be classified into three categories based on their interaction with microwaves: ①transparent, where microwaves pass through without any losses; e.g. fused quartz, several glasses, ceramics, teflon; ②conductors, where microwaves are reflected and cannot penetrate, e.g. copper; and ③ absorbing, where microwaves are absorbed depending on the value of the dielectric loss factor, it is the most important class of materials for microwave synthesis, e.g. aqueous solution and polar solvent.

Fig.4.33 shows how polar molecules in a microwave field attempt to reorient themselves in accordance with the rapidly changing field[76]. When applying microwave irradiation to the material, the electric and magnetic components are changing rapidly (2.4×10^9 per second at a frequency of 2450MHz) and the molecules cannot respond quickly to the change in direction, giving rise to friction and therefore causing them to warm up[77].

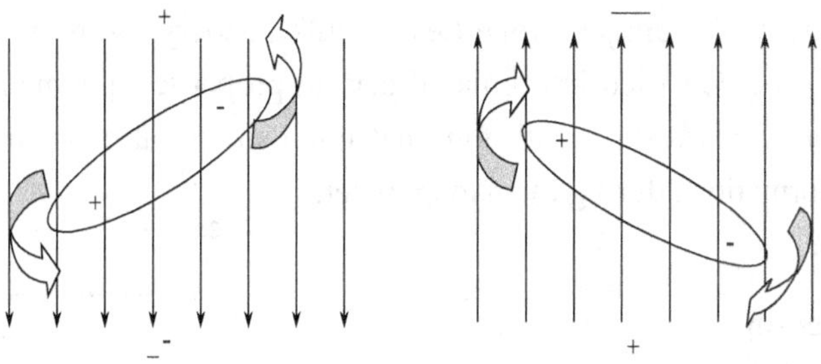

Fig.4.33 Realignment of a dipole in an electromagnetic field[76]

Ionic conduction is another important microwave heating mechanism. When a microwave field is applied to a solution containing ions, they move due to their inherent charge. As a result, ions collide and the collisions cause the conversion of kinetic energy to thermal energy. As the concentration of ions increases in solution, more collisions occur, causing the solution to be heated quickly[78]. In many cases, the remarkable characteristic of microwave synthesis is decrease of reaction time.

Yu et al. prepared 2～4nm colloidal Pt particles by irradiating a mixture of poly(*N*-vinyl-2-pyrrolidone), aqueous H_2PtCl_6, ethylene glycol, and NaOH with 2450MHz microwaves in an open beaker for 30s[79]. Tsuji et al. prepared PVP-stabilized, Ni nanoparticulate in ethylene glycol by a similar method[80].

4.5.7 Ultrasonic wave-assisted synthesis

Depending on the frequency, ultrasonic wave is divided into three regions: power ultrasound (20～100kHz), high frequency ultrasound (100kHz～1MHz), and diagnostic ultrasound (1～500MHz). Ultrasound ranging from 20kHz to 100kHz is used to the systems, in which chemical and physical changes are desired[81].

Ultrasonic wave is mechanical waves that are not absorbed by object, but they can transmit through any substances, such as solid, liquid and gas. The energy of ultrasound is insufficient to cause chemical reactions. But when it travels through media, a series of compressions and

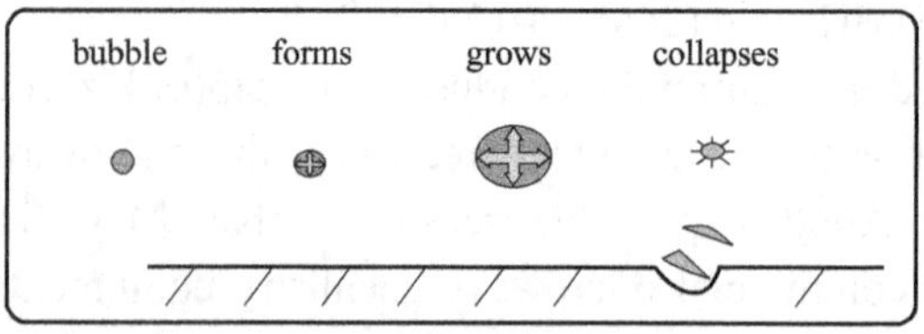

Fig.4.34 Representation of development and collapse of the cavitation bubble and solid erosion[83]

rarefactions are generated. The compression cycles exert a positive pressure on the liquid by pushing the molecules together and the rarefaction cycles exert a negative pressure by pulling the molecules from one another. Because of this excessively large negative pressure, microbubbles (cavitation bubbles) are formed in the rarefaction regions. These microbubbles grow in successive cycles and reach to an unstable diameter so as to collapse violently with producing shock waves (a lifetime of few microseconds)[82].

This process by which the bubbles form, grow and undergo violent collapse is known as cavitation. Representation of development and collapse of the cavitation bubble as well as solid erosion are given in Fig. 4.34[83].

4.6 Synthesis of bulk materials by consolidation of nanopowders

To take advantage of the unique properties of bulk nanocrystalline materials, the nanometer range powders have to be densified into bulk of certain properties, geometry, and size. The most important is to achieve densification with minimal microstructural coarsening and/or undesirable microstructural transformations during the nanopowders consolidation process.

4.6.1 Cold compaction

An initial consolidation, typically cold compaction to generate a solid that can be handled, is called a green body. The usual dependence of green density of nanometals as a function of applied pressure shows in Fig.4.35. At lower pressures, there is a linear increase of density with pressure. At larger pressures, the slope is considerably less. This region is attributed to work hardening for the samples of 15nm Ni and 26nm Fe[84]. As shown in Fig.4.35, little work hardening with flattened compaction curves at high pressures in 50nm Ni is experimentally observed[84, 85].

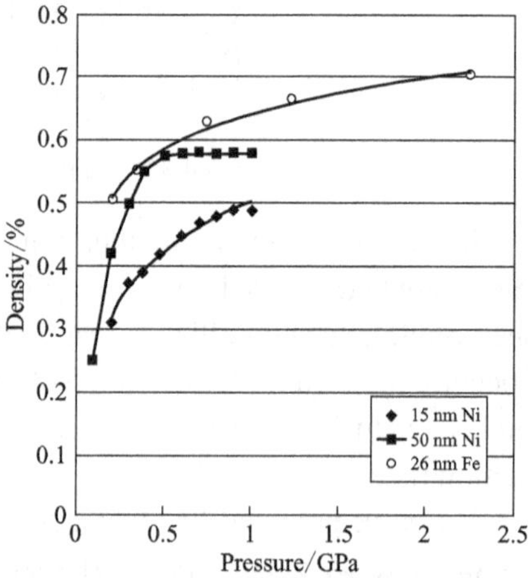

Fig. 4.35 Pressure effects on the cold compaction of 15nm and 50nm Ni powders and 26nm Fe powders[84, 85]

4.6.2 Warm compaction

To eliminate adsorbates and enhance interparticle bonding, warm compaction at temperatures up to 402℃ has been largely applied to nanopowders consolidation process. For most metal nanopowders, these temperatures are still in the cold working regime[86, 87]. Warm compaction of 70～80nm TiN powders results in densities exceeding 90% at 402℃[86].

The consolidated bulk samples by hot pressing the amorphous precursor powders between 700℃ and 850℃ at 275MPa for 0.5h to 2h comprised a matrix of α-Fe with dispersion clusters of 10nm Mo_2C precipitates. Smaller carbide precipitates within large matrix grains had little effect on preventing grain growth. The some results indicated that increasing pressure and pressing temperature did not have significant effect on reducing the porosity in all samples during the hot pressing. The retention of nanostructures in powder consolidation of multicomponent engineering materials with full density at a practical pressure range is still a challenge.

4.7 Template-assisted self-assembly nanostructured materials

4.7.1 Principles of self-assembly

According to Whitesides, "a self-assembling process is one in which humans are not actively involved, in which atoms, molecules, aggregates of molecules and components arrange themselves into ordered, functioning entities without human intervention"[88].

Self-assembly has become a very effective and promising approach to synthesize a wide range of novel nanoscale materials. Fig.4.36 shows the principles of self-assembly and the emerging structures of amphiphile molecules in water[89]. Due to an anisotropic distribution of precursor molecules at pure water-air interface, a resulting force pointing towards the solvent is shown in Fig.4.36(a). When amphiphiles are added in Fig.4.36(b), they accumulate at this interface. The hydrophobic tail points towards the gas phase, while the hydrophilic part is located in the aqueous domain. The amphiphile is enriched at the interface until it is fully occupied. However, dispersion of amphiphilic molecules in the continuous phase is possible when micelles are formed due to minimization of interface energy. In a micelle, the hydrophilic units point towards the solvent while the hydrophobic chains are separated from the water in a hydrophobic core in Fig.4.36(c). By the addition of more and more amphiphile, more and more micelles are formed. They can assemble to dense packing as fcc or bcc in Fig.4.36(d).

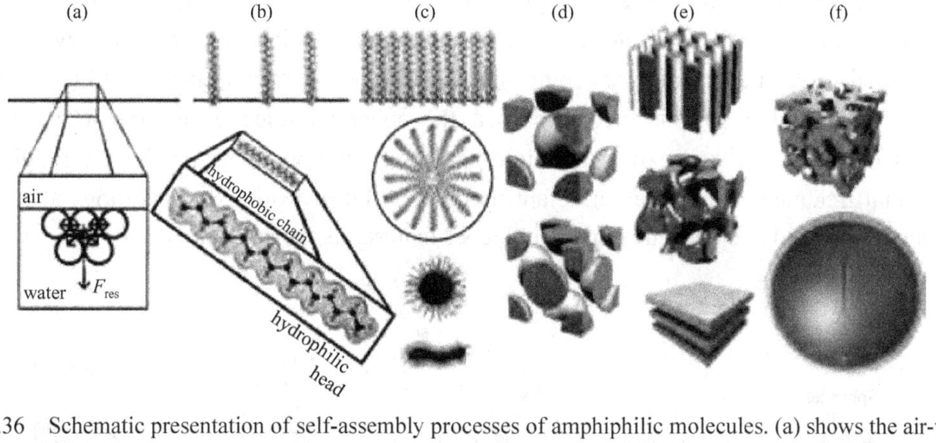

Fig.4.36 Schematic presentation of self-assembly processes of amphiphilic molecules. (a) shows the air-water interface and a force diagram; (b) and (c) shows the enrichment of amphiphiles at the air-water interface and the formation of micelles; (d) and (e) show higher order structures, dense micellar packing (hexagonal and cubic symmetry), and lyotropic phases (hexagonal, gyroid, and lamellar) obtained for higher amphiphile concentration. Even larger structures (f) are obtained when swelling agents or selective solvents are used micro-emulsions and vesicles[89]

Furthermore, the micelles fuse and build continuous structures, so-called lyotropic phases. First, two-dimensional phases as a hexagonal alignment of cylinders are built. Then, at higher concentrations, even the cylinders fuse and build sheets of a lamellar lyotropic liquid crystal in Fig.4.36(e). Bicontinuous three-dimensional structures as the gyroid phase can emerge during the transition of hexagonal to lamellar phases. Micelles and lyotropic phases can be swollen by different solvents to become emulsions[90], for instance, microemulsions or vesicles in Fig.4.36(f). Other self-assembly processes that occur in supra-molecular chemistry are described elsewhere [91, 92]. Self-assembled structures are ideal helpers for nanosynthesis due to the structural control over several length scales.

4.7.2 Self-assembly of MCM-41

In 1992, Mobil scientists demonstrated that surfactant and silicate could co-assemble and form nanocomposites via van der Waals and electrostatic interactions. Surfactant removal results in the formation of mesoporous silica molecular sieves (MCM-41) in Fig.4.37[93].

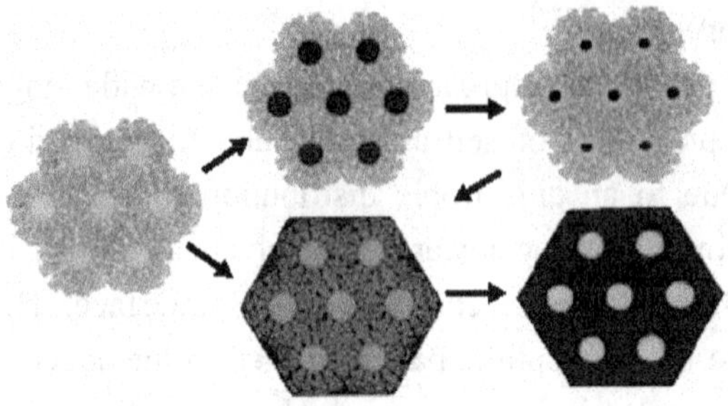

Fig.4.37 Templating assisted self-assembly routes to form mesoporous material[93]

This co-assembling synthesis approach has been extended to create a large variety of mesoporous materials through van der Waals, electrostatic, hydrogen bonding, or other noncovalent interactions including mesoporous SiO_2[94~97], SiO_2 containing interstitial Ti[98] or Al[99], aluminosilicates[100], $AlPO_4$[101~105], Al_2O_3[106], TiO_2[107], Nb_2O_5[108], SnO_2[109, 110], ZrO_2[111], mixed oxides[112] and metal Pt[113].

A widely accepted mechanism for templating assisted self-assembly routes are shown in Fig.4.38[114, 115]. The surfactants firstly assemble into cylindrical micelles, and then the soluble silicate species intercalate into the hydrophilic region between the head groups. These cylindrical micelles produce interacting via electrostatic or hydrogen bonding, or other noncovalent, and a highly ordered inorganic-organic nanocomposite was formed as shown in Fig.4.38 (route B). The addition of the soluble silicate species caused the spherical micelles to transform into order aggregates (route A). A solid of inorganic-organic nanocomposite is obtained through further hydrothermal treatment. After the surfactant template is burned out by calcinations at more than 500℃, a well-ordered mesoporous molecular sieves material is fabricated.

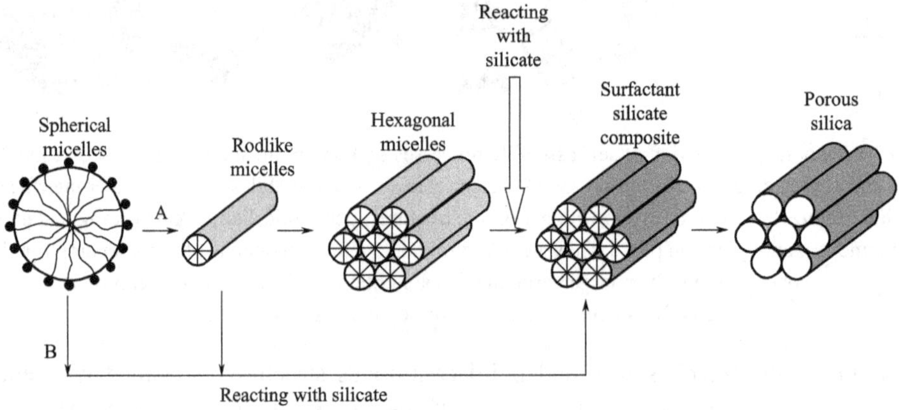

Fig.4.38 Templating mechanisms initially suggested for the formation of ordered nanoscale materials[114, 115]

4.8 Self-assembly of nanocrystals

Like molecular systems, nanocrystals capped with suitable ligands spontaneously assemble into ordered aggregates. The forces that govern the nanocrystal assembly are different in many ways. Surface tension for example, plays an important role because in a nanocrystal, a large fraction of atoms are present at the surface.

Surfactant molecules which self-assemble on solid surfaces have proved to be the best means of obtaining ordered arrays of nanocrystals. The way in which the nanocrystals organize themselves depends critically on the core diameter, the nature of the ligand and the dispersive medium used. Thiolized metal nanocrystals readily arrange into two-dimensional arrays on removal of the solvent.

An example showing the effect of the length of the alkyl chains on 2D self-organisation is provided by Ag nanocrystals deposited on a substrate of graphite[116]. When the length of the alkyl chain $C_nH_{2n+1}SH$ coating the particles is reduced, i.e., n is reduced from 12 to 8, this favors the formation of less extensive monolayers with a fixed distance between particles (see Fig.4.39)[116, 117]. This distance corresponds to the length of a dodecane thiol chain, which is about 2nm [see Fig.4.39(a)]. Therefore this indicates an interpenetration of the chains in the case of nanocrystals

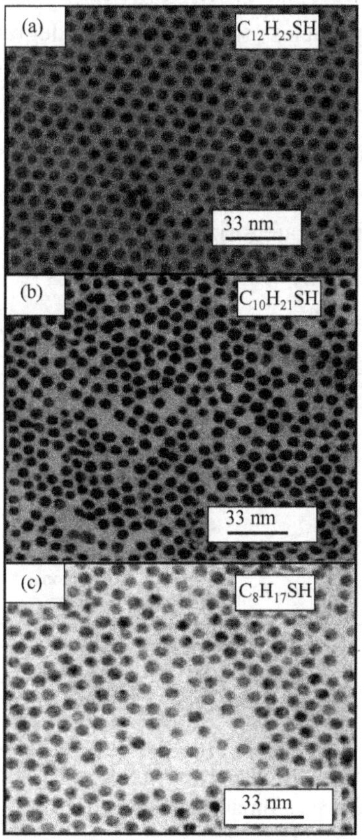

Fig.4.39 TEM images showing the effect of the length of the alkyl chains on 2D organization of Ag nanocrystals (a) dodecane thiol $C_{12}H_{25}SH$; (b) decane thiol $C_{10}H_{21}SH$; (c) octane thiol $C_8H_{17}SH$[116, 117]

coated with dodecane thiol, whereas for nanocrystals coated with shorter chains such as decane thiol or octane thiol, the chains no longer interpenetrate [see Fig.4.39(b) and (c)]. It seems that for shorter chains the interaction with the substrate is stronger. Since this interaction is attractive, the particles cannot move around easily on the substrate, which would explain a looser distance between particles.

4.9 Synthesis and assembly of anisotropic nanoparticles

4.9.1 Anisotropic nanoparticles with feature size

Novel physical or chemical properties of nanoparticles are different from bulk materials because of their feature size at nanometer scale and diverse shape. Thus far, many researchers have focused on developing efficient methods for making nanometer-sized particles. Recently, controlling size distribution or shape has been a new issue in this research field. Therefore, innovative methodology has been required to solve this problem. One candidate is the synthesis of nanoparticles with controllable and directional interaction. Self-assembly of those nanoparticles, or directional self-assembly, will provide cost-effective manufacturing strategy for many useful nanodevices, including sensors, photonic circuits, optoelectronic devices and so on [118].

It is necessary to design, synthesize and functionalize the nanoparticles since the most of nanoparticles developed so far were isotropic particles. Notably, it is important that intrinsic function or properties of nanoparticles should not be degraded during functionalization or assembling process. Recently, Glotzer and Solomon have categorized few types of anisotropic particles at micrometer and nanometer scale which have new structures [119].

However, it is very critical to prepare anisotropic particles with narrow size distribution and then assemble them into three-dimensional structures. The anisotropic nanoparticles with feature size in the range of 1～100 nm can be categorized depending on their shapes as shown in Fig.4.40.

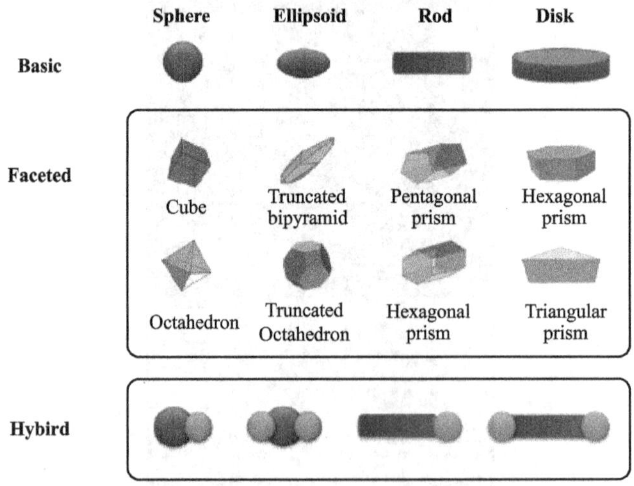

Fig.4.40 Schematic diagram of anisotropic nanoparticles with basic (top), faceted (middle) and hybrid shape (bottom) [120]

4.9.2 Rod-like particles

For crystalline materials, the growth ratio of crystal plane, which is considerably related to interaction between a surface atom and organic molecule, plays a critical factor in determining morphology of a nanostructure. The stronger the binding is of a surfactant to on-growing crystal plane atom, the slower the growth rate of the surface direction become. Consequently, a fast-growing crystal plane disappears gradually as a reaction proceeds and a plane with relatively slow growth becomes the outer surface plane of a final nanoparticle.

When the aspect ratio (a.r.) of nanoparticles is a.r.<5, 5<a.r.<25, or a.r.>25, corresponding particles are classified as short nanorods, long nanorods, or nanowires, respectively[120]. The general strategies for the nanorods in liquid phase are to use intrinsic crystallographic structure, oriented attachment, or selective adhesion of surfactant (Fig.4.41) [121].

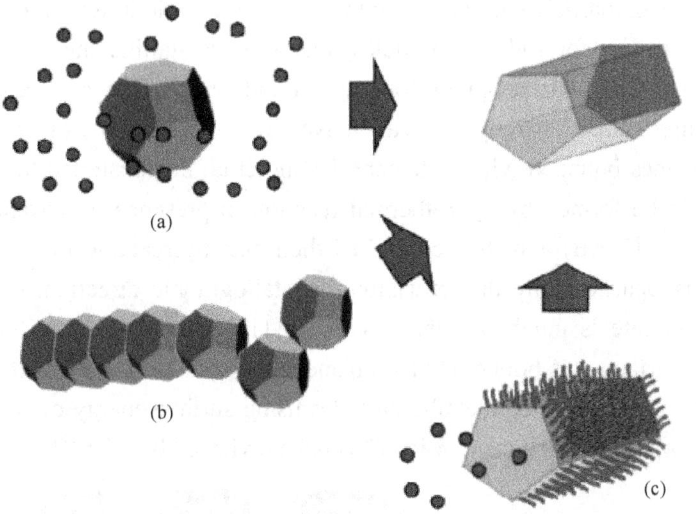

Fig.4.41 Schematic diagram of (a) intrinsic crystallographic structure, (b) oriented attachment, and (c) selective adhesion of surfactant for rod-like nanoparticles[121]

In case of using intrinsic crystallographic structure, an inorganic compound should have asymmetric surface energy. For instance, CdS, CdSe, CdTe, ZnSe, and MnS are a hexagonal wurtzite structure. As shown in Fig.4.42, the surface energy of (001) face in wurtzite structure is higher than that of the other faces because of high packing density [122]. Using this surface energy difference between crystallographic surfaces, the CdSe nanorods were synthesized by simply adding precursors in solvent above the decomposition temperature of precursors, in which 2mL of stock solution [Se:Cd(CH_3)$_2$:tributylphosphine = 1:2:38 by weight] were added into 4g of trioctylphosphine oxide at 360℃ [123]. Agostiano et al. demonstrated that ZnSe nanorods can be synthesized in similar manner, in which the aspect ratio was also controlled by changing injection rate [124].

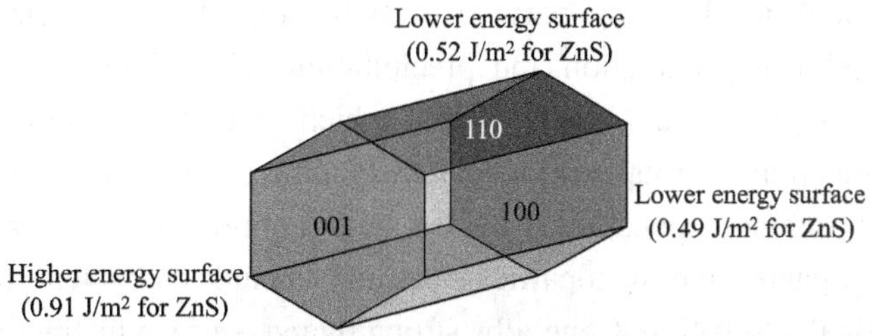

Fig.4.42 Illustration of surface energy for wurtzite hexagonal phase[122]

In case of the oriented attachment for the nanorods, the particles undergo fusion at preferential crystallographic surfaces. For example, cubic ZnS nanorods were formed by the oriented attachment of ZnS nanoparticles[125]. Zn precursor was injected into a solution containing sulfur element and 1-hexadecylamine at 125℃, followed by heating at 300℃. ZnS nanoparticles were gradually transformed into rod during the aging process at 60℃[125]. Recently, oriented attachment of ZnO nanoparticles was also observed by Weller and his colleagues. At first, the quasi-spherical particles were formed by zinc acetate dihydrate dissolved in methanol at about 60℃ and then KOH in methanol was added into ZnO particle suspension. Subsequently, during the evaporation of the solvent and then reflux of the resulting solution, rod-shaped particles were produced [126], in which charges on primary nanoparticle surface may play an important role in formation of ZnO nanorods.

As described earlier for rod-like particles, facets of crystalline metal nanoparticles can be controlled. In most cases, surfactant or ions are bound on specific surfaces preferentially and growth rate is suppressed. Thereby, we can make faceted nanoparticles which have specified crystallographic planes bounded with surfactant. Huang et al. demonstrated that the octahedral Au nanoparticles could be formed by hydrothermal reaction in presence of surfantant (cetyltrimethyl ammonium bromide, CTAB) and they controlled their size by reaction time[127]. Intrinsic growth rates of faceted nanoparticles are different along crystallographic direction. In case of octahedra, when (100) growth rate is much fast than that of (111), octahedal shape forms as depicted in Fig.4.43. If ion or surfactant is bound on (100) plane, cubes can be formed from octahedra. General synthetic methodology for faceted metallic particles using surface energy of crystallographic plane or preferential adsorption of surfactant or ions has been reviewed by Xia[128].

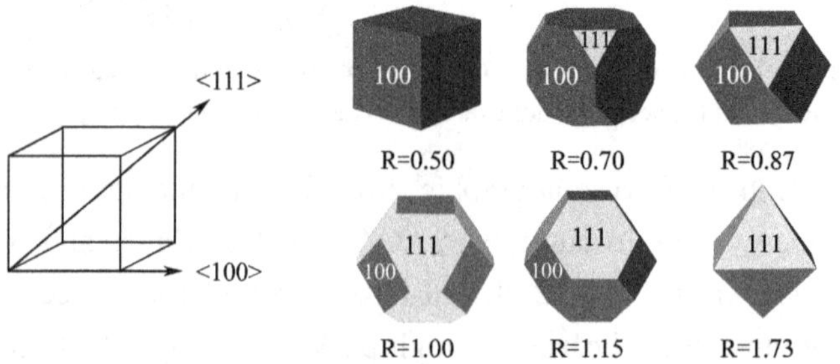

Fig.4.43 Schematic model for the various shapes of a nanostructures depending on the ratio (R) of the growth rate along (100) to (111) [129]

4.9.3 Preparation of various shaped Pt nanoparticles

There are many factors in determining the morphology of final product in colloidal synthesis including reaction precursor, ligands, reaction media (solvents), and reaction environment (temperature and pressure). Ligands, also termed as surfactants, play many important roles in the colloidal synthesis, such as directing particle growth, controlling particle size and stabilizing nanoparticles to prohibit agglomeration and precipitation. The widely used ligands range from organic molecules, polymers and ionic salts, which can bind onto particles through a surface-interacting functional group. Some common functional groups include thiols, amines, carboxylic acids, phosphines, phosphine oxides, and ammonium and carboxylate salts. The interaction between ligands and nanoparticles mainly includes covalent bonding, electrostatic interaction, and physical adsorption. Generally, strong ligand–surface interaction, such as thiols and amines, can restrict the particle growth and thus produces nanoparticles with small size. The multidentate ligands that can increase the surface interaction for weakly bond functional groups can also restrict the particle size. In addition, the concentration of ligands could play an important role in controlling the particle size. For example, higher concentration of ligands gives out small nanoparticles, while solution with diluted ligands produces relatively larger particles. This strategy has been employed to obtain Au nanoparticles with size in the range of 1～4 nm [130~134].

In addition to controlling the particle size, ligands are also essential in determining the particle shape. To minimize the surface area as well as the surface energy, densely packed atomic structures like face-centered-cubic crystals tend to form the so-called Wulff polyhedrons (a truncated octahedron) enclosed by a mix of (111) and (100) facets [135~139]. Since some ligands can selectively bind to certain facets and lower the surface energy, the addition of ligands thus provides an opportunity to prepare other shaped nanoparticles rather than the Wulff polyhedrons. In the case of Pt metal nanoparticles, bromide ions can stabilize the (100) facets as shown in Fig.4.44. When $NaBH_4$ was used as the reducing agent and tetradecylammonium bromide (TTAB) was used as the ligands, Pt nanocubes with size around 15nm were obtained, exposing only (100) facets. However, when molecular hydrogen that favors (111) facets was also introduced into the reaction system, Pt cuboctahedras with size around 15nm were prepared, exposing both (100) and (111) facets. If TTAB was replaced with poly(vinyl)pyrrolidone (PVP) and hydrogen was still being used, Pt tetrahedrons could be obtained with size controllable in the range of 3～10nm[140, 141].

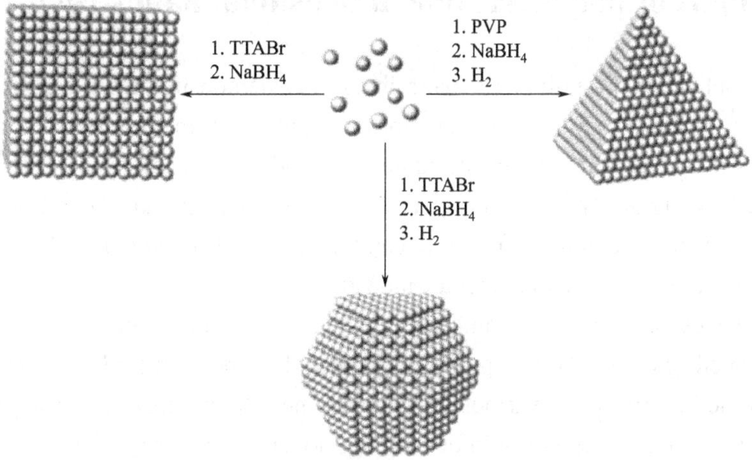

Fig.4.44 Schematic illustration of the preparation of various shaped Pt nanoparticles by varying the synthetic condition[135]

4.9.4 Preparation of various shaped Rh nanoparticles

Humphrey et al. recently reported a seedless polyol synthesis of monodisperse (111)- and (100)-oriented rhodium nanocrystals in sub-10nm (6.5nm) sizes [142]. In this report, ethylene glycol (EG) was used as the solvent as well as the reducing agent, while PVP (poly(vinyl)pyrrolidone) was used as a capping ligand. In a typical process, surfactants were dissolved in a polyol solvent, which was heated up to certain temperature under the protection of an inert gas, e.g., nitrogen or argon. The metal precursor was first dissolved in the polyol solvent and then injected into the reaction media at a high temperature. Upon injection, metal ions were quickly reduced by polyols and protected by ligands from forming large aggregates.

When using $[Rh(Ac)_2]_2$ as the metal precursor, (111)-oriented Rh nanopolyhedra containing 76% (111)-twinned hexagons were obtained; whereas, when employing $RhCl_3$ as the metal precursor in the presence of alkylammonium bromide, such as tetramethylammonium bromide and trimethyl (tetradecyl) ammonium bromide (100)-oriented Rh nanocubes were obtained with 85% selectivity, as shown in Fig.4.45.

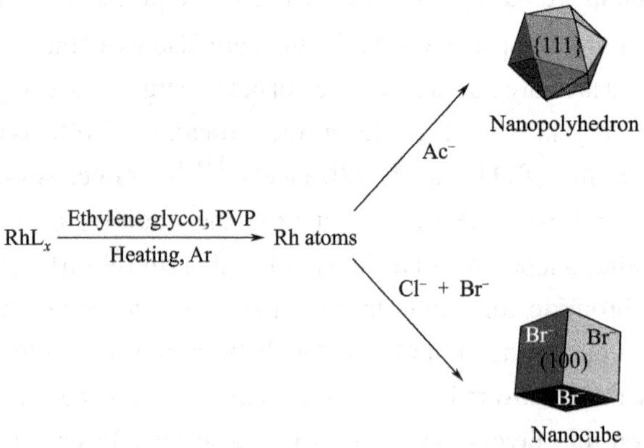

Fig.4.45 Schematic illustration of the reaction pathways responsible for the formation of (111)- and (100)-oriented Rh nanocrystals [142]

4.10 Synthesis of polymeric one dimensional nanostructures (ODNS)

The first application of polymeric nanofibres was demonstrated in a pioneering work by Martin et al.[143,144] where they reported extraordinary increase in the electronic conductivity of polymeric fibers with the confinement of dimension and size in mesoscopic range. Conducting polymers (doped or undoped)[145] science has progressed in leaps and bounds to a stage where scientists are now looking them as a possible candidate to oust the silicon, which has reached the limits of miniaturization, from semiconductor industry.

Polymeric nanotubes, nanowires and nanorods have since then flooded the scene of research with unprecedented growth. This rapid growth rate has been fueled by synthetically novel approaches and the importance to various cross disciplines. Number of scientific groups across the globe have been working successfully in developing novel routes for synthesis of one dimensional polymers such as electrospinning[146], co-electrospinning[147], membrane based[148], tubes by fiber templates (TUFT)[149], electrochemical polymerization within templates and chemical and self-assembly routesand so on[150,151]. All the processes described above are able to yield polymeric ODNS with desired properties and precise aspect ratios.

4.10.1 Electrospinning synthesis of polymer ODNS

Electrospinning is one of the most widely used techniques for synthesis of polymeric fibers. This technique is a novel yet simple and capable of assembling fibrous polymers to diameters from micron level down to less than 100 nm. It is a high voltage process to form ultra-fine solid fibers from a stream of polymeric solution or melt onto a target cathode through a nozzle. As the process is based on electric field, it can be controlled or guided to yield several geometric three dimensional shapes. However, our description will be restricted to the synthesis of nanofibres which are one dimensional ultra-fine solid fibers having large surface to volume/mass ratio (approximately 100 m^2/g for a fiber with a diameter of 100 nm). MacDiarmid et al. were able to electrospin polyaniline-based nanofiber with diameter below 30 nm[152]. However, precise control of these parameters is required for obtaining specificity in the product shape, size or properties. A variety of particles can be added to the polymeric blend, solution or melt to reinforce the fiber matrix or encapsulate special materials in the fibers. One of the advantages of electrospinning is that water can be used as a solvent and hence water soluble polymers such as poly-ethyl oxide (PEO), polyvinyl alcohol (PVA), poly-lactic acid (PLA), etc. can also be electrospun. Thus the entire gamut of polar and non-polar organic polymers can be spun using this technique[153].

The electrospinning technique is a high voltage electrostatic method in which a polymer solution (or melt) maintained at a high positive potential, in a variety of different polar/nonpolar solvents is placed in a hypodermic syringe at a fixed distance from a metal cathode[154,155]. The polymeric solution or melt coming out of the syringe or needle forms a droplet owing to its surface tension, which gets charged to various degrees depending on applied voltage and the concerned polymer. The mutual charge repulsion causes a force directly opposite to the surface tension[156]. As the intensity of applied field is increased the droplet changes its shape from hemisphere to an elongated cone like structure popularly known as Taylor[157]. As the applied voltage crosses a critical value of potential, the surface charge overcomes the surface tension of the droplet on the tip and a jet of fluid is ejected from the tip of Taylor cone. This polymer undergoes a whipping process (or solidification in case of melt) in which the solvent evaporates within the flight distance. The polymer is highly stretched as it lands on the cathode. Dry fibers accumulate on the surface of the cathode resulting in a non-woven mesh of nano- to micron-diameter fibers depending upon experimental parameters as shown in the schematic of the process in Fig.4.46[158].

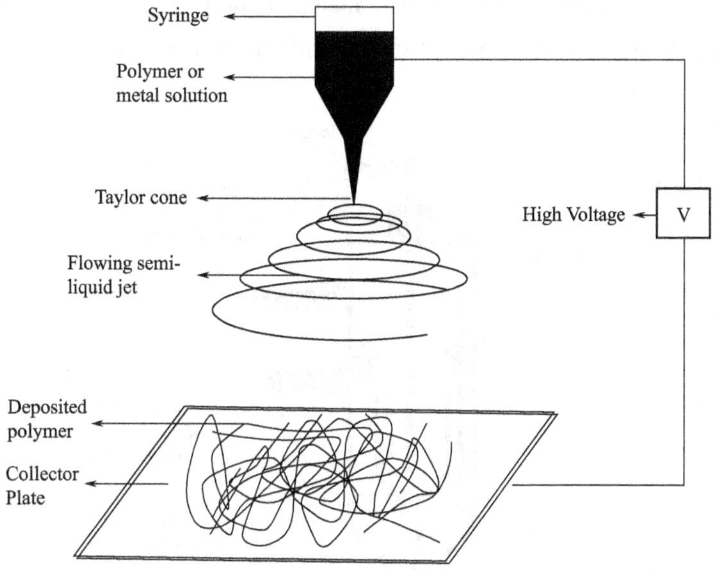

Fig.4.46 A schematic diagram of electrospinning process[158]

As stated earlier the electrospinning process is characterized by rapid evaporation of the solvent or solidification of the melt. The nanostructure formation occurs on a millisecond time scale and hence one would expect that the nucleation and growth of the crystals would be different from their bulk or macro counterparts. However this may not be the case and the mechanism for nucleation has found to be the same as that observed for thicker fibers.

Though the process is very simple in operation but is very complex in actual mechanism of spinning of fibers and is controlled by various types of instabilities such as the Rayleigh instability, an axisymmetric instability and the whipping or bending instability[159~161]. The whipping action is attributed as the major factor governing the elongation and thinning of the electrospun fibers. The other instabilities are responsible for fluctuations of the radius of the jet. Electrospinning process parameters can be divided into (a) Electromechanical: applied potential, flow rate of polymer, distance between the cathode and the anode. (b) Chemical: molecular weight of the polymer, molecular weight distribution, architecture (branched or chain polymer), co- or mono-polymer, conductivity and dielectric constant of the polymer solution (depending on the functional groups), (c) Rheological: viscosity, concentration of solution, surface tension, etc. and (d) External: temperature, humidity, air velocity or vacuum of the chamber.

Recently a new version of electrospinning termed as co-electrospinning has been developed wherein the researchers have produced PEO-PDT type core–shell nanofibers[147]. The experimental set-up is characterized by two air inlets as shown in Fig.4.47. This process is capable of producing core–shell nano/meso fibers. Both liquids flowing from the core and the surrounding concentric annular nozzles form a compound droplet which undergoes a transformation into a compound Taylor cone with a compound jet co-electrospun from its tip. This jet when pulled by electric field undergoes the similar instability effects, as in the traditional process. As the solvent evaporates the compound jet solidifies resulting in compound core–shell nanofibers. The high speed of electrospinning process prevents the mixing of core and shell polymers. Co-electrospinning can also be applied to polymer-metal salt systems as shown by synthesis of PLA-$Pd(OAc)_2$ system[147]. On annealing the coaxial fibers for 2h at 170℃ metallization of $Pd(OAc)_2$ into Pd was observed. Co-electrospinning can also be used to form the polymer nanotubes and hence can be of extensive use in a variety of systems.

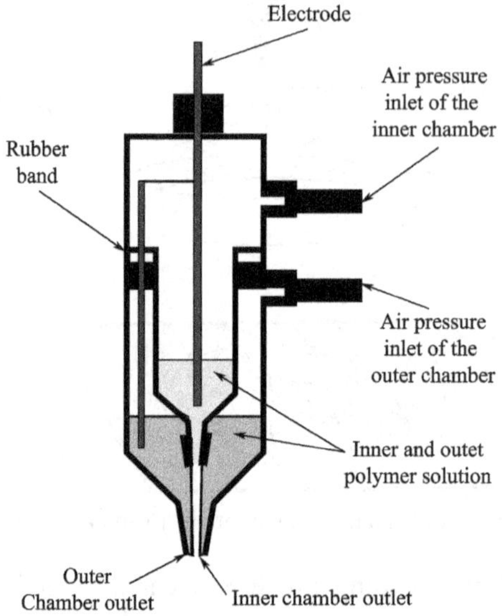

Fig.4.47　Experimental set-up for co-electrospinning of compound core-shell nanofibers[147]

4.10.2 Membrane/template-based synthesis of polymer ODNS

The significance of the template-based method has been emphasized in almost every inorganic system. The same was also found to have a prominence in case of the polymer ODNS, as a result of its reliability and preciseness in the product. However, polymeric solutions or melts have an advantage over metallic or metal oxides structures synthesized using this method because of their inherently high wettability. Thus formation of polymeric ODNS using membrane based or template based synthesis is relatively easy and hence popular. Martin et al. have pioneered the use of this technique to form ordered ODNS. A few more significant details about the membrane or template based synthesis are presented below.

(1) Track-etch polymeric membrane

Most of the polymeric ODNS work in template based synthesis has been limited to two types of membranes, the "track-etch" polymeric membranes and the porous alumina/silica membranes (the latter is less popular in polymeric systems). Track etch membranes infer their name from their synthesis via the track-etch[162] method. These membranes contain cylindrical pores of uniform diameters and the most common material used to prepare these membranes is polycarbonate. Pore diameters down to 10 nm with a pore density of 10^9 pores/cm^2 are commercially available. On the other hand, the porous alumina or silica membranes are prepared by electrochemical etching (or anodization) of the base metals aluminum or silicon to produce ordered nanoporosity on the surface of these materials. Historically, the conductive polymers were synthesized using oxidative polymerization of polymers either electrochemically or using chemical oxidizing agent within the pores of these templates[163].

(2) CVD polymerization methods

CVD has also been one of the major contributors for producing polymeric ODNS from templates. Process is simple and involves placing of the membrane in a high temperature furnace while reactive gases are passed through the furnace. Thermal decomposition of gas occurs in the pores resulting in the decomposition/deposition of polymer along the walls of the pores. Synthesis of PPV nanotubes and nanorods using CVD polymerization methods in the pores of alumina and polycarbonate filters having pore diameters of 10～200nm have been reported[164]. In a similar effort polypyrrole was synthesized using track etched membranes and their further electrochemical polymerization[165]. Different electrolytes were used to see the effect of electrolyte on the morphology and properties of polypyrrole nanotubes.

(3) Surfactant template

Fig.4.48 shows the comb-shaped supramolecules and their hierarchical self-organization with primary and secondary structures. The schemes can, in principle, be used both for flexible and rod like polymers. In the first case, simple hydrogen bonds can be sufficient, but in the latter case a synergistic combination of bondings (recognition) is generally required to oppose macrophase separation tendency. In A through C, the self organized structures allow enhanced processibility due to plastization, and solid films can be obtained after the side chains are cleaved (D). Self-organization of supramolecules obtained by connecting amphiphiles to one of the blocks of a diblock copolymer (E) results in hierarchically structured materials. Functionalizable nanoporous materials (G) are obtained by cleaving the side chains from a lamellae-within-cylinders structure (F). Disk-like objects (H) may be prepared from the same structure by crosslinking slices within the cylinders, whereas nanorods (I) result from cleaving the side chains from a cylinder-within-lamellae structure. In the scheme, (A) is shown as a flexible polymer, whereas (B) and (C) are shown as rod like chains, respectively[166].

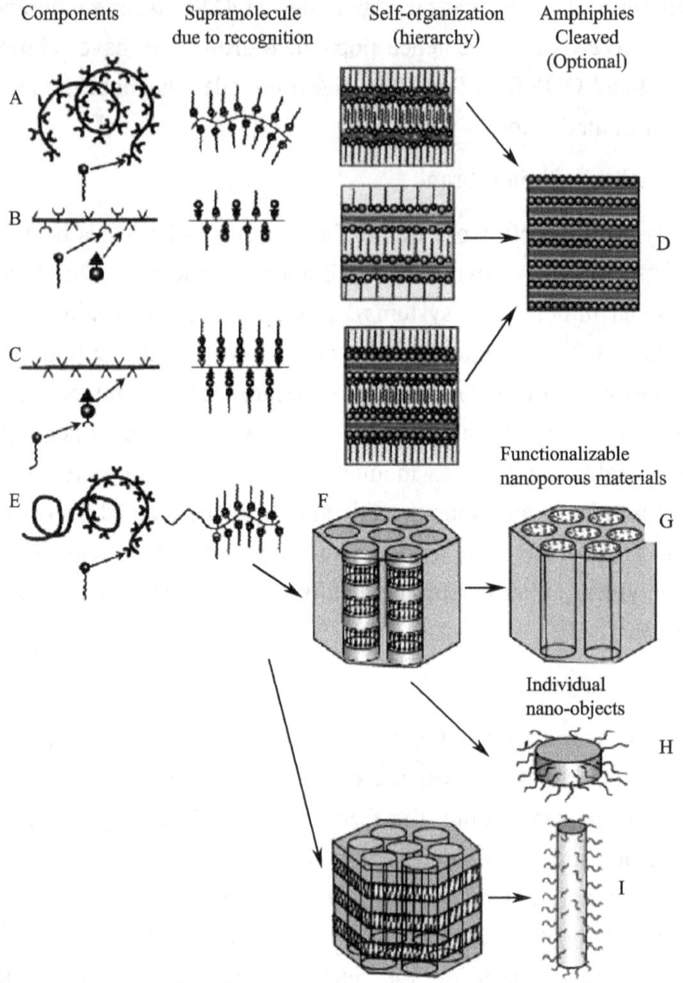

Fig.4.48 Comb-shaped supramolecules and their hierarchical self-organization

4.10.3 Template-free synthesis of polymer ODNS

Researchers across the globe opened various fronts of research out of their own scientific interests for synthesis of polymeric ODNS. These processes, commonly known as "template-free synthesis" or "self-assembly" have successfully been able to produce polymeric ODNS. Their major advantage is the reduction in diameters up to or below 50nm; however, the distribution of diameters of the nanowires, nanotubes or nanorods is not uniform.

Polypyrrole microtubules were synthesized by this method using β-naphthalene sulfonic acid (NSA) and ammonium persulphate (APS) as oxidant. The internal diameter of tubes was found to be 30～100nm. "In situ doping polymerization" technique substitutes the use of templating and detemplating steps with a single step process through simultaneous doping and nanostructure formation. The internal diameter of the microtubules could be controlled by varying the reaction temperature, while the microtubules formed during in situ polymerization at 0℃ have a diameter of 100 nm, the same synthesized at 25℃ have an internal diameter of 400nm. However, the internal diameter for a given temperature was found to increase with increasing NSA concentration. It must be noted that room temperature conductivity of so formed microtubules was found to be slightly less than granular polypyrrole at room temperature[167]. Another set of experiments reported the effect of polymerization time and rate as well as solution's ionic strength on the morphology, conductivity and molecular structure of NSA/polypyrrole tubes[168]. It was observed that the formation of NSA/polypyrrole microtubules via a template-free method was a slow and self-assembled process.

A host of other synthetic methods which are either a modification or completely new processes were explored. For example, synthesis of dendritic polyaniline in a surfactant gel and synthesis of large array of nanowires by a three step electrochemical deposition process[169,170]. The transport properties of these low dimensional systems could be of great interest owing to their potential in future nanoelectronics circuitry. Out of several conducting polymers available, polyacetylene (PA), polyaniline (PANi), poly(3,4-ethylenedioxythiophene) (PEDOT), polypyrrole (PPy) and polyphenylenevinylene (PPV) in their doped or undoped form have attracted most of the attention.

It is evident from the band gap energy that these polymeric materials have very low conductivity and fall in the insulator/semiconductor range. In order to make them applicable as conducting materials, their basic poor intrinsic conductivity must be increased. This can be achieved by doping these polymeric materials with various oxidizing or reducing agents.

4.11 Green nanosynthesis

Nearly all of the principles of green chemistry can be readily applied to the design of nanoscale products[171]. Many preparations of the building blocks in nanotechnology involve hazardous chemicals, low material conversions, high energy requirements and generation of hazardous substances. Thus, there are multiple opportunities to develop greener processes for the manufacture of these materials.

Green chemistry is the utilization of a set of principles that reduces or eliminates generation of hazardous substances in the design, manufacture, and application of chemical products[172]. Green nanoscience and nanotechnology involve the application of green chemistry principles to design nanoscale products[173]. A growing number of applications in nanoscience/nanotechnology are being developed to promise environment-friendly including new catalysts for environmental remediation[174], cheap and efficient photovoltaic technology[175], thermoelectric materials for cooling without refrigerants[176] and nanocomposites for vehicles[177]. Nanoscale sensors[178] can also offer faster response times and lower detection limits, making on-site, real-time detection possible.

4.11.1 Prevent wastes

One subset of these principles is to prevent waste, use safer solvents/alternative reaction media, and reduce derivatives by designing some methods that minimize the process steps and the amount of ancillary materials used to carry out those steps.

An illustrative example involves the fabrication of nanoscale features on a substrate such as a silicon wafer. The traditional strategy for producing these structures is a "top-down" approach that creates nanoscale structures through a lithographic process involving a significant number of depositions, patterning, etching, and cleaning steps. This method contributes to the waste stream. Therefore, a green approach includes self-assembly or "bottom-up" processes by employing self-assembly reactions or "direct" writing deposition to generate and interconnect the structures. Such alternatives eliminate many process steps and minimize the amount of materials and solvents.

Solvent is necessary for the purification and size selection of nanomaterials. Current methods for purification of nanoparticle samples involve washing or extraction to remove impurities. This process typically requires liters of solvent per gram of nanoparticles and is not usually effective in removal of all the impurities. Size selection is essentially purification procedure including extraction or fractional crystallization, or chromatographic methods to separate the different sizes.

4.11.2 Atom economy

Another subset of the principles attempts to maximize materials efficiency, i.e., conversion of raw materials into desired products by enhancing reaction selectivity and efficiency. The concept of atom economy readily applies to wet-chemical nanomaterial preparations. However, the concept also applies to the fabrication of extended nanoscale structures that uses "bottom-up" approaches such as self-assembly of molecules or nanoscale subunits into more complex structures. Because these approaches can be able to transform much more raw materials into products than corresponding "top-down" methods, they have higher atom economy.

4.11.3 Using safer solvents

The safer solvents/alternative reaction media and renewable feedstock can be employed to enhance the process safety or reduce the hazards associated with a process. This is an area of active investigations within nanoscience. There are already a few examples that illustrate the application of these principles to enhance process safety by replacing or reducing the use of hazardous solvents. In some cases, the use of benign feedstock derived from renewable sources may prove a successful strategy for enhancing safety in nanomaterial production.

4.11.4 Enhance energy efficiency

The last subset of the principles involves enhancing energy efficiency. "Bottom-up" assembly of nanodevices greatly reduced the number of processes can be accomplished under very mild conditions and could reduce the chances of particle contamination. The development of catalysts that can accelerate these reactions may be a useful strategy. In most cases, in-situ monitoring of reaction conditions will enhance energy efficiency and save cost.

References

[1] R. Feymann, There's plenty of room at the bottom: an invitation to enter a new field of physics, Lecture at California Institute of Technology, 29(1959).

[2] P. Sanguansri, M.A. Augustin, Trends in Food Science & Technology, 17 (2006) 547-556.

[3] W.A. Goddard, D.W. Brenner, S.E. Lyshevski, G.J. Iafrate, Handbook of nanoscience, engineering, and technology, Second edition, (2007) Taylor & Francis Group, LLC, P1-1.

[4] M.V. Tirrell, A. Katz, Self-assembly in materials synthesis, MRS Bull, 30 (2005).
[5] Z. Hassan, C.H. Lai,Ideals and realities: selected essays of Abdus Salam. (1984) World Scientific, Singapore.
[6] C. Brechignac, P. Houdy, M. Lahmani, Nanomaterials and Nanochemistry, (2007) Springer-Verlag Berlin Heidelberg, ISBN: 978-3-540-72992-1, P457.
[7] C.K. Carl, Processing, Properties and Potential Applications, (2002) Noyes Publications, ISBN: 0-8155-1451-4.
[8] C. Suryanarayana, Nanocrystalline Materials, Int. Mat. Rev., 40 (1995) 41-64.
[9] B.S. Murty, S. Ranganathan, Novel Materials Synthesis By Mechanical Alloying/Milling, Int. Mat. Rev., 43 (1999) 101-141.
[10] P.S. Gilman, J.S. Benjamin, Mechanical Alloying, (1983) Annual Review of Materials Science, Annual Reviews, Palo Alto, CA P279-300.
[11] J.S. Benjamin, T.E. Volin, The Mechanism of Mechanical Alloying, Metall. Trans., 5 (1974) 1929-1934
[12] M.L. Lau, H.G. Jiang, W. Nuchter, E.J. Lavernia, Thermal Spraying of Nanocrystalline Ni Coatings, Phys. Stat. Sol. (A), 166 (1998) 257-268.
[13] H.S. Nalwa, Encyclopedia of Nanoscience and Nanotechnology, 1 (2004) American Scientific Publishers, ISBN: 1-58883-057-8, P687-726.
[14] P.P. Phule, S.H. Risbud, J.Mater .Sci., 25 (1990) 1169.
[15] J. C. Mutin, J.C. Niepce, J.Mater .Sci.Lett., 18 (1983) 304.
[16] H. Tagwa, J. Ohashi, Denki Kagaku Ky Okao, 52 (1984) 485.
[17] H.S. Nalwa, Encyclopedia of Nanoscience and Nanotechnology, 1 (1) (2004) American Scientific Publishers, ISBN: 1-58883-057-8, P719.
[18] K. Robert, H. Ian, G. Mark, Nanoscale Science and Technology, (2007) ISBN: 978-7-03-018257-9, P38-41.
[19] L.W. Zhong, Mater. Today, 7 (6) (2004) 26.
[20] M. Kohfer, W. Fritzsche, (2004) Wiley-VCH Verlag GmbH & Co. KGaA, Weinheim, ISBN: 3-527-30750-8, P43.
[21] M. Anpo, S. Dohshi, M. Kitano, Y. Hu, M. Takeuchi, M. Matsuoka, Annu. Rev. Mater. Res., 35 (2005) 1.
[22] H. Hahn, R.S. Averback, J. Appl. Phys., 67 (1990) 1113.
[23] T.S. Yoshida, Y. Takeyama, Yamada, K. Mutoh, Appl. Phys. Lett., 68 (1996) 1772.
[24] T.T. Kodas, M.J. Hampden-Smith, Aerosol Processing of Materials, (1999) Wiley-VCH, New York.
[25] K. Kimura, Fine Particles: Synthesis, Characterization, and Mechanisms of Growth, 92nd ed, (2000) P513-550.
[26] H. Muhlenweg, A. Schild, A. Gutsch, C. Klasen, S. Pratsinis, J.Metastable Nanocryst.Mater,8 (2000) 941.
[27] H.S. Nalwa, Encyclopedia of Nanoscience and Nanotechnology, 1 (1) (2004) American Scientific Publishers, P691.
[28] B.L. Cushing, V.L. Kolesnichenko, C.J. O'Connor, Recent Advances in the Liquid-Phase Syntheses of Inorganic Nanoparticles, Chem. Rev., 104 (2004) 3893-3946.
[29] G.N. Glavee, K.J. Klabunde, C.M. Sorensen, Hadjipanayis, G.C. Langmuir, 9 (1993) 162.
[30] G.N. Glavee, K.J. Klabunde, C.M. Sorensen, Hadjipanayis, G. C. Inorg. Chem., 34 (1995) 28.
[31] G.N. Glavee, K.J. Klabunde, C.M. Sorensen, Hadjipanayis, G.C. Tang, Z.X. Yiping, L. Nanostruct. Mater., 3 (1993) 391.
[32] G.N. Glavee, K.J. Klabunde, C.M. Sorensen, Hadjipanayis, G. C. Langmuir, 10 (1994) 4726.
[33] I.D. Dragieva, Z.B. Stoynov, K.J. Klabunde, Sci. Mater., 44 (2001) 2187.
[34] Y. Tan, X. Dai, Y. Li, D. Zhu, J. Mater. Chem., 13 (2003) 1069.
[35] I.Pastoriza-Santos, L. M.Liz-Marzan, Langmuir, 15 (1999) 948.
[36] Z.S. Pillai, P.V. Kamat, J. Phys. Chem. B, 108 (2004) 945.
[37] N.B. Vargaftik, Table on the Thermophysical Properties of Liquids and Gases, (1975) Hemisphere Publishing Corporation.
[38] B.L.S. Komarneni, Chemical processing of ceramics, (2005) Taylor & Francis Group, LLC, ISBN-10: 1-57444-648-7 (Hardback), ISBN-13: 978-1-57444-648-7 (Electronic), P32.

[39] R.I. Walton, Chem. Soc. Rev., 31 (2002) 230.
[40] W.J. Dawson, Hydrothermal synthesis of advanced ceramic powders, Am. Ceram. Soc. Bull. 67 (1988) 1673.
[41] E. Matijevic, W.P. Hsu, Preparation and properties of monodispersed colloidal particles of lanthanide compounds. I. Gadolinium, europium, terbium, samarium, and cerium(III), J. Colloid Interface Sci. 118 (1987) 506.
[42] A. Cabanas, J.A. Darr, E. Lester, M. Poliakoff, Continuous hydrothermal synthesis of inorganic materials in a near-critical water flow reactor; the one-step synthesis of nano-particulate $Ce_{1-x}Zr_xO_2$ (x = 0–1) solid solutions, J. Mater. Chem. 11 (2001) 561.
[43] A. Cabanas, J.A. Darr, M. Poliakoff, E. Lester, A continuous and clean one-step synthesis of nano-particulate $Ce_{1-x}Zr_xO_2$ solid solutions in near-critical water, Chem. Commun. (2000) 901.
[44] E. Reverchon, R. Adami, Nanomaterials and supercritical fluids, Journal of Supercritical Fluids 37 (2006) 1-22.
[45] A. Landsberg, T.T. Campbell, J. Metals, 856 (1965).
[46] Y.D. Tretyakov, O.A. Shlyakhtin, J.Mater .Chem., 9 (1999) 19.
[47] B.L.S. Komarneni, Chemical processing of ceramics, (2005) Taylor & Francis Group, LLC, ISBN-10: 1-57444-648-7 (Hardback), ISBN-13: 978-1-57444-648-7 (Electronic), P94.
[48] B. Huang, M. Cao, F. Nie, H. Huang, C. Hu, Construction and properties of structure- and size-controlled micro/nano-energetic materials, Defence Technology 9 (2013) 59-79.
[49] Y.F. Maa, P.A. Nguyen, T. Sweeney, S.J. Shire, C.C. Hsu, Protein inhalation powders: spray drying vs spray freeze drying, Pharm. Res. 16(2) (1999) 249-254.
[50] S. Kaluza, M. Muhler, On the precipitation mechanism and the role of the post-precipitation steps during the synthesis of binary ZnO-Al_2O_3 composites with high specific surface area, J. Mater Chem 19 (2009) 3914-3922.
[51] J.D. Engstrom, D.T. Simpson, E.S. Lai, III R.O. Williams, K.P. Johnston. Morphology of protein particles produced by spray freezing of concentrated solutions, Eur. J. Pharm. Biopharm. 65(2) (2007)149-162.
[52] P.J. McGrath, R.M. Laine, Theoretical process development for freezedrying spray-frozen aerosols, J. Am. Ceram. Soc. 75(5) (1992) 1223-1228.
[53] M.D. Rigterink, Advances in technology of the cryochemical process, Am. Ceram. Soc. Bull. 51(2) (1972) 158-161.
[54] B. Huang, Z.Q. Qiao, F.D. Nie, M.H. Cao, J. Su, H. Huang et al., Fabrication of FOX-7 quasi-three-dimensional grids of one-dimensional nanostructures via a spray freeze-drying technique and size-dependence of thermal properties. J. Hazard. Mater. 184(1-3) (2010) 561-566.
[55] M. Ebelmen, Ann. Chim. Phys., 16 (1846) 129.
[56] C.J. Blinker, G.W. Scherrer, Sol-Gel Science: The Physics and Chemistry of Sol-Gel Processing, (1990) San Diego, CA: Academic Press.
[57] S. Sakka, et al., J. Non-Cryst. Solids, 63 (1984) 223-235.
[58] D. Gallagher, L. Klein, J. Colloid Interface Sci., 109 (1986) 40-45.
[59] H. Kozuka, H. Kuroki, S. Sakka, J. Non-Cryst. Solids, 100 (1-3) (1988) 226-230.
[60] T.P. Hoar, J.H. Schulman, Nature, 152 (1943) 102.
[61] J.H. Schulman, D.P. Riley, J. Colloid Sci., 3 (1948) 383.
[62] H.S. Nalwa, Encyclopedia of Nanoscience and Nanotechnology, 2 (1) (2004) American Scientific Publishers, P539.
[63] I. Capek, Preparation of metal nanoparticles in water-in-oil (w/o) microemulsions, Adv. Colloid Interface Sci., 110 (2004) 49-74.
[64] M.P. Pileni, Catal. Today, 58 (2000) 151-166.
[65] M.P. Pileni, Langmuir, 13 (1997) 3266-3276.
[66] M.P. Pileni, Langmuir, 17 (2001) 7476-7486.
[67] I. Lisiecki, et al., Phys. Rev. B, 61 (2000) 4968-4974.

[68] C. Brechignac, P. Houdy, M. Lahmani, Nanomaterials and Nanochemistry, (2007) Springer-Verlag Berlin Heidelberg, ISBN 978-3-540-72992-1, P449.
[69] J. Tanori, M.P. Pileni, Change in the shape of copper nanoparticles in ordered phases, Adv. Mat., 7 (1995) 862-864.
[70] A. Filankembo, M.P. Pileni, J. Phys. Chem. B, 104 (2000) 5865-5868.
[71] C. Brechignac, P. Houdy, M. Lahmani, Nanomaterials and Nanochemistry, (2007) Springer-Verlag Berlin Heidelberg, ISBN 978-3-540-72992-1, P448-450.
[72] C. Brechignac, P. Houdy, M. Lahmani, Nanomaterials and Nanochemistry, (2007) Springer-Verlag Berlin Heidelberg, ISBN 978-3-540-72992-1, P450-451.
[73] A. Filankembo,et al., J. Phys. Chem. B, 107 (2003) 97.
[74] M. Oghbaei, O. Mirzaee, Microwave versus conventional sintering: A review of fundamentals, advantages and applications, J. Alloys Compd., 494 (2010) 175-189.
[75] Y.S Li, W.S Yang, Microwave synthesis of zeolite membranes: A review, J. Membr. Sci., 316 (2008) 3-17.
[76] M. Al-Harahsheh, S.W. Kingman, Microwave-assisted leaching—a review, Hydrometallurgy, 73 (2004) 189-203.
[77] S.A. Galema, Microwave chemistry, Chem. Soc. Rev., 26 (1997) 233-238.
[78] D.M.P. Mingos, D.R. Baghurst, Application of microwave dielectric heating effects to synthetic problems in chemistry. Chem. Soc. Rev., 20 (1991) 1-47.
[79] W. Yu, W. Tu, H. Liu, Langmuir, 15 (1999) 6.
[80] M. Tsuji, M. Hashimoto, T. Tsuji, Chem. Lett., (2002) 1232.
[81] G. Ruecroft, D. Hipkiss, T. Ly, N. Maxted, P.W. Cains, Sonocrystallization: the use of ultrasound for improved industrial crystallization, Org. Process Res. Dev., 9 (2005) 923-932.
[82] S. Pilli, P. Bhunia, S. Yan, R.J. LeBlanc, R.D. Tyagi, R.Y. Surampalli, Ultrasonic pretreatment of sludge: A review, Ultrasonics Sonochemistry, (2010) doi:10.1016/j.ultsonch.2010.02.014.
[83] B. Viswanathan, National Centre for Catalysis Research (NCCR), Madras (2008).
[84] O. Dominguez, M. Phillippot, J. Bigot, The Relationship Between Consolidation Behavior and Particle Size in Fe Nanometric Powders, Scr. Met. Mater., 32 (1995) 13-17.
[85] R.A. Andrievski, Compaction and sintering of ultrafine powders, Intern.Powder Metall., 30 (1994) 59-66.
[86] R.A. Andrievski, V.I. Torbov, E.N. Kurkin, Nanocrystalline titanium nitride, Proceedings of the 13th International Plansee Seminar, (H. Bildstein and R. Eck, eds.), Metallwerk Plansee, Reutte, 3 (1993) 649-655.
[87] E.Y. Gutmanas, L.I. Trusov, I. Gotman, Consolidation, microstructure and mechanical properties of nanocrystalline metal powders, Nanostr. Mater., 8 (1994) 893-901.
[88] G. M. Whitesides, Self-assembling materials, Sci. American, 273(3) (1995) 146.
[89] H. S. Nalwa, Encyclopedia of Nanoscience and Nanotechnology, 6 (1) 2004 American Scientific Publishers, P182.
[90] D.F. Evans, H. Wennerstrom, The Colloidal Domain, (1999) Wiley-VCH, Weinheim.
[91] J.M. Lehn, Angew. Chem.-Int. Edit. Engl., 29 (1990) 1304.
[92] L.A. Cuccia, J. Lehn, J. Homo, M. Schmutz, Angew. Chem., 112 (2000) 239.
[93] S. Forster, T. plantenberg, Angew. Chem. Int. Ed., 41 (2002) 688.
[94] Z. Luan, H. He, W. Zhou, J. Klinowski, J. Chem. Soc., Faraday Trans., 94 (1998) 979.
[95] H. Yang, G. Vovk, N. Coombs, I. Sokolov, G.A. Ozin, J. Mater. Chem., 8 (1998) 743.
[96] X. Chen, L. Huang, Q. Li, J. Phys. Chem. B, 101 (1997) 8460.
[97] M. T. Anderson, P.S. Sawyer, T. Rieker, Microporous Mesoporous Mater., 20 (1998) 53.
[98] A. Corma, Q. Kan, F. Rey, Chem. Commun., (1998) 579-580.
[99] A.A. Romero, M.D. Alba, J. Klinowski, J. Phys. Chem. B, 102 (1998) 123.
[100] M. Templin, A. Franck, A.D. Chesne, H. Leist, Y. Zhang, R. Ulrich, V. Schadler, U. Wiesner, Science, 278 (1997) 1795.

[101] T. Kimura, Y. Sugahara, K. Kuroda, Chem. Commun., (1998) 559.
[102] Z. Luan, D. Zhao, H. He, J. Klinowski, L. Kevan, J. Phys. Chem. B, 102 (1998) 1250.
[103] A. Firouzi, F. Atef, A.G. Oertli, G.D. Stucky, B.F. Chmelka, J. Am. Chem. Soc., 119 (1997) 3596.
[104] S. Oliver, A. Kuperman, G.A. Ozin, Angew. Chem. Int. Ed., 37 (1998) 46.
[105] S.R.J. Oliver, G.A. Ozin, J. Mater. Chem., 8 (1998) 1081.
[106] M. Yada, H. Kitamura, M. Machida, T. Kijima, Langmuir, 13 (1997) 5252.
[107] S. Cabrera, J.E. Haskouri, A. Beltran-Porter, D. Beltran-Porter, M.D. Marcos, P. Amoros, Solid State Sci., 2 (2000) 513.
[108] V.F. Stone, Jr. Davis, R.J. Davis, Chem. Mater., 10 (1998) 1468.
[109] G.J. Li, S. Kawi, Talanta, 45 (1998) 759.
[110] L. Qi, J. Ma, H. Cheng, Z. Zhao, Langmuir, 14 (1998) 2579.
[111] G. Pacheco, E. Zhao, A. Garcia, A. Sklyarov, J.J. Fripiat, J. Mater. Chem., 8 (1998) 219.
[112] K.C. Kwiatkowski, C.M. Lukehart, Synthesis and Processing, Handbook of Nanostructured Materials and Nanotech, 1 (2000) P387-421.
[113] G.S. Attard, P.N. Bartlett, N.R.B. Coleman, J.M. Elliott, J.R. Owen, J.H. Wang, Science, 278 (1997) 838.
[114] J.S. Beck, J.C. Vartuli, W.J. Roth, M.E. Leonowicz, C.T. Kresgc, K.D. Schmitt, C.T.-W Chu, D.H. Olson, E.W. Sheppard, S.B. McCullen, J.B. Higgins, J.L. Schlenker, A new family of mesoporous molecular sieves prepared with liquid crystalline templates, J. Am. Chem. Soc., 114 (1992) 10834-10843.
[115] C.T. Kresge, M.E. Leonowicz, W.J. Roth, J.C. Vartuli, J.S. Beck, Ordered mesoporous molecular-sieves synthesized by a liquid-crystal template mechanism, Nature, 359 (1992) 710-712.
[116] L. Motte, E. Lacaze, M. Maillard, M.P. Pileni, Langmuir, 16 (2000) 3803.
[117] C. Brechignac, P. Houdy, M. Lahmani, Nanomaterials and Nanochemistry, (2007) Springer-Verlag Berlin Heidelberg, ISBN 978-3-540-72992-1, P506.
[118] G. Lee, Y. Cho, S. Park, G. Yi, Synthesis and assembly of anisotropic nanoparticles, Korean J. Chem. Eng., 28(8) (2011) 1641-1650.
[119] S. C. Glotzer, M. J. Solomon, Nat. Mater. 6 (2007) 557.
[120] C. J. Murphy, A. M. Gole, S. E. Hunyadi, C. J. Orendorff, Inorg. Chem. 45 (2006) 7544.
[121] Y. Xia, P. Yang, Y. Sun, Y. Wu, B. Mayers, B. Gates, Y. Yin, F. Kim, H. Yan, Adv. Mater. 15 (2003) 353.
[122] Y.-W. Jun, J.-H. Lee, J.-S. Choi, J. Cheon, J. Phys. Chem. B, 109 (2005) 14795.
[123] X. Peng, L. Manna, W. Yang, J. Wickham, E. Scher, A. Kadavanich, A. P. Alivisatos, Nature 404 (2000) 59.
[124] P. D. Cozzoli, L. Manna, M. L. Curri, S. Kudera, C. Giannini, M. Striccoli, A. Agostiano, Chem. Mater. 17 (2005) 1296.
[125] J. H. Yu, J. Joo, H. M. Park, S.-I. Baik, Y.W. Kim, S. C. Kim, T. Hyeon, J. Am. Chem. Soc. 127 (2005) 5662.
[126] C. Pacholski, A. Kornowski, H. Weller, Angew. Chem. Int. Ed. 41 (2002) 1188.
[127] C.-C. Chang, H.-L. Wu, C.-H. Kuo, M. H. Huang, Chem. Mater.20 (2008) 7570.
[128] X, Lu, M. Rycenga, S. E. Skrabalak, B. Wiley, Y. Xia, Annu. Rev. Phys. Chem. 60 (2009) 167.
[129] K.-S. Choi, *Dalton Trans.* 5432 (2008).
[130] K. Na, Q. Zhang, G.A. Somorjai, Colloidal metal nanocatalysts: synthesis, characterization, and catalytic applications, J. Clust Sci. 25 (2014) 83-114.
[131] G. Mpourmpakis, S. Caratzoulas, D. G. Vlachos, Nano. Lett. 10 (2010) 3408.
[132] M. C. Daniel, D. Astruc, Chem. Rev. 104 (2004) 293.
[133] X. M. Lin, H. M. Jaeger, C. M. Sorensen, K. J. Klabunde, J. Phys. Chem. B 105 (2001) 3353.
[134] Z. X. Wang, B. Tan, I. Hussain, N. Schaeffer, M. F. Wyatt, M. Brust, A. I. Cooper Langmuir 23 (2007) 885.
[135] K. An, S. Alayoglu, T. Ewers, and G. A. Somorjai, J. Colloid Interface Sci. 373 (2012) 1.
[136] G. Wulff, Z. Kristallogr, 34 (1901) 449.
[137] J. L. Elechiguerra, J. Reyes-Gasga, M. J. Yacaman, J. Mater. Chem. 16 (2006) 3906.
[138] E. Leontidis, K. Kleitou, T. Kyprianidou-Leodidou, V. Bekiari, P. Lianos, Langmuir 18 (2002) 3659.

[139] Q. Zhang, N. Li, J. Goebl, Z. D. Lu, Y. D. Yin, J. Am. Chem. Soc. 133 (2011) 18931.
[140] Y. T. Yu, B. Q. Xu, Appl. Organomet. Chem. 20 (2006) 638.
[141] J. M. Petroski, Z. L. Wang, T. C. Green, M. A. El-Sayed, J. Phys. Chem. B 102 (1998) 3316.
[142] Y. W. Zhang, M. E. Grass, W. Y. Huang, G. A. Somorjai, Langmuir 26 (2010) 16463.
[143] S.V.N.T. Kuchibhatla, A.S. Karakoti, D. Bera, S. Seal, One dimensional nanostructured materials, Prog Mater Sci 52 (2007) 699-913.
[144] Z. Cai, C.R. Martin, J Am Chem Soc111 (1989) 4138.
[145] A.G. MacDiarmid, Rev Mod Phys 73 (2001) 701.
[146] I. Chun, D.H. Reneker, Nanotechnology 7 (1996)216.
[147] Z. Sun et al., Adv Mater 15 (2003)1929.
[148] C.R. Martin, Chem Mater 8 (1996)1739.
[149] M. Bognitzki et al. Adv Mater 12 (2000) 637.
[150] X.M. Yang, Z.X. Zhu, T.Y. Dai, Y. Lu, Macromol Rapid Commun 26 (2005) 1736.
[151] Y. Shen et al., J Polym Sci Part A: Polym Chem 37 (1999) 1443.
[152] Y. Zhou et al., Appl Phys Lett 83 (2003) 3800.
[153] R. Dersch et al., Polym Adv Technol 16 (2005) 276.
[154] M. Bognitzki et al. Polym Eng Sci 41 (2001) 982.
[155] D.H. Reneker, A.L. Yarin, H. Fong, S.J. Koombhongse, Appl Phys 87 (2000) 4531.
[156] J. Doshi, D.H.Reneker, J Electrostat 35 (1995) 151.
[157] G.Taylor, Proc R Soc London Ser A – Math Phys Eng Sci 313 (1969) 453.
[158] Y.X Li. Adv Mater 16 (2004) 1151.
[159] M.M. Hohman, M. Shin, G. Rutledge, M.P.Brenner, Phys Fluids13 (2001) 2201.
[160] H..R Darrell, L.Y. Alexander, F. Hao, K. Sureeporn, J Appl Phys 87 (2000) 4531.
[161] M.M. Hohman, M. Shin, G. Rutledge, M.P. Brenner, Phys Fluids 13 (2001) 2221.
[162] R.L. Fleischer, Price PB, Walker RM. Nuclear tracks in solids. Berkley, CA: University of California Press; 1975.
[163] C.R. Martin, Accounts Chem Res 28 (1995) 61.
[164] K. Kim, J.I.Jin, Nano Lett 1 (2001) 631.
[165] S. Demoustier-Champagne, P.Y.Stavaux, Chem Mater 11 (1999) 829.
[166] O. Ikkala, G. ten Brinke, Science 295 (2002) 2407.
[167] D.E.A. Mott, N.F. Davis, Electronic processes in non-crystalline materials. Oxford, England: Clarendon Press; 1979.
[168] M.W. Jing Liu, J Polym Sci Part A: Polym Chem 39 (2001) 997.
[169] Liang Liang et al., Angew Chem Int Ed 41 (2002) 3665.
[170] G. Li, Z. Zhang, Macromolecules 37 (2004) 2683.
[171] J.A. Dahl, B.L.S. Maddux, J.E. Hutchison, Chem. Rev., 107 (2007) 2228-2269.
[172] P. Anastas, J. Warner, Theory and Practice, (1998) Green Chem.; Oxford University Press.
[173] L.C. McKenzie, J.E. Hutchison, Chim. Oggi., 22 (2004) 30.
[174] P.V. Kamat, R. Huehn, R. Nicolaescu, J. Phys. Chem. B, 106 (2002) 788.
[175] T. Hasobe, H. Imahori, S. F ukuzumi, P.V. Kamat, J. Phys. Chem. B, 107 (2003) 12105.
[176] R. Venkatasubramanian, E. Siivola, T. Colpitts, B. O'Quinn, Nature, 413 (2001) 597.
[177] S.M. Lloyd, L.B. Lave, Environ. Sci. Technol., 37 (2003) 3458.
[178] J.I. Hahm, C.M. Lieber, Nano Lett., 4 (2004) 51.

Review questions

1. What does "there is plenty of room at the bottom" means?
2. What are the "top-down" and "bottom-up" approaches?

3. Do you describe the work principle of horizontal mill?

4. What are the differences between high energy ball mill and traditional ball mill?

5. What is the physical vapor deposition method (PVD)?

6. Please explain the principle of sputtering technique of direct current (DC) glow discharge.

7. What is the principle of magnetron sputtering?

8. What is plasma?

9. What does the "laser ablation" means?

10. What is the chemical vapor deposition (CVD)?

11. What is the solvethermal method?

12. What are the supercritical temperature and supercritical pressure of water? Do you think the properties of solvent at the critical point are same those of solvent at ambient pressure and temperature or not? Why? What kind of vessel can be used in the hydryothermal synthesis?

13. What is the freeze-drying method (cryochemical synthesis method)?

14. What is sol-gel method?

15. What is the characteristic of surfactant? What is reverse micelle?

16. Can you depict diagrams of multiple emulsions about "oil-in-water-in-oil" or "water-in-oil-in-water"?

17. Please describe the principle of microemulsions method.

18. What kind of wavelength can we utilize for microwave synthsis? What kinds of materials can we be used as reactants, solvents and reactors? What is the distinguishing feature of microwave?

19. What is the ultrasonic-assisted synthesis?

20. What is self-assembly?

Vocabulary

2-pyrrolidone n. [医] 2-吡咯烷酮

ablation [ə'bleɪʃ(ə)n] n. 消融（烧蚀，风化，切开，脱离）

acetate ['æsɪteɪt] n. 醋酸盐

additive ['ædɪtɪv] n. 添加剂；adj. 加添的

adhesion [əd'hiːʒ(ə)n] n. 黏附；支持；固守

aerosol ['eərəsɒl] n. 气溶胶；气雾剂；喷雾器；浮质；adj. 喷雾的；喷雾器的

agglomeration [əglɒmə'reɪʃ(ə)n] n. 凝聚；结块；附聚

aggregate ['ægrɪgət] vi. 集合；聚集；合计；vt. 集合；聚集；n. 合计；总计；adj. 聚合的

alkoxide [æl'kɔksaid,-sid] n. [化] 醇盐；[化] 酚盐

aluminosilicate [əˌljʊməno'sɪləkɪt] n. [化] 铝硅酸盐；铝矽酸盐

amine [ə'miːn] n. [有化] 胺（等于amin）

ammonia [ə'məʊnɪə] n. [化] 氨，阿摩尼亚

ammonium [ə'məʊnɪəm] n. [无化] 铵；氨盐基

amorphous [ə'mɔːfəs] adj. 无定形的；无组织的；非晶形的

amphiphilic [ˌæmfə'fɪlɪk] adj. [生化] 两亲的；[生化] 两性分子的

amplitude ['æmplɪtjuːd] n. 振幅；丰富，充足；广阔

anisotropic [ˌænaɪsə(ʊ)'trɒpɪk] adj. [植] 各向异性的；非均质的

annealing [ə'niːl] vt. 使退火，韧炼；锻炼；n. 退火；锻炼，磨练

annular ['ænjʊlə] adj. 环形的，有环纹的

anode ['ænəʊd] n. 阳极

AOT (sodium sulfosuccinate) 二-(2-乙基己基)琥珀酸酯磺酸钠

apparatus [ˌæpəˈreɪtəs] n. 装置，设备；仪器；器官

aqueous ['eɪkwɪəs] adj. 水的，水般的

asymmetric [ˌæsɪ'metrɪk] adj. 不对称的；非对称的

attachment [ə'tætʃm(ə)nt] n. 附件；依恋；连接物；扣押财产

attrition [ə'trɪʃ(ə)n] n. 摩擦；磨损；消耗

autoclave ['ɔːtə(ʊ)kleɪv] n. 高压反应釜；高压锅；vt.

用高压锅烹饪
axisymmetric [ˌæksisɪ'mɛtrɪk] adj. 轴对称的
barium ['beərɪəm] n. 钡
batch [bætʃ] n. 一批；一炉；一次所制之量；vt. 分批处理
bitartrate [baɪ'tɑ:treɪt] n. 酒石酸氢盐；酸式酒石酸盐
bombard [bɒm'bɑ:d] vt. 轰炸；炮击 n. 射石炮
borohydride [ˌbɔ:rəu'haidraid] n. [化] 氢硼化物
bottom-up ['bɔtəm'ʌp] adj. 自底向上的；从细节到总体的
bromide ['brəʊmaɪd] n. [无化] 溴化物；庸俗的人；陈词滥调
bromine ['brəʊmi:n] n. [化] 溴
Brownian motion 布朗运动
bubble ['bʌb(ə)l] n. 气泡，泡沫，泡状物；透明圆形罩，圆形顶；vi. 冒泡；vt. 使冒泡
bulky ['bʌlkɪ] adj. 庞大的；体积大的；笨重的
calcination [ˌkælsɪ'neʃən] n. 煅烧；[化工] 焙烧；烧矿法
candidate ['kændɪdeɪt; -dət] n. 候选人，候补者；应试者
carboxylate [kɑ:'bɒksɪleɪt] n. 羧酸盐；羧化物
carburized ['kɑ:bjəraɪz] adj. 渗碳的；v. 使渗碳
cathode ['kæθəʊd] n. 阴极
cavitation [ˌkævi'teiʃən] n. 气穴现象；空穴作用；成穴
centrifugal [ˌsentrɪ'fju:g(ə)l; sen'trɪfjʊg(ə)l] adj. 离心的
cerium ['sɪərɪəm] n. [化学] 铈
cetyltrimethyl ammonium bromide 十六烷基三甲基溴化铵
chalcogenide ['kælkədʒənaɪd] n. 氧属化物；硫族化物
chemical vapor deposition 化学气相沉积
chlorine ['klɔ:ri:n] n. 氯，氯气
chromatographic adj. 色谱法的；层析法的
citrate ['sɪtreɪt] n. 柠檬酸盐
coarsening n. 粗化；结晶粒粗大化；变粗 v. 变粗（coarsen 的 ing 形式）
coaxial [kəʊ'æksɪəl] adj. 同轴的，共轴的
comminution [ˌkɑmə'njʊʃən] n. 粉碎；[医] 捣碎
compound ['kɒmpaʊnd] vt. 合成；混合；vi. 和解；妥协；n. 化合物；adj. 复合的；混合的
concentric [kən'sentrɪk] adj. 同轴的；同中心的
cone [kəʊn] n. 圆锥体，圆锥形；球果；vt. 使成锥形
convention [kən'venʃ(ə)n] n. 大会，协定，惯例
coprecipitation ['kəupriˌsipi'teiʃən] n. 共同沉淀
cosurfactant [kəʊsɜ:'fæktænt] n. 助表面活性剂；面活性剂；活性剂
coulombic [ku:'lɒmbɪk] adj. [电] 库仑的；库仑定律的
critical ['krɪtɪk(ə)l] adj. 临界的；批评的，爱挑剔的；决定性的；危险的；鉴定的
crucible ['kru:sɪb(ə)l] n. 坩埚
cryogen ['kraɪə(ʊ)dʒ(ə)n] n. 冷冻剂；[制冷] 致冷剂
crystallographic [ˌkrɪstəlɑ'græfɪk] adj. 结晶的
CTAB (cetyltrimethyl ammonium bromide) abbr.十六烷基三甲基溴化铵
cylindrical [sɪ'lɪndrɪkəl] adj. 圆柱形的；圆柱体的
decomposition [ˌdi:kɒmpə'zɪʃn] n. 分解，腐败，变质
deflected [dɪ'flektɪd] adj. 偏离的，偏向的；v. 打歪
dehydrate [di:haɪ'dreɪt; di:'haɪdreɪt] n. [无化] 二水合物；二水物
dehydration [ˌdi:haɪ'dreɪʃən] n. 脱水
derivative[dɪ'rɪvətɪv] adj. 引出的，系出的；n. 引出之物，衍生字；n. 导数，微商
dielectric [ˌdaɪɪ'lektrɪk] adj. 非传导性的；诱电性的；n. 电介质；绝缘体
dielectric loss factor 介质（电）损耗因子
dihydrochloride [daiˌhaidrə'klɔ:raid] n. 二盐酸化物；二氢氯化物
dimer ['daɪmə] n. 二聚物；二量体
disproportionation ['disprəˌpɔ:ʃə'neiʃən] n. 歧化作用；不相均
dramatically [drə'mætɪkəlɪ] adv. 戏剧地，引人注目地，从戏剧角度，显著地
droplet ['drɒplɪt] n. 小滴
eject [ɪ'dʒekt] v. 喷射，放逐，驱逐
electrochemical [ɪˌlɛktro'kɛmɪkl] adj. 电气化学的
electrode [ɪ'lektrəʊd] n. 电极
electrospinning 静电纺丝；电纺丝；电纺丝法
element ['elɪm(ə)nt] n. 元素；要素；原理；成分；自然环境
elongated ['i:lɒŋgeitid] adj. 瘦长的，[植] 细长的
enclosed [ɪn'klozd] adj. 被附上的；与世隔绝的；v. 附上
endothermic [ˌendəʊ'θɜ:mɪk] adj. [化] 吸热的；[动] 温血的
entrap [ɪn'træp; en-] vt. 使陷入；使陷罗网；欺骗
equiaxed ['i:kwiækst] adj. 各向等大的，等轴的
equilibrium [ˌi:kwɪ'lɪbrɪəm; ˌekwɪ-] n. 均衡；平静；保持平衡的能力
etch [etʃ] vt. 蚀刻；鲜明地描述；铭记；vi. 蚀刻；

n. 刻蚀；腐蚀剂
etching ['etʃɪŋ] n. 蚀刻术；蚀刻版画；v. 蚀刻；刻画；adj. 蚀刻的
ethylene ['eθɪliːn; -θ(ə)l-] n. [化] 乙烯
ethylene glycol 乙二醇，乙烯乙二醇
eutectic [juː'tektɪk] n. 共熔合金；adj. 共熔的；容易溶解的
evaporation [ɪˌvæpə'reɪʃən] n. 蒸发，脱水，干燥
extraction [ɪk'strækʃ(ə)n; ek-] n. 抽出；拔出；取出；抽出物；出身
facet ['fæsɪt; -et] n. 面；方面；小平面 vt. 在…上琢面
feedstock ['fiːdstɒk] n. 原料；给料
filter ['fɪltə] vi. 滤过；渗入；n. 滤波器；[化工] 过滤器；滤光器；vt. 过滤；渗透
flocculation [flɔkju'leiʃən] n. 絮凝；絮结产物
flow [fləʊ] vi. 流动，飘扬；vt. 淹没，溢过；n. 流动；泛滥
fluctuations [ˌflʌktju'eiʃəns] n. [物] 波动（fluctuation 的复数）；变动；起伏现象
functionalize ['fʌŋkʃənəlaiz] vt. （美）使职能化；根据工作人员职能分派
fusion ['fjuːʒ(ə)n] n. 融合；熔化；熔接；融合物；[物] 核聚变
gamut ['fjuːʒ(ə)n] n. 全音阶；全音域；整个范围
gaseous ['gæsɪəs; 'geɪsɪəs] adj. 气体的，气态的
geometric [ˌdʒɪə'metrɪk] adj. 几何学的，几何学图形的
geothermal [dʒiːə(ʊ)'θɜːm(ə)l] adj. [地物] 地热的；[地物] 地温的
glycol ['glaɪkɒl] n. [化]乙二醇；甘醇；二羟基醇
granular ['grænjʊlə] adj. 粒状的；颗粒的
grinding ['graɪndɪŋ] adj. 刺耳的；研磨的；令人难以忍受的
gyroid ['paɪrɔɪd] n. 螺旋二十四面体
halide ['heɪlaɪd] n. 卤化物；adj. 卤化物的
hard-tech 硬技术
hazardous ['hæzədəs] adj. 危险的
helium ['hiːlɪəm] n. 氦
helix ['hiːlɪks] n. 螺旋，螺旋状物，耳轮
heterogeneous [ˌhet(ə)rə(ʊ)'dʒiːnɪəs; -'dʒen-] adj. 异种的，异质的，由不同成分形成的
hexadecylamine n. 十六烷基胺
hexagonal [hek'sægənəl] adj. 六边的，六角形的
hexane ['hekseɪn] n. [化] 己烷
hierarchical [haɪə'rɑːkɪk(ə)l] adj. 分层的；等级体系的
homogeneity [ˌhɒmə(ʊ)dʒɪ'neɪɪtɪ; -dʒɪ'niːɪtɪ; ˌhəʊm-] n. 同种；同质；同次性
homogeneous [ˌhɒmə(ʊ)'dʒiːnɪəs; -'dʒen-] adj. 均匀的；齐次的；同种的
hopping ['hɒpɪŋ] adj. 忙碌的；卖力的；n. 跳跃；v. 跳跃(hop 的 ing 形式)；使…兴奋
hydrate ['haɪdreɪt] n. [化] 水合物；[化] 氢氧化物；vt. 使成水化合物；vi. 与水化合
hydrazine ['haɪdrəziːn] n. [化] 肼；[化] 联氨；[化] 酰肼
hydrocarbon [ˌhaɪdrə(ʊ)'kɑːb(ə)n] n. 碳氢化合物
hydrolysis [haɪ'drɒlɪsɪs] n. 水解
hydrolytic [ˌhaidrə'litik] adj. 水解的，水解作用的
hydrophilic [ˌhaɪdrə(ʊ)'fɪlɪk] n. 软性接触透镜吸水隐形眼镜片；adj. [化] 亲水的
hydrophobic [haɪdrə(ʊ)'fəʊbɪk] adj.疏水的
hydrothermal [haɪdrə(ʊ)'θɜːm(ə)l] adj. 水热的；热液的
hydroxide [haɪ'drɒksaɪd] n. [无化] 氢氧化物；羟化物
hypodermic [haɪpə(ʊ)'dɜːmɪk] n. 皮下注射；皮下注射器；adj. 皮下的
igniter [ɪg'naɪtər] n. 点火器；点火剂；发火器
ilmenite ['ɪlmənaɪt] n. [矿] 钛铁矿
imitate ['ɪmɪteɪt] vt. 模仿，仿效；仿造，仿制
impinging [ɪm'pɪndʒ] n. 冲击；碰撞；v. 冲击，撞击
inorganic [ɪnɔː'gænɪk] adj. [无化] 无机的；无生物的
insulator ['ɪnsjʊleɪtə]n. [物] 绝缘体；从事绝缘工作的工人
intercalate [ɪn'tɜːkəleɪt; ˌɪntəkə'leɪt] vt. 设置；插入
interdiffusion n. 相互扩散
intermediate [ˌɪntə'miːdɪət] vi. 起媒介作用；adj. 中间的，中级的；n. 中间物；媒介
interstitial [ˌɪntə'stɪʃ(ə)l] adj. 空隙的；[物] 填隙的；n. [物] 填隙原子
intricate ['ɪntrɪkət] adj. 复杂的，错综的，缠结的，难懂的
intriguing ['ɪntriːgɪŋ; ɪn'triːgɪŋ] adj. 有趣的；迷人的；v. 引起…的兴趣；策划阴谋
intrinsic [ɪn'trɪnsɪk] adj. 本质的，固有的
ionized ['aɪənˌaɪz] adj. 电离的；已电离的；v. 使电离，电离
irregular [ɪ'regjʊlə] n. 不规则物；不合规格的产品；adj. 不规则的；非正规的；不合法的
isooctane [ˌaɪsəʊ'ɒkteɪn] n. 异辛烷
jet [dʒet] n. 喷射，喷出；v. 射出，迸出，喷射
judiciously [dʒuː'diʃəsli] adv. 明智而审慎地，有见地

kinetic [kɪ'netɪk; kaɪ-] adj. 运动的，动力学的
lamellae [lə'mɛli] n. 片晶；薄片
laser ablation 激光消融；激光剥蚀；雷射钻孔
ligand ['lɪg(ə)nd] n. 配合基（向心配合价体）
lipophilic [ˌlɪpə(ʊ)'fɪlɪk; ˌlaɪ-] adj. 亲脂性的，亲脂的
lithography [lɪ'θɒgrəfɪ] n. 平版印刷术，光刻
lyotropic [ljət'rɒpɪk] adj. 易溶的；感胶离子的
magnetron ['mægnɪtrɒn] n. 磁控管
magnetron sputtering 磁控溅射法；磁控管溅射
membrane ['membreɪn] n. 膜；薄膜；羊皮纸
mesh [meʃ] n. 网眼；网丝；圈套；vi. 相啮合；vt. [机] 啮合；以网捕捉
metastable [ˌmetə'steɪb(ə)l] adj. 亚稳的；相对稳定的
methanol ['meθənɒl] n. 甲醇（methyl alcohol）
methodology [meθə'dɒlədʒɪ] n. 方法学，方法论
micell [maɪ'sɛl] n. [物]胶束；胶态离子；微胞
microemulsion [ˌmaikrəui'mʌlʃən] n. 微乳液；微型乳剂
milling ['mɪlɪŋ] n. 磨；制粉；轧齿边；v. 碾磨；磨成粉；滚（碾轧）金属
miniaturize ['mɪnɪtʃəraɪz]vt. 使小型化；使微型化(等于 miniaturise)
miscible ['mɪsɪb(ə)l] adj. 易混合的；可溶混的；能混溶的
monodisperse [ˌmɔnəudis'pə:s] adj. [化] 单分散的
monomer ['mɒnəmə] n. 单体；单元结构
morphology [mɔ:'fɒlədʒɪ] n. 形态学，语形论，形态论
morphology [mɔ:'fɒlədʒɪ] n. 形态学，形态论；[语] 词法，[语] 词态学
motion ['məʊʃ(ə)n] v. 运动，向…打手势；n. 打手势，示意，移动，动作，提议
multidentate [ˌmʌltɪ'dɛntet] adj. 多齿状突起的，多齿的
mutual ['mju:tʃʊəl; -tjʊəl] adj. 共同的；相互的，彼此的
nebulizer ['nɛbjəˌlaɪzɚ] n. 喷雾器
nitrate ['naɪtreɪt] n. 硝酸盐；vt. 用硝酸处理
nitrogen ['naɪtrədʒ(ə)n] n. 氮
nonequilibrium n. 非平衡态；adj. 不平衡的
nozzle ['nɒz(ə)l] n. 喷嘴；管口；鼻
nucleophilic [ˌnju:kliəu'filik] adj. 亲核的；亲质子的
octahedral [ˌɑktə'hidrəl] adj. 八面体的；有八面的
octahedron [ˌɒktə'hi:drən; -'hed-] n. 八面体
on-site [ɒn saɪt] adj. 现场的
optoelectronic [ˌɒptəʊɪˌlek'trɒnɪk] adj. 光电子的
oriented ['ɔ:rɪentɪd] adj. 导向的；定向的；以…为方向的
orthotitanate n. 原钛酸盐
oscillatory ['ɒsɪlətərɪ] adj. 振动的，动摇的，变动的
oust [aʊst] vt. 驱逐；剥夺；取代
oxidize ['ɒksɪdaɪz] vt. 使氧化；使生锈；vi. 氧化
oxometalate n. 含氧金属盐
penetrate ['penɪtreɪt] vt. 渗透；穿透；洞察；vi. 渗透；刺入；看透
phosphide ['fɒsfaɪd] n. [化] 磷化物
phosphine ['fɒsfi:n] n. 磷化氢；三氢化磷
photonic [ˌfoʊ'tɑnɪk] adj. 光激性的
photovoltaic [ˌfəʊtəʊvɒl'teɪɪk] adj. 光电伏打的，光电的，光伏的
physical vapor deposition method 物理气相沉积法
pioneering [paɪə'nɪərɪŋ] adj. 首创的；先驱的
plasma ['plæzmə] n. 等离子体
polarons ['pəulərən] n. [物][遗] 极化子
poly(*N*-vinyl-2-pyrrolidone) 聚(*N*-乙烯-2-吡咯烷酮)
polyaniline [pɒli':ænɪli:n] n. 聚苯胺
polycarbonate [ˌpɒlɪ'kɑ:bəneɪt] n. 聚碳酸酯
polycrystalline [ˌpɒlɪ'krɪstə'laɪn] adj. [物] 多晶的
polymerization [ˌpɒlɪməraɪ'zeɪʃn] n. 聚合；[化] 聚合作用
polyol ['pɔliɔlˌ -ɔul] n. 多元醇；多羟基化合物
polypyrrole [pɒ'laɪpɪrəʊl] n.聚吡咯
potassium [pə'tæsɪəm] n. 钾
precautionary [prɪ'kɔ:ʃ(ə)n(ə)rɪ] adj. 预防的；留心的；预先警戒的
precursor [prɪ'kɜ:sə] n. 前躯体，先驱者，先进者
preferential [ˌprefə'renʃ(ə)l] adj. 优先的；选择的；特惠的；先取的
prism ['prɪz(ə)m] n. 棱镜
projection [prə'dʒekʃ(ə)n] n. 投射；规划；突出；发射；推测
propeller [prə'pelə] n. 螺旋桨，推进器
propose [prə'pəʊz] vt. 建议；打算，计划；求婚；vi. 建议；求婚；打算
pulsed laser deposition 脉冲激光沉积
purification [ˌpjʊərɪfɪ'keɪʃən]n. 净化；提纯；涤罪
pyrolysis [paɪ'rɒlɪsɪs] [paiə'rɔlisis, ˌpi-] n. [化] 热解；[化] 高温分解
pyrrolidone [pɪ'rɒlɪdəʊn] n. 吡咯烷酮
quarternary ['kwɔ:tənəri] adj. 四元的；四进制的；四级的
quasi ['kweɪzaɪ] adj. 准的；类似的；外表的；adv. 似是；有如

quasi-one-dimensional 准一维的
quench [kwen(t)ʃ] vt. 熄灭，[机] 淬火；解渴；结束；冷浸；vi. 熄灭；平息
radius ['reɪdɪəs] n. 半径，半径范围
reactant [rɪ'ækt(ə)nt] n. 反应物
refinement [rɪ'faɪnm(ə)nt] n. 精制；提纯；细微
reflux ['riːflʌks] n. 逆流；退潮
refrigerant [rɪ'frɪdʒ(ə)r(ə)nt] n. 制冷剂；adj. 制冷的；冷却的
reorient [riː'ɔːrɪent; -'ɒr-] vt. 使适应；再调整
repulsion [rɪ'pʌlʃ(ə)n] n. 排斥；反驳；反感；厌恶
reticular [rɪ'tɪkjʊlə] adj. 网状的；错综的
reverse micelle 反向胶团，反胶态分子团
rheological [riə'ladʒikəl] adj. 流变学的；液流学的
self-assembly [ˌselfə'sembli] n. 自组装 adj. 自行组装的
sensor ['sensə] n. 传感器
simultaneous [ˌsɪm(ə)l'teɪnɪəs] adj. 同时的；联立的；同时发生的；n. 同时译员
sinter ['sɪntə] n. 烧结物；熔渣；泉华；vt. 使烧结；使熔结；vi. 烧结；熔结
sintering ['sɪntərɪŋ] n. 烧结；adj. 烧结的；v. 烧结；使熔结
sodium ['səʊdɪəm] n. 钠
soft-tech 软技术
sol-gel method 溶胶-凝胶法
solid-state ['sɒlɪdˌsteɪt] adj. 固态的
solvent ['sɒlv(ə)nt] n. 溶剂；解决方法
solvethermal 溶剂热的
spectrometry [spek'trɒmɪtrɪ] n. 光谱测定法
spherical ['sferɪk(ə)l] adj. 球形的，球面的；天体的
spray [spreɪ] n. 水沫；喷雾；喷雾器；vt. 喷射；vi. 喷
sputtering ['spʌtərɪŋ] n. 溅射；喷溅涂覆法；v. 喷溅；发出爆裂声
stable ['steɪb(ə)l] n. 马厩；牛棚；adj. 稳定的；牢固的；坚定的；vi. 被关在马厩；vt. 赶入马房
stoichiometric [ˌstɒɪkɪə(ʊ)'metrɪk] adj. 化学计量的；化学计算的
sublimation [ˌsʌblɪ'meɪʃən] n. 升华，升华作用；高尚化；升华物
subsequent ['sʌbsɪkw(ə)nt] adj. 后来的，随后的
subset ['sʌbset] n. 子集；子设备；小团体
substitution [sʌbstɪ'tjuːʃn] n. 代替；置换；代替物
substrate ['sʌbstreɪt] n. 基质；衬底
sulfur ['sʌlfɚ] vt. 用硫磺处理；n. 硫磺；硫磺色
supercritical [suːpə'krɪtɪk(ə)l; sjuː-] adj. [物][物化] 超临界的；吹毛求疵的
supersaturation [ˌsʊpɚsætʃə'reʃən] n. [化] 过饱和
surfactant [sə'fækt(ə)nt] n. [化] 表面活性剂 adj. [化] 表面活性剂的
suspension [sə'spenʃ(ə)n] n. 悬浮；暂停；停职
synthetic [sɪn'θetɪk] adj. 合成的，人造的；n. 人工制品
syringe [sɪ'rɪn(d)ʒ; 'sɪ-] vt. 注射，冲洗；n. 注射器；洗涤器
teflon ['tɛflɑn] n. 铁氟龙（商标名）；聚四氟乙烯
template ['templeɪt; -plɪt] n. 模板，样板
ternary ['tɜːnərɪ] adj. 三元的，三重的；三进制的；第三的
tetrahedron [ˌtetrə'hiːdrən; -'hed-] n. [晶体] 四面体
thermodynamic [ˌθɜːməʊdaɪ'næmɪk] adj. 热力学的；使用热动力的
thiols [θi'olz] n. [有化] 硫醇（等于mercaptan）
titanium [taɪ'teɪnɪəm; tɪ-] n. 钛
titanium tetraethoxide 四乙氧基钛
titanium tetraisopropoxide 四异丙氧基钛
top-down [ˌtɒp'daʊn] adj. 自上向下；组织管理严密的
transesterification ['trænsəsˌterəfi'keiʃən] n. [化] 酯基转移；酯基转移作用
trialkylsilyl 三烷基甲硅烷基
triangular [traɪ'æŋgjʊlə] adj. 三角形的，三人间的
trimethoxy silane 三甲氧基硅烷
truncate [trʌŋ'keɪt; 'trʌŋ-] adj. 截短的；被删节的；vt. 把…截短；缩短；使成平面
tungsten ['tʌŋst(ə)n] n. 钨，钨锰铁矿
ultrasonic [ʌltrə'sɒnɪk] adj. 超声的；超音速的；n. 超声波
uniformity [juːnɪ'fɔːmɪtɪ] n. 同样，一致(性)
vapor phase deposition 气相沉积法；汽相淀积
velocity [və'lɒsəti] n. 速率，速度；周转率；迅速；高速，快速
via ['vaɪə] prep. 取道，通过；经由
vinyl ['vaɪnl] n. 乙烯基
volatile ['vɒlətaɪl] adj. 反复无常的，挥发性的；n. 挥发物
weld [weld] vt. 焊接；使结合；使成整体；vi. 焊牢；n. 焊接；焊接点
whipping ['wɪpɪŋ] n. 鞭打；笞刑；用来缚扎的绳索
wurtzite ['wɚtˌsaɪt] n. [矿物] 纤维锌矿
xerogel ['ziərədʒel] n. 干凝胶
zinc [zɪŋk] vt. 镀锌于…；涂锌于…；用锌处理；n. 锌
zirconium [zɜː'kəʊnɪəm] n. [化学] 锆

4. 纳米材料制备

在过去数十年里，科学家已经研究了至少有一维处于纳米量级的许多新材料的合成与表征方法。如：纳米粒子、纳米膜和纳米管。然而，设计和制备具有可控性能的纳米材料仍然是纳米科技的一项重大的和长期的挑战。最常见的制备过程包括物理方法，例如：电弧放电法、热蒸发法、激光/电子束加热法、溅射法等；化学方法，例如：化学气相沉积法、固相反应法、溶胶-凝胶法、沉淀法以及水热法等。

Feymann[1]在他的“底部有大量的空间”的著名演讲中提到，对于硅技术的摩尔规律而言，随着尺寸减小的同时，运算速度亦随之提高。国际科技路线图指出，至少在未来五年，硅技术还可以继续缩小器件的尺寸。为了超越目前的技术，我们不得不思考可替代的技术。我们已经发现了一些可替代的选择，例如碳纳米管器件和分子开关，这就给试图亲自制造这些器件和开关的技术人员提出了新的挑战。这些新技术不仅提出了材料方面的挑战，而且也预示着体系水平的挑战。

在某种程度上，硅技术是纳米技术奇迹的开始。近年来，许多半导体公司已经制造出尺寸在 90nm 左右的器件，一些公司正准备将器件尺寸降至 65nm。有些公司甚至设想出如何将尺寸降至更小。当缩减器件尺寸时，就会增加其数量。纳米材料的制备有多种途径。了解纳米材料制备过程中的一些工艺特征是非常重要的，这是因为制备的工艺路线通常决定了所制备的材料的性能。

4.1 “自上而下”和“自下而上”的合成方法

通过“自上而下”或“自下而上”的方法都可以制备出新型纳米材料。

“自上而下”的方法是采用粉碎、研磨、蚀刻及光刻等工艺，将块体材料制备成纳米材料。因此，“自上而下”的方法是通过施加一种“力”将材料的尺寸减小。在材料的尺寸减小的过程中，控制和颗粒变小的程度直接地影响着所制备材料的性质[2]。目前，纳米材料的商业规模化生产主要采用“自上而下”的方法。

在“自下而上”的方法中，原子、分子、甚至纳米粒子本身也可以作为构建单元，并通过改变构建单元的尺寸，控制其表面和内部结构，然后控制它们的组织和组装，从而制造出复杂的纳米结构。

在整个“自下而上”的研发路线图中，有两种常用的方法。第一种是硬技术的方法[3]，该方法是利用大型复杂的设备诱导操纵一个一个的原子，或一个一个的分子，然后将相对简单的构建单元组建成可应用的纳米结构器件。第二种是软技术的方法，该方法是通过进行宏观尺度上的操控，经由自组装路线进行构建单元的复杂设计[4]。与硬技术相反[5]，该方法利用的是相对廉价的和常规的设备。如图 4.1 所示，就尺度而言，“自上向下”的方法与“自下而上”的方法相反，而生物过程实质上是“自上而下”和“自下而上”两种方法的中间过程。

本章介绍的合成方法主要包括固相方法、物理气相沉积法（PVD）、化学气相沉积法（CVD）、液相合成方法、通过固结纳米粉合成块体纳米材料、模板辅助自组装纳米材料以及绿色合成方法。

4.2 固相方法

4.2.1 机械磨

机械磨方法实际上属于材料的固态加工过程，是“自上而下”方法的一个实例。高能球磨法是具有重要工业意义的一种纳米制备方法，也称为机械研磨或机械合金化。图 4.2 是几种不同磨机的示意图[6]。

众所周知，搅拌磨机亦称为粉碎机或搅拌球磨机。将连接有次级臂的中心轴旋转而使球体处于运动状态，磨机的筒体被固定。粗颗粒材料在旋转的筒体中被硬钢球或碳化钨球机械地挤压成粉体，通常在控制的气氛条件下，以避免不必要的反应发生，如：氧化反应。这种在颗粒内部经由晶界的形成与重组，颗粒反复的变形就会引起粒度的大幅度减小。

水平磨就是磨机的筒体绕着它的水平轴旋转。旋转产生的离心力和重力的共同作用，从而引起钢球在粉体中的升起与落下。

一维振动磨通常被称为振动磨。磨机筒体在垂直摆动。在这种作用下，1kg 的钢球在粉体中升起与落下。

行星磨是磨机筒体被固定在旋转的台面上。在现代化的磨机中，筒体自身也在旋转，旋转的筒体与旋转台面可以是同向旋转，也可以是不同向旋转。

三维振动磨也被称为三轴振动磨。它与一维振动磨的工作原理相同，只不过该振动磨要在三维空间自由振动，其振动的方式更复杂。球体不仅与侧壁发生碰撞（摩擦和冲击），而且与底板和顶板发生碰撞。

高能球磨力是通过高振动频率、小振幅来获取的。球体长期处于相对运动状态的高能球磨机，与传统工业中粉磨材料的干法或湿法磨机不同。低能磨机通过机械合成可以制备某些材料，但粉磨时间过长，需要几百乃至上千小时，与只需数小时或数十小时的高能球磨机相比，大大地降低了其工业潜力。然而，高能球磨机的生产能力应满足工业需求。

值得注意的是，在机械粉磨过程中，来自磨具（Fe）和气氛（惰性气体中含有微量的 O_2、N_2）的污染，已成为引起被制备材料的表面和界面污染的主要问题。机械粉磨和机械合金化已经用于大规模的商业制造[7, 8]。

机械合金化是一种高能球磨过程，在该过程中，自然混合的粉体不断地被焊接和断裂，达到在原子水平上的合金化[9]。机械合金化/粉磨过程的影响因素包括粉磨时间、配料比、粉磨环境以及颗粒的内部力学特性。最近一项关于机械加工制备 Ni 纳米晶的微结构进展研究表明，粉磨环境和温度强烈地影响着粉末的变形行为。在机械粉磨过程中，粒子的连续焊接及断裂促进了不规则的薄片大量形成[10]。结构的磨细化程度取决于输入的机械能量和材料的加工硬化程度[11]。当在液氮（代替甲醇）的条件下粉磨时，Ni 的晶粒尺寸已变得足够细小，如表 4.1 所示[12]。

4.2.2 固相反应[13]

通过控制相变或固体材料的反应，固相反应为生产纳米粒子提供了可能性。这种方法的优点是生产工艺简单。固相反应技术是用于高产率地生产单分散的纳米氧化物纳米粒子的一种便捷的、廉价的及高效的制备方法。

这是制备许多陶瓷粉末的最传统的方式，如：$BaTiO_3$。它涉及将 $BaCO_3$ 和 TiO_2 颗粒进行粉磨以进行粉碎和混合。然后，将混合物在 900～1200℃下煅烧使其生成 $BaTiO_3$。反应机理如下[14~16]。

（1）Ba 通过扩散进入 TiO_2 中发生界面反应，在 TiO_2 周围形成 $BaTiO_3$ 壳(层)；生成的壳限制了 Ba 向 TiO_2 中进一步扩散。

（2）形成原钛酸盐相。

（3）最终形成 $BaTiO_3$。

高的焙烧温度以促进 Ba 和 Ti 的相互扩散，导致固态反应形成 $BaTiO_3$，如图 4.3 所示[17]。

烧结被定义为在热的作用下离散的材料的固结过程。在致密化过程中，如果材料的一部分达到熔点，则称为液相烧结；反之，称为固相烧结。此外，通过烧结过程进行材料的化学合成，就称为反应性烧结。

4.3 物理气相沉积法（PVD）

物理气相沉积法（PVD）涉及固体材料通过物理过程转换成气相，然后进行冷却、气相材料在基质上沉积，在此过程中，材料会与系统中的气体发生反应，其性能或许发生某些变化。物理气相沉积转化过程中的实例有热蒸发法、激光消融法或脉冲激光沉积法（激光器的纳秒脉冲聚焦于块体靶材的表面）及溅射法(通过用原子或离子轰击靶材而溅射出靶材粒子)。PVD 可用于制备薄膜、多层膜、纳米管、纳米丝或纳米粒子。

4.3.1 热蒸发 PVD 法

（1）炉中气相膨胀

物理气相沉积法的一个实例是气相膨胀，是高压气相通过一个喷嘴向低压环境背景膨胀，以生产过饱和的气体，气体流速可达到超音速，该过程可以导致极小的、数个原子以上的原子簇成核，其成核过程可用质谱进行分析。虽然这种簇的数量很少，但这对小金属簇的物理研究很有用，由于簇具有非常稳定的几何和电子结构，它们常常是由原子的幻数所组成。这种系统如图 4.4 所示[18]。

（2）惰性气体蒸发-冷凝法

另一种 PVD 过程是惰性气体蒸发-冷凝法，纳米颗粒在惰性气氛下由蒸发的金属再冷凝所形成。虽然这项技术比较古老，但近期该技术已用于真正的纳米粉体的制备。这里所说的材料，通常是一种金属材料放入一个温控的坩埚中，使其蒸发进入低压的惰性气体环境[超高真空腔（UHV）]。这种金属蒸气与惰性气体发生碰撞而被冷却，形成过饱和蒸气，然后均匀成核。通常采用不同的惰性气体压力，可将粒径控制在 1～100nm 范围内。采用液氮冷却，粒子在指形冷凝器上沉积，然后将其刮下，并压制成致密的纳米材料。图 4.5 显示了一个典型的通过惰性气体蒸发-冷凝技术，生产纳米粉体的实验装置。该方法是使用置于装置腔体中的金属电阻加热源（或以射频加热、电子束或激光束作为热源）对金属源加热使其蒸发，先将腔体抽真空至 10^{-7}Torr，再注入惰性气体使腔体处于低压。蒸发的金属原子蒸气从热源迁移到冷凝器过程中，与腔体中的惰性气体原子发生碰撞，从而失去了动能。最终粒子沉积于冷凝器的表面。惰性气体冷凝技术中主要利用液氮作为冷却源以提高沉淀效率。

（3）固相粉末的热蒸发

在高温条件下，通过直接蒸发市售的金属氧化物粉末，可以制备带状的准一维纳米结构（称为纳米带）半导体氧化物。合成的带状氧化物是纯的、结构均匀的单晶，大部分无位错现象。

在控制生长动力学、控制局部生长温度、控制原材料的化学组成的实验条件下，利用固态热升华工艺合成了多种纳米结构的 ZnO。图 4.6 给出了 ZnO 的纳米结构，如：纳米螺旋管、纳米弓、纳米桨、纳米线和纳米笼[19]。

4.3.2 等离子体辅助 PVD 法

下面将介绍几种不同类型的等离子体辅助 PVD 法。

（1）直流（DC）辉光放电溅射技术

最简单的溅射过程是阴极溅射，如图 4.7 所示。在充有较低压力（1.5～15Pa）的惰性气体（通常是 Ar 气）的腔体内的阴极和阳极之间施加一直流电压（0.3～1.5kV），就会产生等离子体。

通过阳极和阴极之间的辉光放电产生一种具有高能量（> 30eV）的 Ar^+等离子体。这种 Ar^+等离子体被电场所加速，穿过腔体轰击靶材，使靶材原子从表面逃逸出来，溅射到基质的表面上（阳极），这些原子在惰性气体中冷却和凝结导致在基质上形成纳米粒子或薄膜。然而，只有很少一小部分（约 0.01%）的惰性气体原子被电离。轰击靶材的离子流相当小，在基质上的沉积率（约 0.1μm/h）也非常小。溅射法的优点如下：不需要坩埚；可制备高熔点的金属粒子；使用反应性气体可合成合金粒子；能制备粒状薄膜；可控制合成窄粒度分布的粒子；如果采用数个不同材料同时作为靶材，可制备纳米复合材料等。同时，电压、电流、气体压力及靶材是制备纳米粒子最重要的影响因素。

溅射是利用机械活化替代热活化过程从靶材上离解出粒子[20]。

直流辉光放电也涉及通过一个热丝发射电子使气体原子电离，如图 4.8 所示。等离子体中气体离子被加速后产生定向的离子束。如果气体是一种反应前躯体气体，该离子束可直接沉积在基质上。另外，若使用的是惰性气体，则惰性气体离子束将撞击靶材，将溅射出的中性原子沉积到相邻的基质上，如图 4.8 所示[18]。

（2）磁控溅射法

磁控溅射法是通过在两块平行的极板间，施加一个高的直流电压使其产生等离子体，如图 4.9 所示[18]。在溅射靶材的附近施加一个静磁场，使惰性气体电离产生正等离子体。该等离子体被磁场加速以轰击靶材，从而，靶材上溅射出的中性原子沉积到基质的表面上。磁场的另一优点是防止在靶材料上生成二次电子并冲击基质而引起基质的发热或损坏。该方法的沉积速率足够高（约 1μm/min），可应用于工业化制备。若选用多种旋转的靶材，可以在基质上生成多层涂层。

最近，Anpo 等人[21] 采用射频磁控溅射法成功地制备出具有可见光响应的 TiO_2 薄膜，该薄膜用于分离在可见光辐照下光分解水产生的 H_2 和 O_2，如图 4.10 所示。该系统使用的基质（石英或钛箔）与加热器相连，靶材和基质的距离固定在 70～90mm 之间。通过在 Ar 气氛中引入功率为 300W 的射频电源，以产生轰击 TiO_2 靶材的 Ar^+ 等离子体，从靶材上溅射出的中性原子沉积在基质表面，基质温度保持在 200～600℃。当基质的温度为 25℃时，采用射频磁控溅射法可完成 Pt 的沉积。

（3）真空电弧沉积法

真空电弧沉积法中，阴极由靶材组成，阳极与一个点火器相连，当阴极接触点火器时，就会产生一个低压和强电流的自支持电弧。

电弧从阴极上驱逐出大量的离子，同时，从阴极斑落下小颗粒。电弧中的离子被加速冲向基质，如果有需求，可以利用磁场使其偏移（改变运动方向）；在沉积之前，任何大颗粒都会被过滤掉，如图 4.11 所示[18]。在高真空度条件下，阴极电弧在没有背景气体的状况下，仍然可以工作。然而，通过引入背景气体如氮气，就能实现沉积。即使基质的温度较低，高能量的离子也能生成致密的膜。因此，电弧技术通常用于硬涂层的沉积。直流源的等离子体也可以由射频、微波源而产生。这种方法的优点在于提供更高的电流密度和更高的沉积速率。

（4）用等离子枪进行的等离子体溅射

在等离子体溅射过程中，惰性气体电离的高能离子（Ar^+或 Kr^+）轰击金属靶材。高能离子源，即等离子体，可由直流电源或射频电源（等离子体枪）产生。等离子体中的高能离子被加速，轰击靶材，引起靶材料表面逸出大量的中性原子和少量的离子[22]。

等离子体是经由等离子体枪直流放电产生的阳离子、阴离子、电子以及中性物种等组成的离子化的气体。

从靶材表面驱逐出的材料（中性原子）具有高能量，这些原子在冷的基质上其能量不足以形成薄膜时，则以粒子形式收集在冷的基质上。从靶材表面驱逐出的粒子被惰性气体带走离开溅射源，然后被收集，如图 4.12 所示。

4.3.3 激光消融法

激光消融法的原理是在一个真空腔中，具有高能量的脉冲激光束，经由一个外部透镜聚焦在旋转的靶材上，靶材受热蒸发。真空腔保持恒定的氦气压力，蒸发材料的纳米颗粒在基质上沉积之前，得到冷却和凝结。激光驱动的沉积过程为零污染。激光消融源的基本实验装置是由 Yoshida 等人设计的，如图 4.13 所示[23]。

4.4 化学气相沉积法（CVD）

CVD 是一种化学气相反应过程，该过程是利用易挥发性金属的蒸发，并进行化学反应以形成理想的化合物，在气相中迅速冷却生成纳米粒子。CVD 法生成的纳米粒子具有纯度高、均匀性好、粒径小、分布窄、分散性好以及化学反应活性高等许多优点。

这种成熟的方法可以在较低的处理温度和一般真空条件下，制备单层膜、多层膜、复合材料、纳米结构、功能涂层材料、多晶材料及无定形材料。CVD 过程涉及在活化环境下（采用热、光或等离子体活化）的气体化合物的各种化学反应，最终形成稳定的固态产物。由于 CVD 的通用属性，可涉及各种类型的化学反应，气相（均相）反应和表面（非均相）反应交织在一起。随着反应物温度及分压的增加，气相反应变得日趋重要。一个极高的反应物浓度主要发生气相反应，导致均匀成核。根据使用的前驱体和施加的沉积条件，可以把各种各样的化学反应划分为以下几种形式：热分解、还原、氧化、复合反应、歧化反应和可逆反应。以上化学反应类型的实例如下。

气相分解法是所谓的单一化合物加热分解的方法。一种化合物加热进行蒸发和分解，反应如下。

① 高温分解或热分解　下面是喷雾热解法的一个实例。

一个喷雾热分解过程通常也被称作气溶胶热分解或蒸发热分解法[24]。它由三部分组成：气溶胶的生成、炉内（焙烧）形成固体粒子、粒子的收集，如图 4.14 所示。使用空气、氮气或混合型气体，将含有反应物的均匀溶液（有机物、水或混合物）转化成气溶胶；在气溶胶中形成的液滴尺寸依赖于所使用的气溶胶发生器、载气的体积以及液体的属性[25, 26]。一个超声波发生器通常产生很细的液滴，其适用于制造纳米颗粒[24]。然后由气体将气溶胶带入预热炉；在炉内，液滴经过溶剂蒸发、溶质沉淀、溶质分解和氧化烧结，最终形成粒子。最终采用气体对粒子进行冷却，如图 4.14 所示[24, 27]。

② 还原反应。

③ 氧化反应。

④ 复合反应。

⑤ 歧化反应。

⑥ 可逆反应。

CVD 方法通用的化学属性可以通过沉积膜的事实进一步得到验证，许多不同的反应物或前躯体可用于不同的化学反应。例如，使用各种反应物，通过下列任何一种化学反应都可获得氧化硅膜。

当改变不同的反应物比率和沉积条件，可沉积出不同的薄膜。例如，根据反应物 Si_2Cl_6 和 N_2O 的比率和沉积条件，可生成氧化硅薄膜或氮化硅薄膜。

虽然 CVD 是一个通过化学动力学和传递现象所控制的非平衡过程，平衡分析仍然有利于对 CVD 过程的了解。

4.5　液相合成方法

为了制备不同形态的纳米材料，液相合成方法总是作为首选。这些合成反应在相对较低的温度下进行，且需要较低的能量。

4.5.1　沉淀法

沉淀法是合成纳米粒子和纳米胶囊的一种方便的和奏效的方法。早期许多纳米粒子的制备是将水溶液中的可溶性产物共沉淀，然后使产物热分解生成氧化物。共沉淀反应涉及同时出现成核、生长、晶粒长大（粗化）和凝聚过程，要把每个过程隔离，并进行独立研究是相当困难的，共沉淀的基本机理至今还不完全清楚[28]。

（1）从水溶液中合成金属

从水溶液或非水溶液中沉积金属，通常需要化学还原金属阳离子。还原剂有多种形式，如 H_2、溶剂化的 ABH_4（A 是碱金属）、水合肼（$N_2H_4 \cdot H_2O$）、盐酸肼（$N_2H_4 \cdot 2HCl$）。

通常情况下，氧化-还原反应过程的可能性可用标准电极电位 $E^{\ominus}$ 判断，众所周知，$E^{\ominus}$ 对应于电化学半反应。

所有反应的 $E^{\ominus}$ 值都是相对于 H_2 而言的，规定为标准温度和标准压力（STP）下的 H_2 的半反应和 $E^{\ominus}$ 值。

在 H_2 存在的情况下，适当地调节 pH 值可以使水溶液中许多金属离子被还原成金属。

硼氢化物离子的电化学半反应和 $E^{\ominus}$ 如下。

然而，选择硼氢化物离子作为一种还原剂是明智之举，尤其是在水溶液系统中[29~32]。最近已有相关采用硼氢化物专门用于制备金属纳米粒子的综述性文献报道[33]。

水合肼以任意比例溶于水，但是 N_2H_4 显碱性，具有化学活性的自由离子通常表示为 $N_2H_5^+$。

或在盐酸肼的情况下。

盐酸肼离子 $N_2H_5^+$ 的标准还原电势如下。

理论上，在过量的还原剂及控制合适的 pH 值时，凡 $E^{\ominus}$ 值大于−0.481V 或−0.23V 的任何一种金属就可能被还原。关于从溶液中沉淀金属，很显然包括第一周期的过渡金属离子，如 Fe^{2+}、Fe^{3+}、Co^{2+}、Ni^{2+} 及 Cu^{2+}，也包括一些第二和第三周期的过渡金属，以及大部分后过渡元素和少数的非金属。把金阳离子还原成金属金是系统地研究金属沉淀反应的一个实例。

通常是以 $AuCl_4^-$ 的形式存在的金阳离子，很容易被 H_2 气还原。虽然 $AuCl_4^-$ 具有很强的氧化性（$E^{\ominus}$ = +1.002V），但弱的还原剂如羧酸盐或醇就足以将其还原。

Tan[34]等人报道了采用酒石酸氢钾还原的方法合成 Au、Pt、Pd 和 Ag 纳米粒子；在加入合适的稳定剂时，所有的产物形成了稳定的胶体。

（2）在非水溶剂中合成金属纳米材料

一个类似的反应，3-氨丙基三甲氧基硅烷作为稳定剂，采用 *N,N*-二甲基甲酰胺（DMF）还原 $AgNO_3$ 或 $AgClO_4$ 制备 Ag 纳米粒子[35]，反应过程中，DMF 被氧化成羧酸。

在这种情况下，通过调节温度和[DMF]/[Ag]摩尔比，Ag 纳米粒子尺寸可控制在 6～20nm。硅烷基稳定剂对水解和缩合反应是敏感的，尤其是在升高温度和酸性条件下。尽管没有明显的结块出现，但一些 Ag 粒子已经包裹上了薄的 SiO_2 层，如图 4.15 所示。

采用柠檬酸还原金离子可形成球形粒子。但采用类似的反应还原银离子时（图 4.16），由于柠檬酸起到配位剂的作用，依据反应条件，可生产不同形貌分布的较大颗粒（60～200nm）的 Ag 粒子[35]。形成柠檬酸-银配合物影响晶体的生长，甚至容易发生光化学反应，使球形的纳米晶转变为三角锥形的纳米晶。研究结果表明：当银离子保持不变时，过量的柠檬酸盐离子使银纳米粒子的生长速度急剧降低。一系列的脉冲实验表明：在反应的早前阶段，柠檬酸盐离子与 Ag_2^+ 二聚体形成配合物[36]。

4.5.2 溶剂热法

“溶剂热”是指在一种液体或在温度高于介质沸点的超临界介质中的反应。水热反应是溶剂热反应的一种类型。为了实现温度高于反应介质沸点的反应，压力容器（高压反应釜）通常是必不可少的。

必须指出的是，高于或低于溶剂的沸点时，溶剂的液体结构实质上是不变的，这是因为液体的可压缩性很小的缘故。但是，当溶剂接近其临界点时，由于溶剂的密度发生变化，所以溶剂的结构发生了急剧改变。

通过溶剂热反应可以制备出各种纳米材料，例如：金属、金属氧化物、硫化物、氮化物、磷化物、金属含氧酸盐簇状物、有机-无机杂化材料以及碳纳米管。

大部分溶剂热反应的产物是具有一定形态的纳米颗粒或微米颗粒。通常报道的产物的粒径分布范围很窄，并且可形成单分散粒子。在溶剂热反应中，当溶剂分子或添加剂分子优先被吸附在产物的某一表面（或有一种特殊的相互作用），该表面的生长被阻止，因此，在溶剂热反应中，具有独特形貌的产物就会形成。通过溶剂热反应可获得各种类型的产物：纳米棒，纳米线，纳米管和纳米片状物。

（1）反应性溶剂

① 水　水[沸点（bp），100℃；临界温度（T_c），374℃；临界压力（P_c），221bar（1bar=10^5Pa，下同）]是溶剂热反应中最广泛使用的反应介质。

就水的相图而言，只有对水的液-气相线进行深入的研究是必不可少的（如图 4.17）。低于 100℃时，液态水的平衡蒸气压低于 1bar；在沸点以上，是可用于溶剂热反应的压力范围。根据图 4.17 所示的平衡线，可得到封闭系统中的温度对应的压力，374.1℃对应的压力是 221bar，在该点（临界点）时，气相和液相再也无法区分开来，在临界点附近，流体的性质发生了本质上的改变。超临界流体变得比气体更易压缩、比液体更加致密。这些流体可以通过压力和温度的改变来调整其气态或液态的性质。在恒温下，压力的增加可以导致流体的密度（ρ）增加，流体增浓。因此，在水热方法中，温度、水压力和反应时间是三个主要的物理参数。

② 液氨　液氨（bp，78℃；T_c，132.4℃；P_c，112.9bar）通常用于溶剂热合成氮化物。有报道称通过此方法可制得亚稳或难以得到的氮化物材料。

③ 有机介质　多种有机溶剂已被广泛应用于合成无机材料。从热力学的角度来看，所有的有机化合物在高温下都有可能分解成碳、氢（或水，如果有机化合物含有氧原子）、氮等。

在醇类中的溶剂热反应有时被称为“醇热反应”；类似于“水热”和“水解”，这个词来源于醇解。

在乙二醇中的溶剂热反应被称为“乙二醇热”反应。1,2-乙二醇是最简单的乙二醇。这种化合物在高温下很稳定。表 4.2 给出了几个常见化合物的临界温度和临界压力[37]。

（2）溶剂热合成方法

采用各种溶剂热方法能够合成金属氧化物，如：金属氢氧化物的溶剂热脱水、金属醇盐的溶剂热分解、混合氧化物的溶剂热合成、无定形氧化物的溶剂热结晶、溶剂热离子交换或溶剂热夹层和金属的溶剂热氧化。金属氧化物结晶粒子的溶剂热合成如图 4.18 所示[38]。

（3）水热合成

“水热”这个词来自地球科学，它是指高温和高水压的一种体系。在一个典型的水热研究中，被称为高压反应釜或弹式反应釜的高温、高压设备是必不可少的。在高温（大于 100℃）和压力（数个大气压之上）的合成中，水热合成所用的水既可作为催化剂，偶尔又可作为固相成分。

实验室进行亚临界溶剂热合成时，一个常用的高压反应釜是由不锈钢外壳和聚四氟乙烯内衬组成，如图 4.19 所示[39]。目前，有各种不同范围的压力-温度、不同体积的高压反应釜。首先，选择合适的反应器的最重要的参数是实验温度和压力条件。

钛铁矿（$FeTiO_3$）是一种非常稳定的矿物。从这种矿物中提取二氧化钛（TiO_2）是有可能的。在 500℃和 300kgf/cm^2（压力）反应条件下，用 10mol/L KOH 或 10mol/L NaOH 与钛酸铁以 5:3 比例混合（钛酸铁:水），63h 后，钛铁矿被完全分解。在同样条件下，如果将比例改为 5:4（钛酸铁:水），钛铁矿分解需要 39h。采用 KOH 溶液的反应式如下。

（4）超临界水热合成

水热合成（HTS）是通过模仿天然地热过程制备合成材料[40,41]。金属盐水溶液的反应平衡随温度而变化，导致形成金属氢氧化物或金属氧化物。超临界水(SCW) 为水热合成提供了一个极好的反应介质，因为水的介电常数与溶剂密度随着压力和温度的变化而发生改变，从而改变反应速率和平衡。预期的益处之一是反应速率更高，获得的粒子更小。反应产物必须是不溶于超临界水中。超临界水热合成法（HTS-SCW）通常操作如下：制备金属盐的水溶液，并对其加压加热。加压的金属盐溶液与超临界水在一个混合点结合在一起，这就导致了快速的加热和随后的反应。在溶液离开反应器之后，急冷并在线过滤除去较大的颗粒。冷却水被直接加入到反应器中以终止反应，如图 4.20 所示。目前，提出了两种不同的反应模型：第一种是使用一个间歇式反应器，其特征是反应时间长；第二种是使用流动式反应器，以确保其连续操作。

Cabanas 等人[42,43]采用硝酸铈铵和醋酸锆的混合物作为前驱体，通过一个小的连续反应器的超临界水热合成法，制备出了 $Ce_{1-x}Zr_xO_2$。根据这些作者的观点，与间歇式过程相比，连续式反应器能够更好地控制实验条件。

超临界水热合成法（HTS-SCW）可以制备非常小的纳米颗粒；然而，这个过程需要具有挑战性的操作条件，因为水的超临界点位于 374℃和 220bar（22.0MPa）。由于在超临界水（SCW）中操作，同时是高压和高温，具有特殊性能的不锈钢材质是必需的。此外，超临界水（SCW）是一种强氧化剂，不仅用于材料加工，而且用于与超临界水接触的植物成分的提取。超临界水热合成法只适用于在高温下稳定的化合物。

4.5.3 冷冻干燥法（低温化学合成法）

各种各样的低温工艺已经应用于材料科学和技术。其中，Landsberg 和 Campbell 第一次采用低温化学合成法制备出 W 和 W-Re 合金合成纳米晶粉体[45]。

冷冻干燥方法包括下列步骤：①溶液制备；②溶液的快速冻结；③冷冻-干燥去除水；④热分解和热处理。Tretyakov 和 Shlyakhtin[46] 综述了目前采用低温化学合成法制备氧化物材料的研究进展。

（1）冷冻过程

在低温化学合成中，水溶液的冷冻是通过快速冷却完成的。这个过程通常是在强烈的机械搅拌下，经由一个喷嘴或超声波喷雾器将溶液喷入液氮中来实现的。冷冻产物是由 0.01～0.5mm 的蓬松的聚集球形颗粒所组成，并经过蒸发将其与制冷剂分离。

液氮容易蒸发而减缓了冷冻液滴的冷却速率，因此，有些作者建议使用冷的己烷代替液氮，如图 4.20 所示。在冷冻过程中，盐溶液的低温结晶产物可能是由以下三种形式所组成：①冰的六边晶体与结晶盐或盐的较大的水合物晶体进行充分的机械混合物；②晶体冰和无定形的固体溶液的混合物；③一种无定形的化学均匀的固体。

（2）干燥过程

合成过程中下一个至关重要的阶段是冷冻-干燥工艺。选择较低的温度和降低压力达到缩短干燥时间，同时防止通过冰-盐共溶体的溶化形成大量液相。相对于其它工艺，该方法的优势在于可以使用各种工业型号的冷冻-干燥机。图 4.20 是特别适合实验室小型制备的冷冻-干燥示意图。

从冷冻产物中除去溶剂的主要技术是把固体冰直接转化为蒸汽，从而避免液相的形成。

通常液相的形成会降低快速冷冻获得的低温结晶产物的形态均匀性。

升华是一个强烈的吸热过程。若使冷冻-干燥过程保持一个合适的速率，则需要一个持续的外部供热。冷冻-干燥工艺条件可以根据压力-温度相图给予解释。根据吉布斯定律，水的相图中的三相点（三种形态共存：固态、液态和气态），在 $T = 273.15\text{K}$ 和 $p = 610.5\text{Pa}$ 时可以实现（图 4.17）。所以，在冰升华过程中，冰和水蒸气的共存是必不可少的。在外部供热条件下，如果系统压力不超过 610.5Pa（4.58Torr），冰就可以蒸发（升华）而不溶化。根据 100～610Pa 的平衡水蒸气压力值[47]，冷冻-干燥过程的通常温度是 253～273K。

（3）热分解

在大多数情况下，冷冻-干燥不是合成工艺的最终阶段。其产品需要进一步加工，使其转化为氧化物或金属粉末。这些后续的热分解和粉体加工技术与其它粉末的化学合成方法相同或相似。由于形成固溶体的缘故，盐混合物中每一种组分的分解温度通常都低于每一种纯盐的分解温度。尽管氧化物的形成温度为 200～900℃，但复合氧化物的生成，通常发生在盐前躯体热分解的最后阶段。低温结晶产物是由不同的化合物形成的高分散的机械混合物所组成。所以，在合成的最后阶段，热加过程中形成均匀的复合氧化物。

（4）喷雾冷冻干燥技术

喷雾冷冻干燥技术是制备纳米粒子的先进方法之一，已经广泛用于合成具有纳米结构的生物药物和无机材料[48~51]。通过该方法得到的粒子通常具有引人注目的优点。比如：由于在冷冻剂中快速冷冻（闪冻）可形成分子尺度的均一性；也因为在低温真空干燥条件下冰的升华，可得到最小的颗粒团聚[52,53]。迄今为止，该体系已被用来合成纳米结构的含能材料。

具有不同单元尺寸的三维网格结构的 1,1-二氨基-2,2-二硝基乙烯(FOX-7)也可以通过冷冻-干燥技术制备[54]。三维网格结构的 FOX-7 的制备如图 4.22(a)所示。图 4.22(b)和(c)中 FOX-7 典型的扫描电镜图片表明：三维 FOX-7 结构是由一维的纳米结构所组成。纳米 FOX-7 产物的尺寸大小和网格结构可以很容易地通过调变原料水溶液的浓度来控制。通过对这种结构的可能形成机理的深入研究揭示了在液氮的条件下，骤冷导致高的成核速率，由于水溶液和致冷剂之间快速的热传递，小颗粒和冰迅速形成。冰升华后，具有相对较高能量的小粒子开始自组装聚集，从而形成三维的网格结构，这一发展过程的框架如图 4.22(d)所示。更重要的是，这种简便的合成方式代表了制备其它水溶性含能化合物纳米结构材料的一种通用方法。

4.5.4 溶胶-凝胶法

溶胶-凝胶法是由溶液、溶胶、凝胶、固化、金属有机化合物或无机化合物的热处理等所组成的一种方法。该方法可用于制备金属氧化物和其它化合物。

传统上，溶胶-凝胶法涉及金属醇盐的水解和缩聚。金属醇盐的通式是 $M(OR)_x$，而通式中烷氧基（—OR）与醇相关。通常金属醇盐的合成涉及金属物种（金属、金属氢氧化物、金属氧化物或金属卤化物）和醇的反应。例如，正如 Ebelmen 最早报道的那样，由 $SiCl_4$ 和合适的醇很容易制得 $Si(OR)_4$[55]。

在无水条件下，形成稳定的 $Si(OR)_4$ 溶胶，其中，R 代表 CH_3，C_2H_5 或者是 C_3H_7。

将水引入系统之后，水解就开始了，也就是来自水的一个羟基取代一个烷氧基，同时生成一个自由的醇分子，随后，发生缩合反应。

这些缩合反应导致了凝胶的形成。缩聚反应过程涉及两个水解的小分子结合在一起，释放出一个醇分子或水分子。

醇盐前躯体的水解程度强烈地影响着 Si—O—Si 的网络结构。因为 OH^-是比—OR 更容易离去的基团，缩聚反应可以被裁剪成所需要的二聚体、链状或三维聚集体，如图 4.23 所示。其结果，在聚合过程的早期阶段有利于形成 Si—O—Si 链，在老化过程中，随后生成支链和交联。这些凝胶是由直径 0.5～5nm 的聚合物小颗粒组成，依据水解-聚合的反应速率、醇盐-溶剂和醇盐-水的最初比例，形成的凝胶或多或少是处于凝聚状态，并且较密实。

溶胶-凝胶路线可用于制备纯的、化学计量的、致密的、等轴的以及单分散颗粒。例如，通过控制钛酸四异丙酯的水解制备平均粒径 70～300nm 的无定形 TiO_2 粒子。采用钛酸四乙酯合成单分散的非晶粉体时，仅出现均匀成核而没有絮凝产生。根据反应动力学和颗粒的老化（后合成）条件，合成的纳米氧化物粒子可以是非晶体或晶体。由于这些纳米结构的氧化物粒子具有大量的高活性表面，可以通过渗碳或氮化以形成非氧化物粉体。

在溶胶-凝胶过程中，需要考虑的因素是溶剂、温度、前躯体、催化剂、pH、添加剂和机械搅拌。这些因素可以影响反应动力学、生长、水解和缩聚反应，溶剂影响前躯体的反应动力学和构型，pH 值影响水解及缩聚反应。酸性条件有利于水解，这就意味着在缩聚反应之前所有或几乎所有的水解产物都已形成。在酸性条件下，有较低的交联度。通过改变影响水解、缩聚反应速率的因素，凝胶的结构和性能可以被剪裁。在传统的溶胶-凝胶法中，缩聚反应可能是可逆的，反应过程中需要尽量彻底地移走生成的水或生成的极性溶剂。

Blinker 和 Scherrer[56] 对溶胶-凝胶法及产物进行了分类，如图 4.24 所示。溶胶-凝胶法的主要原理是通过溶胶的水解和缩聚反应，溶胶就自发地转变成了凝胶。

溶剂的蒸发促进了凝胶相的进一步转变，随后形成干凝胶相。溶胶-凝胶转换可能发生在大量的溶液中。然而，当溶液覆盖于固体表面上，可提高凝胶化效率，通过溶液的浸涂、旋转涂及喷涂，在固体基质上可形成 100nm 范围内的干凝胶膜。加热干凝胶可完全去除溶剂分子导致无机簇状物的聚集，形成薄膜状固体材料。然后，重复该过程使其生成厚的多层膜。在高温下进行快速的超临界干燥会导致形成气凝胶，即一种孔隙率大于 75%的多孔材料。另一个极端条件是在环境条件下，一个非常缓慢的溶剂蒸发会引起固相沉淀，最终生成细小的、均匀的粒子。溶胶-凝胶法已用于制造片材、管材，以及纤维状陶瓷材料的工业化生产过程中，如图 4.24 所示[56~59]。

在非水解的溶胶-凝胶法中，传统的水解、缩聚反应将被直接的缩聚反应或酯交换反应所取代。金属卤化物与醇盐可直接发生缩聚反应，或金属醇盐与金属羧酸酯发生缩聚反应。酯交换反应涉及一个三烃基硅基醋酸酯和一个金属醇盐之间的反应。非水解反应不涉及水或极性溶剂。

4.5.5 微乳液方法

（1）微乳液基本原理

将水、油、表面活性剂和一种醇或一种胺共表面活性剂混合，生成的一种清澈、似乎均匀的分散体系被称为“微乳液”[60, 61]。在这种情况下，油相是简单的长链烃，表面活性剂是含有一个亲水头（通常是磺酸根离子或季铵盐）和一个亲油尾的长链有机分子。目前，广泛使用的表面活性剂是十六烷基三甲基溴化铵（CTAB），如图 4.25 所示。

表面活性剂的双亲属性，如 CTAB，使得它与烃和水很容易混合。表面活性剂通过与极性的共表面活性剂发生离子偶极相互作用，形成了球形聚集体。在聚集体中，表面活性剂分子的极性端（离子端）指向中心，共表面活性剂承担着一个带负电荷的“间隔体”的角色，使带正电荷的表面活性剂的头（极性端）的斥力达到最小化，如图 4.26 所示[60]。

表面活性剂分子有一个亲水的极性头和一个亲油的烃基链。这种分子倾向于自发地聚集于不能混溶的两种介质的界面处，例如在水和油的界面处表面活性剂分子形成不同的聚集体，这些分子的特定组成引起油与水之间的表面张力减小。表面活性剂的形状对最终产物的形态起着很重要的作用。当达到一个临界浓度（称为临界胶束浓度或 CMC）时，表面活性剂分子也可以自组装成胶束和其它结构。

如果表面活性剂的短的烃基链上连有一个很大的极性头，其分子有一个锥体形状。这些链将聚集一起形成叫做胶束的球状聚集体（如图 4.27[50]）。当在低浓度下形成胶束时，胶束呈球形，胶束的直径是由它们的烃链的长度和极性头来决定。在不同的条件下，可以形成单体、胶束和反相胶束。

一般而言，两种或两种以上不相溶的液体热力学不稳定分散系，在液-液界面处通过形成水包油或油包水型的乳液而成为动力学稳定的分散系。复杂的乳液，例如：油包水或水包油-油包水的乳液也可能形成，如图 4.28 所示。通常认为纳米乳液是伴随有 1～100nm 分散相液滴的一个连续相的乳液。

（2）反向胶束（球形纳米反应器）

反向胶束（有时称为油包水型的微乳液）是在一种烃溶剂中，表面活性剂分子的一个单层，包裹住可溶性的水池，如图 4.28 所示。在表面活性剂的浓度一定时，水池的大小取决于包裹水的量。这些液滴（水池）可作为纳米反应器。事实上，水池提供了可控制体积在 1～500nm^3 范围内的一个纳米反应介质。

图 4.29 是这个过程的示意图[63]。反向胶束作为尺寸可变的纳米反应器，因为它有尺寸可控和交换水量的两种性质，即使在室温下，反向胶束的小尺寸能够进行永不停息的布朗运动。当把两种溶液混合，由于布朗运动的诱导，水芯（在反相胶束中）进行交换，使两种反应物金属盐和还原剂相互接触，从而促使它们之间的反应，如图 4.29 所示[63]。

许多实验结果表明，反向胶束作为纳米反应器可以制备许多纳米尺寸的化合物，如纳米半导体、纳米金属、纳米氧化物、纳米合金。事实上，对于Ⅱ～Ⅵ族的半导体，如制备的 CdS、CdTe、CdMnS 或 ZnS[64]，颗粒尺寸从 2～4nm 变化，而对于制备的金属[65]，Ag 的粒径在 2～6nm，Cu 的粒径在 2～10nm，纳米粒子的尺寸控制来自油中最小的水池（0.6～6nm），而对于最大的模板（6～12nm），生成的纳米颗粒尺寸没有发生变化，这可由包裹在胶束中水的结构给予解释[53]。事实上，对于低含水量（w=10%），在胶束核中的水分子与抗衡离子强烈地作用。当含水量增加（w>10%），水分子不再参与水化过程，可以认为是自由水。

控制纳米晶体的形状是一个严峻的挑战。一些已发表的结果表明，以三元体系（水-表面活性剂-油）作为模板会导致表面活性剂自组装形成不同的形状。由功能化的表面活性剂组成的胶体模板的形状，可能决定着纳米晶体的形状[66, 67]。图 4.30 是一个内部连通的圆柱形模板制备出了球状的 Cu 纳米粒子和一些圆柱状的 Cu 纳米颗粒[68]。

在 $Cu(AOT)_2$-水-异辛烷［AOT：琥珀酸二（2-乙基己基）酯磺酸钠］三元系统中，让我们思考卤化钠的浓度对制备铜的纳米晶体形状的影响。在相互连通的圆柱状区域，形成了细小的柱状纳米晶［如图 4.31（a）］。当向胶体溶液中加入浓度小于 2×10^{-3}mol/L 的氯离子时，就会生成长条状的铜棒，如图 4.31（b）和（c）所示[69~71]。这些长条状的铜棒与细小的柱状纳米晶具有相同的晶体结构。

如果掺入的是 NaBr 溶液，而不是 NaCl 溶液，就可得到立方纳米晶（如图 4.32[72]），而且，产物的尺寸不会随溴离子浓度的变化而变化。当加入 NO_3^-溶液时，可以得到各种形状和尺寸的纳米晶体，当加入 F^-溶液，则可制备出小的立方晶体[73]。

4.5.6 微波辅助合成

微波是介于红外线和无线电波之间的电磁光谱，其波长在 1mm～1m 之间，频率在 300MHz～30GHz 之间，应用于工业的典型波段是 (915±15)MHz 和 (2450±50)MHz[74,75]。

几乎所有目前涉及的微波合成的相关报道使用的都是 2450MHz（对应的波长为 12.24cm）的微波。原因之一是液态水在接近该频率时吸收微波能量是最大的。另一个可能的原因是 2450MHz 的磁控管大多用于商业微波化学设备中。一般来说，基于材料与微波的相互作用，将材料可分为三个类型：①透过，微波通过时没有任何损失，例如熔融石英、数种玻璃、陶瓷、聚四氟乙烯；②导体，微波只能反射，而不能透过，如铜；③吸收，根据介电损耗因子值，微波可以被吸收，在微波合成法中，它是最重要的一类材料，如水溶液和极性溶剂。

图 4.33 展示了在一个瞬息变化的微波场中[76]，极性分子是如何试图调整自身以适应变化的微波场。当利用微波辐照材料时，电场和磁场迅速地发生变化（在 2450MHz 的频率下，变化速率为每秒 2.4×10^9 次），而分子不能迅速响应到这些方向上的改变，导致分子间发生摩擦并变热[77]。

离子的传导是另一个重要的微波加热机理。当将一种含有离子的溶液置于一个微波场中，离子由于带有电荷而发生运动，其结果是离子之间发生碰撞，碰撞引起了动能转换成热能。当溶液中离子浓度增加时，会出现更多的碰撞，导致溶液很快被加热[78]。在许多情况下，微波合成的显著特点是减少了反应时间。

Yu 等人报道了在敞口的烧杯中利用 2450MHz 的微波对聚（*N*-乙烯-2-吡咯烷酮）、H_2PtCl_6 水溶液、乙二醇和 NaOH 的混合液辐照 30s，制备出粒径为 2～4nm 的 Pt 胶体粒子[79]。Tsuji 等人利用相似的合成方法，在乙二醇中制备出聚乙烯吡咯烷酮（PVP）包裹的 Ni 纳米粒子[80]。

4.5.7 超声波辅助合成

依据频率，可将超声波分为三类：功率超声（20～100kHz）、高频超声（100kHz～1MHz）、诊断超声（1～500MHz）。频率为 20～100kHz 的超声波可用于化学和物理变化较理想的系统[81]。

超声波是机械波，不会被物质吸收，但是它可以通过任何固体、液体或气体物质进行传输。超声波的能量不足以引起化学反应，但当它通过介质传播时，就会产生一系列的压缩波和稀疏波。压缩循环通过推动分子给液体施加一个正压，稀薄循环通过反复拉回分子给液体施加一个负压。因为这种负压过大，在稀薄区会形成微泡（气穴现象），这些微泡在连续的循环中生长，达到一个不稳定的直径时，以至于发生剧烈崩溃产生冲击波（寿命几微秒）[82]。

在这个过程中，气泡的形成、生长和经历剧烈崩溃称为气穴现象。气穴泡的发展和崩溃，以及固态的侵蚀如图 4.34 所示[83]。

4.6 通过固化纳米粉合成块材

为了利用块体纳米晶材料的独特性质，纳米量级的粉体可以被压制成具有一定性能、几何构型及尺寸的块材。最关键的是在实现其致密化的过程中，尽可能地减小微观结构长大/或不良的微结构的转变。

4.6.1 冷压

冷压实是形成一种成型的固体，这种最初的固化被称为一种坯体。纳米金属的坯体密度依赖于施加的压力，如图 4.35 所示。在低压区，密度随压力呈线性增加。在高压区，曲线的斜率大大减小。该区域表明了 15nm 的 Ni 和 26nm 的 Fe 样品的加工硬化过程[84]，如图 4.35 所示。从实验中观察到，在高压区，50nm 的 Ni 呈现出平缓的压实曲线，表明具有一个很小的加工硬化[84, 85]。

4.6.2 热压

为了消除吸附物，并提高粒子间作用力，温度达到 402℃的热压广泛用于纳米粉的固化过程。对于大多数金属纳米粉，这些温度仍处于冷加工区域[86, 87]。在 402℃时，热压 70～80nm 的 TiN 粉其密度比原密度高出 90%[88]。

通过对无定形的前驱体粉末进行热压，温度 700～850℃、压力 275MPa，时间 0.5～2h，形成的固化块体样品，是由α-Fe 的矩阵中分散 10nm 的 Mo_2C 分子簇所组成。而在一个大的材料（矩阵）中更小颗粒的碳化物对防止颗粒的生长几乎没有什么影响。在热压过程中，一些结果表明，增加压力和压制温度对减小所有样品的孔隙率没有明显的影响。在一个实际压力范围内，多组分工程材料的粉末固化以达到高密度，要保持其纳米结构仍然是一个挑战。

4.7 模板辅助自组装纳米结构材料

4.7.1 自组装原理

根据 Whitesides 的观点：“一个自组装过程是一个不涉及人的参与过程，在自组装过程中，原子、分子、分子的聚集体以及组成部分自身排列成有序的功能化的实体，而没有人的干预”[88]。

自组装已成为一个非常有效的和有前景的合成多种新型纳米材料的方法。图 4.36 显示的是自组装原理以及双亲分子在水中的结构[89]。由于前驱体分子在纯水-空气界面处呈现各向异性分布，其合力指向溶剂，如图 4.36（a）所示。当在水中加入两性分子时，如图 4.36（b）所示，两性分子在界面处聚集；疏水端的尾朝向气相，而亲水端插入水中，双亲分子在界面处富集直到界面被完全覆盖。然而，由于界面能的最小化而形成了胶束，双亲分子在连续相中的分散成为可能。在胶束中，亲水基指向溶剂，而疏水基链被疏水核中的水所隔开，如图 4.36（c）所示。通过加入越来越多的双亲分子，就形成了越来越多的胶束。它们能够组装成密堆积的面心立方或体心立方，如图 4.36（d）所示。

此外，胶束融合并且建立连续的结构，即所谓的溶致相。首先，生成了柱状六方排列的二维相。然后，柱状体融合并生成溶致液晶片状结构，如图 4.36（e）所示。在六方相转化为层相的过程中，可能出现螺旋相的双连三维结构。胶束和溶致相能够被不同的溶剂所溶胀而变成乳液[90]，如图 4.36（f）中的微乳液或泡囊。出现在超分子化学过程中的其它自组装过程另当别论[91, 92]。由于多尺度的结构控制，自组装结构是纳米合成中理想的辅助手段。

4.7.2 MCM-41 自组装

1992 年，Mobil 公司的科学家揭示了采用表面活性剂和硅酸盐可以进行共组装，通过范德华力和静电相互作用而形成纳米复合材料。除去表面活性剂就得到介孔硅分子筛（MCM-41），如图 4.37 所示[93]。

利用范德华力、静电力、氢键或其它非共价键之间的相互作用力，这种共组装合成路线已经延伸到制备大量的介孔材料，其中包括 SiO_2[94~97]、掺有 Ti 的 SiO_2[98]或掺有 Al 的 SiO_2[99]、铝硅酸盐[100]、$AlPO_4$[101~105]、Al_2O_3[106]、TiO_2[107]、Nb_2O_5[108]、SnO_2[109, 110]、ZrO_2[111]、混合氧化物[112]和金属 Pt[113]。

一个可以被广泛接受的模板辅助自组装合成路线的机理如图 4.38 所示[114, 115]。首先，表面活性剂组装成圆柱状胶束。然后，可溶性的硅酸盐物种插入胶束头部的亲水区。这些柱状胶束之间通过静电、氢键或其它非共价键产生相互作用，形成高度有序的无机-有机纳米复合物，如图 4.38（路线 B）所示。加入可溶性硅酸盐物种引起球形胶束转变成高度有序的聚集体（路线 A），通过进一步的水热处理，从而获得一种固体的无机-有机纳米复合材料。当在高于 500℃的温度下烧掉表面活性剂模板，就形成了一个高度有序的介孔分子筛材料。

4.8 自组装纳米晶

像分子体系那样，包裹合适配位体的纳米晶可以自发地组装成有序的聚集体。支配纳米晶组装的力有许多不同的方式。例如张力起着重要的作用，因为纳米晶的表面原子占有很大的份额。

在固体表面上自组装的表面活性剂分子证明是获得纳米晶有序排列的最佳手段。纳米晶自组织的方式主要是依靠核的直径、配位体的属性、所使用的分散介质等。硫醇化的纳米晶，在除去溶剂后容易排列成二维阵列。

Ag 纳米晶沉淀在石墨基质上，是烷基链的长度对 2D 自组织的影响一个实例。当包裹粒子的烷基链 $C_nH_{2n+1}SH$ 变短，即当 n 从 12 变到 8 时，就不利于颗粒之间保持一定距离的单层的形成（见图 4.39）[116, 117]。颗粒之间的距离对应于大约 2nm 的十二烷基硫醇的链长［见图 4.39(a)］。因此，这就表明在十二烷基硫醇包裹纳米晶的情形下，硫醇链是相互贯通的，而对于包裹有较短烷基链硫醇的纳米晶，如十烷基硫醇或辛烷基硫醇，其硫醇链不再相互贯通［见图 4.39(b) 和 (c)］。这似乎是短链与基质的相互作用更强，因为这种相互作用是吸引力，颗粒不可能在基质上自由运动，这就解释了颗粒之间具有较大距离的原因。

4.9 各向异性纳米粒子的合成和组装

4.9.1 具有特征尺寸的各向异性纳米粒子

由于纳米粒子独特的纳米尺度以及各种形貌，其新颖的物理或化学性质不同于块体材料。迄今，许多研究者专注于发展制备纳米尺度粒子的有效方法。最近，控制粒径分布和形态已经成为该研究领域的一个新问题。因此，需要创新性的方法来解决这一难题。一种方法是可控制备和定向生长合成纳米粒子。纳米粒子的自组装或者定向自组装，对于许多有用的纳米器件，包括传感器、光子电路、光电器件等提供了尚佳的成本-效益生产策略[118]。

截至目前，合成的大部分纳米粒子都是各向同性的粒子，因此，设计、合成以及对纳米粒子功能化就显得尤为重要。值得注意的是，在功能化或组装过程中，保持纳米粒子的内在功能或性质是十分必要的。最近，Glotzer 和 Solomon 把具有新结构的各向异性的微米及纳米尺度的粒子分为数个类型 [119]。

然而，制备具有窄的尺寸分布的各向异性纳米粒子，并将其组装成三维结构是非常关键的。对特征尺寸在 1～100nm 范围内的各向异性纳米粒子，可依据它们的形貌进行分类，如图 4.40 所示。

4.9.2 棒状粒子

对于晶体材料，晶面的生长比率、表面原子与有机分子之间的相互作用，对纳米结构形貌的形成起关键性作用。表面活性剂与生长着的晶面原子键合得越强，该晶面的生长速率就越慢，因此，在反应进程中，快速增长的晶面逐渐消失，而相对缓慢增长的晶面成为最终纳米材料的外表面。

当纳米粒子的长径比 a.r. < 5，5 < a.r. < 25 或 a.r. > 25 时，则相对应的粒子分别被归类为短纳米棒、长纳米棒及纳米线[120]。对于在液相中形成的纳米棒，一般策略是利用其内禀的晶体结构，采用表面活性剂的定向附着或选择性吸附，如图 4.41 所示[121]。

如果利用其内禀的晶体结构，无机化合物应该具有不对称的表面能，例如硫化镉、硒化镉、碲化镉、硒化锌和硫化锰都具有六角纤锌矿结构，如图 4.42 所示，在纤锌矿结构中，由于（001）晶面高的堆积密度，其（001）晶面的表面能高于其它晶面的表面能[122]。利用晶面之间表面能的差别，在前驱体分解温度以上，通过在溶剂中简单地加入前驱体以合成 CdSe 纳米棒；即：在 360℃，将 2mL 的原液［$Se:Cd(CH_3)_2:C_{12}H_{27}P$=1:2:38，质量比］加入到 4g 氧代三辛基膦中[123]，Agostiano 等人指出可采用类似的方法合成 ZnSe 纳米棒，通过改变滴加速度来控制其长径比[124]。

对于定向附着形成纳米棒的情形，颗粒与优先晶面进行融合。例如，立方 ZnS 纳米棒是由 ZnS 纳米颗粒的定向附着而形成的[125]。在 125℃，将 Zn 前驱体滴加到含有硫元素和十六胺的溶液中，然后加热到 300℃，在 60℃老化时，ZnS 纳米颗粒逐渐转变为纳米棒[125]。最近，Weller 和他的同事也观察到了 ZnO 纳米颗粒的定向附着。首先，在 60℃左右，将二水合乙酸锌溶解在甲醇中，形成准球形颗粒；然后，将 KOH 的甲醇溶液加入到 ZnO 颗粒的悬浮液中。随后，蒸发溶剂，对剩余的溶液进行回流，从而制备出棒状颗粒[126]，起初纳米颗粒表面的电荷在 ZnO 纳米棒的形成过程中可能起到重要的作用。

如前所述的棒状颗粒，晶体金属纳米颗粒的晶面能够控制。在多数情况下，表面活性剂或离子优先结合在特定的晶面上，该晶面的增长率就受到了抑制。因此，我们可以制备出与表面活性剂结合的、具有特殊晶面的纳米粒子。Huang 等人研究发现，在表面活性剂（十六烷基三甲基溴化铵，CTAB）存在时，通过水热反应可以制备出八面体 Au 纳米颗粒，并且可以通过反应时间控制其颗粒大小[127]。沿着晶面方向，晶面纳米颗粒的内禀增长率是不同的。当（100）晶面的增长率远大于（111）晶面时，将会形成正八面体，如图 4.43 所示[129]。如果离子或表面活性剂键合在（100）晶面，则八面体就会转化为立方体。Xia 对于通常利用晶面表面能或利用表面活性剂或离子的选择性吸附制备晶面金属颗粒的合成方法进行了综述[128]。

4.9.3 各种形貌 Pt 纳米粒子的制备

在胶体合成反应过程中，有许多因素决定着最终产物的形貌，其中包括反应前驱体，配体，反应介质(溶剂)以及反应环境(温度和压力)。配体也称为表面活性剂，在胶体合成中扮演着重要的角色，例如粒子的定向生长，控制粒子尺寸，稳定纳米粒子以阻止其团聚和沉淀。目前广泛使用的配体包括有机分子、聚合物和离子盐，可通过表面相互作用的官能团与粒子结合。一些常见的官能团包括硫醇、胺类、羧酸、膦类化合物、氧代膦、铵盐和羧酸盐。配体与纳米颗粒之间的相互作用主要包括共价键、静电相互作用以及物理吸附。通常，强的配体-表面相互作用，如硫醇和胺类，能够限制粒子生长，这样可以制备小尺寸的纳米粒子。能够增强弱键合的官能团的表面相互作用的多齿配体也能限制颗粒的尺寸。此外，配体的浓度在控制粒径方面发挥着重要的作用。例如，较高浓度的配体可以生成小的纳米粒子，而稀的配体溶液则制备出相对较大的粒子。这一方法已用来制备 1～4nm 范围内的纳米 Au 粒子[130~134]。

除了控制粒子大小以外，配体也决定着粒子的形貌。为了减小表面积以及表面能，密堆积的原子结构，如面心立方晶体易于形成包括（111）面和（100）面在内的所谓的伍尔夫多面体（切去顶尖的八面体）[135~139]。由于一些配体可以选择性的与某一晶面键合，并降低其表面能，因此，加入配体，除了制备伍尔夫多面体外，也提供了制备不同形貌的纳米粒子的可能性。在制备 Pt 金属纳米粒子时，溴化物离子能够稳定（100）晶面，如图 4.44 所示。当 $NaBH_4$ 作为还原剂，十四烷基溴化铵（TTAB）作为配体时，可以制备出只暴露（100）面，尺寸在 15nm 的 Pt 纳米立方体粒子。然而，当将有利于（111）面的氢分子引入到反应系统中时，可以制备出（100）面和（111）面暴露在外，尺寸在 15nm 的 Pt 十四面体。如果用聚（乙烯）吡咯烷酮（PVP）替代 TTAB，氢仍然作为还原剂，可以制备出尺寸可控的，范围在 3～10nm 的 Pt 四面体[140, 141]。

4.9.4 制备各种形貌的铑（Rh）纳米粒子

Humphrey 等人报道了一种采用无晶种多元醇合成单分散（111）面和（100）面取向的尺寸分布小于 10nm（及 6.5nm）的 Rh 纳米晶粒子的方法[142]。在这个报道中，乙二醇(EG)既作为溶剂又作为还原剂，而聚(乙烯)吡咯烷酮（PVP）作为封端剂（覆盖剂）。在一个典型的制备过程中，表面活性剂溶解在多元醇溶剂中，然后在惰性气体，如氮气或者氩气保护下加热到某一温度。金属前驱体首先溶解在多元醇溶剂中，然后在高温下注入反应介质。在滴加过程中，金属离子被多元醇快速还原成金属粒子，并被封端剂（覆盖剂）所保护以免形成大的金属粒子聚集体。

当使用醋酸铑$[Rh(Ac)_2]_2$作为金属前驱体时，可以制备出（111）取向的，其中含有 76%（111）孪生六边形的铑（Rh）纳米多面体。然而，在烷基溴化铵存在下，例如四甲基溴化铵或三甲基（十四烷基）溴化铵，使用 $RhCl_3$ 作为金属前驱体时，可以制备出（100）取向的铑（Rh）纳米立方体粒子，其选择性可达 85%，如图 4.45 所示。

4.10 一维纳米结构（ODNS）聚合物的合成

Martin 等人开创性地展示了聚合物纳米纤维的首次应用[143, 144]，他们报道了当把聚合物纤维的维度和尺寸限定在介观范围内，其聚合物纤维的电导率急剧增大。导电聚合物(掺杂或无掺杂)科学飞速地发展到一个阶段[145]，即科学家们现在寻找导电聚合物作为一种备用材料以替代在半导体工业上已经达到微型化极限的硅材料。

从那时起，聚合物纳米管、纳米线和纳米棒的研究如火如荼，达到前所未有的增长。这种飞快的增长速度一直受新型的合成方法及各种交叉学科重要性的推动。在全球范围内很多的研究小组已经成功地开发出合成一维聚合物的新路线，例如：电纺丝[146]，共-电纺丝[147]，薄膜基的[148]，通过纤维模板制备的纳米管[149]，在模板内，通过化学的及自组装路线的电化学聚合[150, 151]等。上述的所有工艺都能够得到具有理想性能和精确长径比的一维纳米结构（ODNS）的聚合物。

4.10.1 电纺丝法合成一维纳米结构（ODNS）聚合物

电纺丝法是制备聚合物纤维最广泛使用的技术之一。这种技术新颖而简单，能够组装直径从微米到小于 100nm 的纤维状聚合物。该方法是经过喷嘴喷射出的一细流聚合物溶液或熔融到靶的阴极上，形成超细固体纤维的高压过程。由于该过程是在电场中进行，因此，可以控制或获得不同三维几何形状的纤维状聚合物。然而，此方法仅限于合成具有大的表面与体积之比（直径 100nm 的纤维其表面积约为 $100m^2/g$）的一维超细固体纳米纤维。Mac Diarmid 等人用静电纺丝制备了直径小于 30nm 的聚苯胺基纳米纤维[152]。然而，只要精确地控制一些参数就能获得形状、尺寸或性能独特的产品。可以添加各种粒子到聚合的混合物、溶液或熔体中，以增强纤维基质或在纤维中封装特殊材料。静电纺丝的优点之一是可以用水作为溶剂，因此，水溶性的聚合物，如：聚环氧乙烷(PEO)、聚乙烯醇(PVA)、聚乳酸(PLA)等也可以用静电纺丝制备。因此，所有极性的和非极性的有机聚合物都可以采用该技术进行静电纺丝[153]。

静电纺丝技术是一种高电压的静电方法，一种聚合物溶液(或熔体)保持较高正电压，各种不同的极性或非极性的溶剂放置在与金属阴极保持固定距离的注射器中[154, 155]。从注射器或针头上流出的聚合物溶液或熔体由于表面张力而形成液滴，依据施加的电压和所采用的聚合物，液滴会带不同程度的电荷。相同电荷相互排斥造成的斥力与表面张力相反[156]。施加的电场强度增加时，液滴从半球形转变成通常被称为 Taylor 锥的一个细长的锥结构[157]。当外加电压超过一个临界值，在注射器针尖上的液滴克服了表面张力，从 Taylor 锥的尖端喷射出一细流液体。这种聚合物在拉丝的距离之内要经历一个抖动的过程以蒸发溶剂（或熔体固化），当聚合物流体落到阴极板上时，被高度拉伸（成丝）。依据实验参数，可获得在阴极板表面收集的直径为纳米或微米量级的非织造布状的干纤维丝，如图 4.46 所示[158]。

如前所述的静电纺丝过程的特点是快速蒸发溶剂或熔体固化过程。纳米结构的形成发生在毫秒范围，因此，人们会认为晶体的成核和增长是不同于块体或宏观物体的。然而，情况并非如此，人们已经发现成核机制与观察到的较厚纤维的成核机理相同。

虽然静电纺丝的操作过程非常简单，但实际上静电纺丝纤维的机理相当复杂，并且要控制各种不稳定性，如瑞利不稳定，轴对称不稳定性，抖动或弯曲的不稳定性[159~161]。支配电纺丝纤维的延伸和变薄的主要因素应归因于抖动作用。其它不稳定性是由喷射半径的波动造成的。电纺丝的工艺参数可分为 (a)机电的：包括施加的电势，聚合物的流速，阴极和阳极之间的距离；(b)化学的：聚合物的分子量，分子量分布，结构（支链或链状聚合物），共聚合物或单一聚合物，聚合物溶液的电导率和介电常数（根据官能团）；(c)流变学的：溶液的黏度、浓度、表面张力等；(d)外部的：温度、湿度、空气流速或腔体中的真空度等。

近来开发了一种新型的电纺丝技术共电纺丝，利用该技术，研究人员已经制备出了聚环氧乙烷-聚十二烷基噻吩（PEO-PDT）型的核-壳（结构的）纳米纤维[147]。实验装置具有两个空气入口，如图 4.47 所示。这种工艺能够制备核-壳结构的纳米纤维或核-壳结构的介尺度纤维。从中心喷嘴以及从周围的同心环状喷嘴流出的两种液体形成一个复合液滴，该复合液滴在共电纺丝的复合喷嘴的尖端经历一种转变，形成一种复合的 Taylor 锥。当受电场的牵制时，这种喷射（出的液体）经受与传统工艺相似的不稳定性影响。随着溶剂的蒸发，复合喷射（液）固化生成复合核-壳（结构的）纳米纤维。电纺工艺的高速度(运转)防止了核-聚合物与壳-聚合物的混合。共电纺丝也可以应用于金属盐聚合物系统，如：合成聚乳酸-醋酸钯（PLA-$Pd(OAc)_2$）系统[147]，同轴纤维在170℃退火2 h，可以观察到$Pd(OAc)_2$金属化成Pd。共电纺丝也可用于制备聚合纳米管，因此，共电纺丝可以广泛应用于各样系统。

4.10.2 基于膜或基于模板的一维纳米结构（ODNS）聚合物的合成

基于模板的方法对几乎所有的无机系统都十分重要。由于合成产品的可靠性和精确性，基于模板的方法用于一维纳米结构（ODNS）聚合物合成，也同样具有显著的特点。然而，由于聚合物固有的高润湿性，使用基于模板方法制备聚合物溶液或熔体优于合成金属或金属氧化物结构。这样，基于膜或基于模板制备一维纳米结构（ODNS）聚合物相对简单，因而很受欢迎。Martin等人已经率先使用这种技术制备出有序的一维纳米结构（ODNS）。一些有关基于膜或基于模板的合成细节描述如下。

（1） 径迹蚀刻聚合物膜

基于模板合成一维纳米结构（ODNS）聚合物的大部分研究局限于两种类型的膜，径迹蚀刻聚合物膜和多孔氧化铝/氧化硅膜（后者在聚合物制备中不大流行）。顾名思义，径迹蚀刻膜经由径迹-蚀刻的方法制备膜[162]。这些膜含有相同直径的圆柱形孔，制备这种膜最常用的材料是聚碳酸酯。商业上常用的是孔密度为10^9个孔/cm^2的孔径小于10 nm的膜。另一方面，通过对金属铝或硅材料进行电化学蚀刻（或阳极处理），在这些材料的表面上制备出有序的纳米孔的氧化铝或氧化硅膜。传统上，常通过电化学方法或通过在模板孔内放入化学氧化剂的方法对聚合物进行氧化聚合反应来制备导电聚合物 [163]。

（2）CVD聚合方法

CVD也是用模板法制备一维纳米结构（ODNS）聚合物的主要方法之一。工艺很简单，即将膜放置在高温炉中，使反应性的气体从炉内通过，气体在孔中发生热分解，生成的聚合物沿着膜的孔壁分解/沉积。使用CVD聚合方法，在孔径为10～200nm的氧化铝及聚碳酸酯模板（过滤器）中合成出的PPV (poly para-phenylenevilylene，聚对苯乙烯撑) 纳米管和纳米棒已有报道[164]。同样，采用径迹蚀刻膜和进一步的电化学聚合制备出了聚吡咯[165]，不同的电解液对聚吡咯纳米管的形貌及性能有一定的影响。

（3）表面活性剂模板

图4.48是梳状超分子及其分层自组织的一级和二级结构。从原理上示意图4.48可适应于柔性的和棒状的聚合物合成。在第一种情况下，简单的氢键足以形成柔性体，但后一种情形，通常需要一种键的协同结合（识别）与宏观相的分隔趋势不同。从A到C，由于塑化剂（表面活性剂）自组织结构可以增强可加工性。当侧链断裂之后可获得固态膜（D）。通过把双亲分子连接到二嵌段共聚物（E）的其中一个嵌段上获得的自组织超分子，会形成分层结构材料。从圆柱内的层状结构（F）中断裂侧链可以获得功能化的纳米孔性材料（G）。在圆柱内通过交联圆片，可以制备出盘状物（H）。而纳米棒（I）则是来自层状的圆柱结构的侧链断裂。图中（A）表示一个柔性的聚合物，而（B）和（C）分别代表棒状的链[166]。

4.10.3 无模板剂的一维纳米结构（ODNS）聚合物的合成

世界各地的研究者根据自己的科研兴趣从不同方面对一维纳米结构（ODNS）聚合物的合成进行了探究。通常，“无模板合成”或“自组装”方法已经可以成功制备一维纳米结构（ODNS）聚合物。它们的主要优点在于可以使一维纳米结构的直径达到50nm，或小于 50nm。然而，其纳米线、纳米管或者纳米棒的直径分布不均匀。

通过这种方法，以β-萘磺酸（NSA）和过硫酸铵（APS）为氧化剂，合成了聚吡咯微管，其中微管的内径在 30～100nm。“原位掺杂聚合”工艺通过掺杂和纳米结构同时形成的单一步骤替代了使用模板和脱除模板的多个步骤。微管的内径可以通过改变反应温度来控制，在 0℃时，原位聚合制备的微管内径达 100nm，在 25℃时，合成的微管内径为 400nm。然而，在给定的某一温度时，内径随着β-萘磺酸（NSA）浓度的增加而增大。必须指出的是，这样合成的微管在室温下电导率略小于室温下合成的粒状聚吡咯的电导率[167]。另一组实验报道了聚合反应的时间、速率以及溶液的离子强度对萘磺酸 (NSA)/聚吡咯管的形貌、电导率及分子结构的影响[168]。可以发现，通过无模板剂法形成的萘磺酸 (NSA)/聚吡咯微管是一个缓慢的和自组装的过程。

人们对许多其它的改良或全新的合成方法进行了探索，例如，在表面活性剂胶体中合成树枝状的聚苯胺[169]，通过三步电化学沉积工艺合成大阵列的纳米线[170]。这些低维系统的传输性能，由于它们在未来的纳电子学电路的潜力而具有更大的吸引力。一些掺杂或未掺杂的导电聚合物，如：聚乙炔（PA），聚苯胺（PANi），聚-3,4-乙撑二氧噻吩（PEDOT），聚吡咯（PPy）以及聚对苯乙烯（PPV）已经受到了很大的关注。

从带隙能来看，很显然，这些高分子材料的带隙能介于绝缘体/半导体之间，具有非常低的导电性。为了使它们适合作为 导电材料，要提升它们弱的本征导电性，可以通过对聚合物材料进行各种氧化剂或还原剂掺杂来实现。

4.11 绿色纳米合成

几乎所有绿色化学的原理都可应用于纳米产品的设计中[171]。在纳米技术中，许多构建单元的制备涉及危险化学品、材料的低转化率、能耗高及产生有害物质，因此，有很多机会开展制造这些材料的绿色工艺。

绿色化学是指在设计、制造和应用化工产品时，利用一系列的原理以减少或消除有害物质的产生[172]。绿色纳米科学和技术是将绿色化学原理应用于纳米产品的设计中[173]。在纳米科学和纳米技术中更多的应用是促进环境友好，其中包括对环境修复的新的催化剂[174]，廉价高效的光伏技术[175]，不加制冷剂而用于冷却的热电材料[176]，交通纳米复合材料[177]。纳米传感器[178]也能提供更快的响应时间、更低的检测限量，使现场实时检测成为可能。

4.11.1 防止废弃物

这些原理之一是通过设计使工艺步骤及辅助原料的用量最小化，以防止废弃物，使用更安全的溶剂/可供选择的反应介质，减少副产品。

一个说明性的例子是涉及在一个基质上，如硅片上制造纳米结构，制备这些结构的传统方法是一个“自上而下”的路线，该路线要制造一个纳米结构，是通过大量的沉积、图案化、蚀刻和清洗步骤的光刻蚀过程，这种工艺造成大量的废弃物。因此，一个绿色方法是涉及自组装工艺或“自下而上”的工艺，该工艺是通过使用自组装反应或“直接地”写沉积物的方式，生成结构或将结构之间连接起来。这样的选择省掉了许多工艺步骤，并使材料和溶剂的用量达到最小化。

在纳米材料的纯化和颗粒尺寸的选择中，溶剂是必不可少的。纳米粒子样品的纯化常用的方法是洗涤或萃取以除去杂质，每克纳米粒子需要数升溶剂，但对除去所有的杂质该方法通常是无效的。颗粒尺寸的选择本质上是一种纯化过程，包括萃取或分步结晶或色谱法以分离不同尺寸的颗粒。

4.11.2 原子经济

这些原理的另外一个方面就是试图使材料的效率达到最大化，就是通过提高反应的选择性和效率，使原料以优化的方式转化为理想的产物。原子经济的概念已经应用于湿化学纳米材料的制备中。而且，原子经济的概念也延伸到纳米结构材料的制备，利用“自下而上”的方法，如分子自组装或将纳米亚结构单元组装成更复杂的结构。与“自上而下”的方法相比，“自下而上”的方法能够将更多的原材料转化为产品，具有更高的原子经济性。

4.11.3 使用更安全的溶剂

使用安全的溶剂/可选择的反应介质和可再生的原料，可以提高过程的安全性或者减少过程的有害性。这是纳米科学中一个活跃的研究领域。已经有一些实例表明，这些原理的应用可以提高过程的安全性，取代或减少有害溶剂的使用。在某些情况下，使用来自于可再生的良性原料，对于提高纳米材料生产的安全性，是一个成功的策略。

4.11.4 提高能源效率

这些原理的最后一个任务是涉及提高能源效率。大大地缩短了流程的“自下而上”的纳米器件组装，能在非常温和的条件下实施，并能减少颗粒污染的机会。加速这些反应的催化剂发展可能是有用的策略。在大多数情况下，反应条件的实时监测能够提高能源效率和节省成本。

复 习 题

1.“底部有很大的空间”的含义是什么？

2. 什么是“自上而下”及“自下而上”的方法？

3. 请描述水平磨的工作原理。

4. 高能球磨与传统的球磨有什么区别？

5. 什么是物理气相沉积法（PVD）？

6. 请解释直流辉光放电溅射技术的原理。

7. 磁控溅射的原理是什么？

8. 什么是等离子体？

9.“激光消融”的含义是什么？

10. 什么是化学气相沉积法（CVD）？

11. 什么是溶剂热方法？

12. 水的超临界温度及超临界压力是多少？你认为在临界点处溶剂的性质与该溶剂在任意压力和温度下的性质相同或不同？为什么？在水热方法中应当使用什么种类的容器？

13. 什么是冷冻-干燥法（低温化学合成法）？

14. 什么是溶胶-凝胶法？

15. 表面活性剂有什么特点？什么是反胶束？

16. 请画出多重乳液“油包水-水包油”和“水包油-油包水”的简图。

17. 请描述微乳液法的原理。

18. 用于微波合成的波长是多少？什么种类的材料能够作为反应物、溶剂和反应器？微波有什么显著的特点？

19. 什么是超声辅助合成法？

20. 什么是自组装？

5. Scanning Tunneling Microscope and Atomic Force Microscope

The scanning tunneling microscope (STM) was invented by Binnig and Rohrer in 1982, for which they were awarded the Nobel Prize in 1986 [1~3]. Later, the atomic force microscope (AFM) based on the principle of the STM was developed, for the research purpose concerned with surface structures of nonconducting and conducting materials [4]. STM and AFM are proved to be powerful tools for obtaining information of atomic morphology on a surface of material. In addition, by use of the nanometer manipulation, one has applied STM and AFM for manufacturing nano-sized objects or devices.

5.1 Scanning tunneling microscope (STM)

5.1.1 Basic principle of STM

STM is based on the principle of electron tunneling to obtain an image of the topography of the surface. When two surfaces of probe and a sample are held close together, their wavefunctions can overlap. When a small bias voltage, V is applied between the tip and the sample, the overlapped electron wavefunction permits quantum mechanical tunneling and produces a tunneling current.

Scanning tunneling microscopy is a process that relies on a very sharp tip connected to a cantilever beam to touch a surface that is composed of atoms and is electrically conductive. The tip is chemically polished or ground and is made of materials of tungsten, iridium, or platinum-iridium. Under an ultra high vacuum condition, a sharp metal tip is brought into extremely close contact (less than 1nm) to a conducting surface as shown in Fig.5.1 [5]. A bias voltage is applied to the junction between the tip and the sample.

The electrons can tunnel quantum-mechanically through the air gap and generate a current. The tunnel current (around 1nA) decreases to one-tenth of its initial value for every 0.1nm increase in gap separation. This current is compared with a reference current and the error signal generated

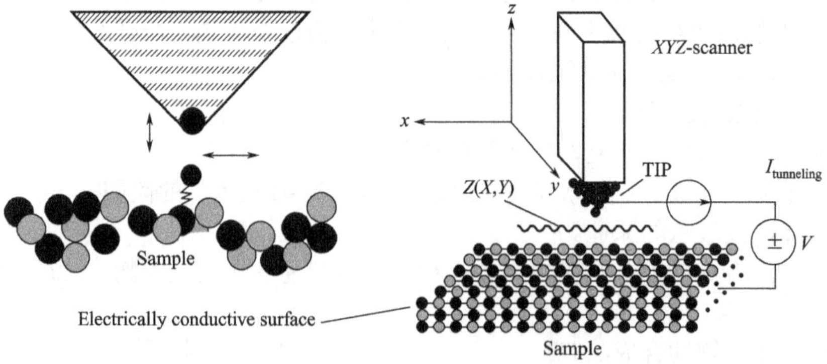

Fig.5.1 Schematic diagram of the basic principle of scanning tunneling microscope [5]

is applied to a gap control z-piezoelectric scanner which moves the tip up or down in order to maintain a constant tunnel current (equal to the reference value) as the tip scans over the sample to record an image.

5.1.2 Operation modes

It can be assumed that electronic states (orbitals) are localized at each atomic site. The response of the tip scanning over the surface can give an image of the surface atomic structure.

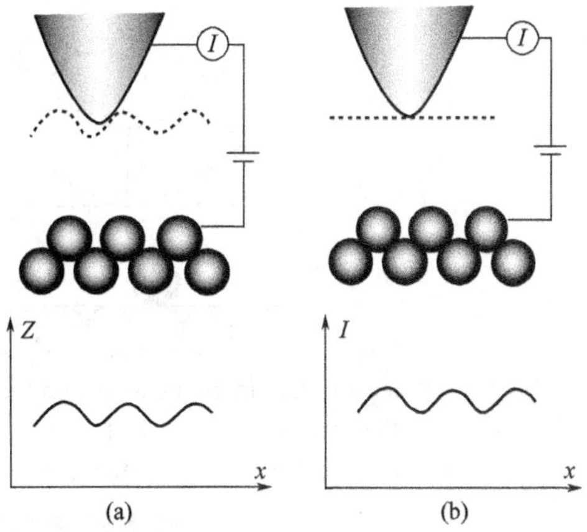

Fig.5.2 Operation modes of STM. (a) constant current mode and (b) constant height mode [6]

To produce images, the STM can be operated in two modes, i.e., constant current mode and constant height mode as shown in Fig.5.2 [6]. Constant current mode means that the tip scans to record the information of feedback-controlled motion of the tip up and down in vertical direction, whereas a constant tunneling current is maintained at each *x-y* position.

An alternative imaging mode is "constant height" operation in which constant height and constant bias applied are simultaneously maintained. When the tip scans on the surface of sample, a very slight difference of topographic structure will lead to a variation in distance between sample and tip, and will generate tunnel current[6].

5.1.3 Application of STM

Nanofabrication is a promising way of producing nano-based electronic components by use of scanning tunneling microscopy. In a normal STM imaging process, the tip-sample interaction is kept small so as to analyze nondestructive. However, if the interaction is made big, the tip can move atoms on the substrate. There is a distinct possibility of manipulating atoms by using STM.

The process of manipulating atoms is technically simple as displayed in Fig.5.3 [7]. The tip is brought just above the atom of sample, and then the tip is lowered to increase the interaction between the tip and the atom. This is done by increasing the tunneling current (to the tune of~ 30nA). Note that the typical tunneling current used for imaging is of the order of 1nA. As a result of this, the interaction between tip and atom is strengthened and the tip can now be moved to the desired location. The interaction potential between the tip and the atom is strong to overcome the energy barrier so that the atom can slide on the surface of sample. After moving atom to the desired location, the tip is withdrawn by reducing the tunneling current. This process can be repeated to obtain the desired structure, atom by atom.

A quantum corral built by arranging 48 Fe atoms on a Cu (111) surface is shown in Fig.5.4 [8], and an electronic interference pattern can be seen inside of the quantum corral. Another research result was reported by Eigler who employed a STM to manipulate Xe atoms on the surface of Ni (110) in a low temperature and ultra high vacuum (UHV) in 1990 [9].

Besides, Tunneling current has been used to break chemical bonds. In the case of organic molecules, the detached fragments have been moved and further recombined in the desired fashion.

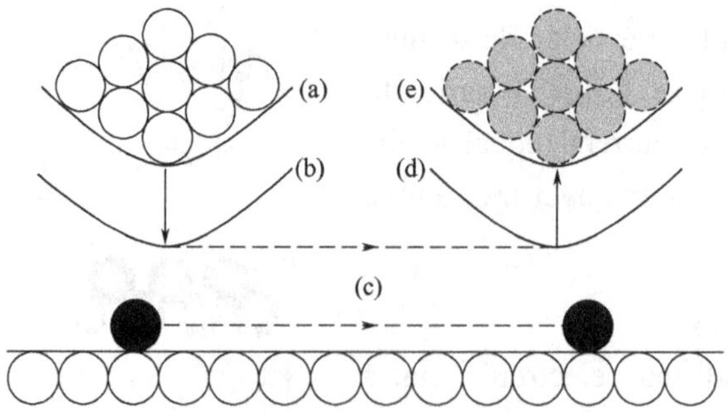

Fig.5.3 Process of manipulating atoms on the surface. An STM tip at position (a) is brought to position (b) by vertical movement, and the tunneling current of position (b) is large. Then the tip is slid over the surface (c) to the desired location (d), and subsequently the tip is brought back to position (e) [7]

Fig.5.4 Circular quantum corral built from 48 Fe atoms on the Cu(111) surface [8]

5.2 Atomic force microscope (AFM)

5.2.1 Basic principle of AFM

The atomic force microscope (AFM) (also called the scanning force microscope) is used in ambient conditions and in ultra high vacuums. The sample may be conductor or insulator. AFM is composed of three main parts: cantilever, sample stage and optical deflection system involving a laser diode and photodetector in Fig.5.5 and Fig.5.6. The laser beam is reflected off the back side of the cantilever. A four quadrant photodetector gives the opportunity to measure both normal bending and torsion of the cantilever, corresponding to normal and lateral forces. The cantilever is usually microfabricated from silicon or silicon nitride with typical dimensions that are 100～300μm in length, 10～30μm in width and 0.5～3μm in thickness. The probe commonly consists of a silicon nitride pyramid with a very sharp tip (the terminal radius of the tip is approximately 10～20nm). This probe is grown on the end of a flexible silicon cantilever.

There are two kinds of design methods. One is that the position of the sample is usually controlled by piezoelectric ceramics which move the sample relative to the cantilever in three dimensions as shown in Fig.5.5. The other is that the cantilever could be mounted on a piezoelectric actuator in order to scan the tip instead of the sample described in Fig.5.6. A sharp tip connected to a cantilever beam is brought into contact with the surface of the sample. When surface is scanned, both normal bending and torsion of the cantilever are measured by a laser from the upper side of the cantilever and a four quadrant photodetector.

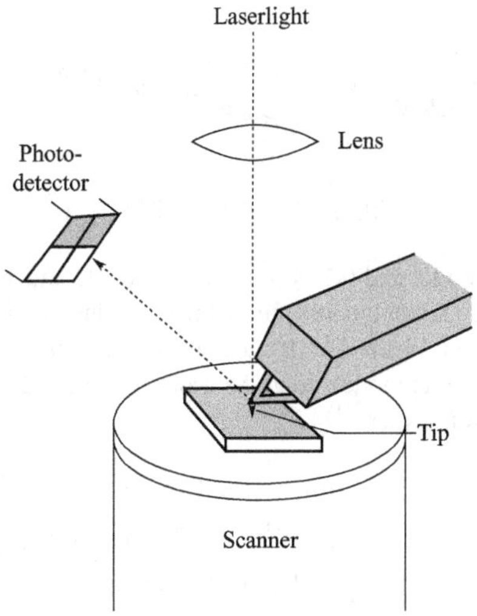

Fig.5.5 Schematic diagram of atomic force microscope

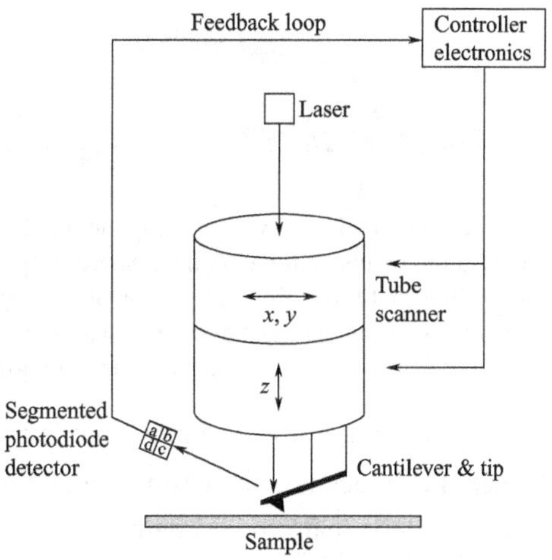

Fig.5.6 Schematic diagram of the basic principle of atomic force microscope [7,10,11]

The cantilever beam and tip are controlled by certain physical law because the cantilever acts as a soft spring of known spring constant (between 0.01N/m and 100N/m) that obeys Hookes Law ($F=-kz$), where F is the force acting on the cantilever; k is the cantilever spring constant and z is the vertical displacement of the end of the cantilever. Therefore, the bending of the cantilever, which is a measure of the tip-sample interaction, is often determined by an optical monitor system (interaction forces between 10^{-7} N and 10^{-15} N can be measured with the AFM).

5.2.2 Mode of operation of AFM

There are a large number of different imaging modes for AFM, providing information about the sample surfaces. The three main imaging modes are illustrated schematically in Fig.5.7. In this figure the interaction forces are sketched when the probe approaches and contacts the surface of sample. Repulsive forces are shown here as being positive while attractive forces are negative by convention [12].

(1) Contact mode

Contact mode implies that the probe remains in contact with the sample at all times so that the force between the tip and the sample becomes repulsive as illustrated in Fig.5.7 and Fig.5.8 (a). This is the simplest operation mode of AFM.

There are two variations on this technique: constant force and variable force. With constant force mode, a feedback mechanism is utilized to keep the deflection, and force applied to cantilever is constant. As the surface of sample is unevenness, the z-axis height is altered to cause a return to the original deflection or "set-point". The change in z-axis position is monitored and this information is used to create a topographical image of the sample surface. For variable force imaging the feedback mechanisms are switched off so that the z-axis height remains constant and the deflection is monitored to produce a topographic image [11,12].

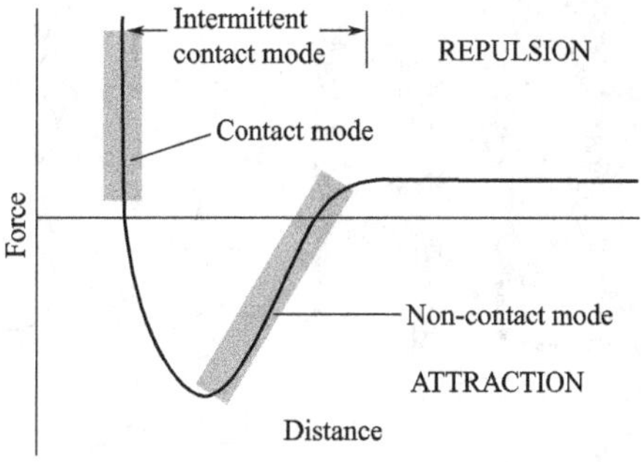

Fig. 5.7 Interaction forces of AFM imaging modes [12]

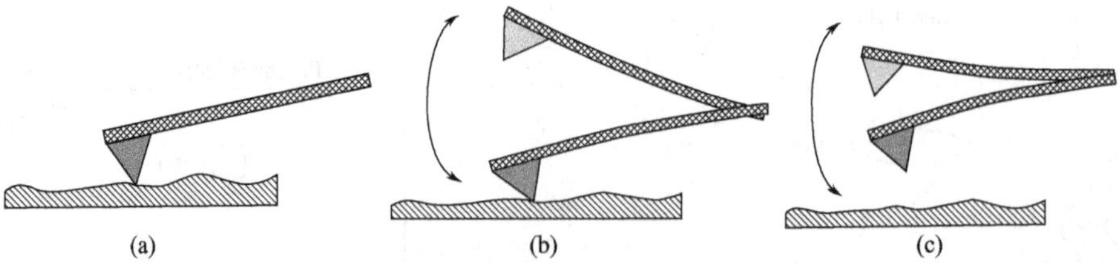

Fig.5.8 Operation modes of AFM. (a) Contact mode; (b) Tapping mode; and (c) Non-contact mode. Contact mode operation is in the repulsive force regime, where the probe is pressed against the sample surface, causing an upwards deflection of the cantilever. Intermittent contact, or tapping mode, the probe is oscillated close to the surface where it repeatedly comes into and out of contact with the surface. Non-contact mode is that the probe is attracted to the surface due to long-range interaction forces [11]

The vertical resolution of contact mode is in the range of 2～10nm, while the lateral resolution could reach the values less than 1nm, depending mostly on the tip radius [13,14]. The disadvantage of the contact mode is that it could damage both the sample and the tip. The tip could leave scratches in the surface of soft materials, and the tip could be broken on hard rough surfaces. Consequently, the scan speed should be kept very low (usually less than 1Hz), depending on the studied surface.

(2) Intermittent contact (tapping) mode

In order to overcome the limitations of contact mode imaging as mentioned above, the intermittent, or tapping, mode of imaging was developed in Fig.5.8(b) [15~17]. Here the cantilever is allowed to oscillate at frequencies in the range from 100kHz to 500kHz, with amplitude of approximately 20nm. Tip contacts sample for a short time in each cycle. The oscillatory amplitude of the cantilever will change as tip encounters different topography of sample. By using a feedback mechanism to alter the *z*-axis height of the piezo and maintain constant amplitude, an image of the topography may be obtained in a similar manner as with contact mode imaging. The resolution of tapping mode is almost the same as that of contact mode, but it is faster and much less damaging than contact AFM.

(3) Non-contact mode

If the set-point is chosen in a range from 5nm to 10nm away from the surface, the long-range interactions such as van der Waals and electrostatic forces occur between the probe and the surface atom, and the tip will be attracted to the surface causing the cantilever to bend towards it. The surface profile can be recorded by maintaining the same value of the cantilever deflection via a feedback system. This is called the noncontact mode.

The resolution of the noncontact AFM is much lower in comparison to the contact mode. However, the scanning is much faster, making this method suitable for studying large and rough surfaces of various natures.

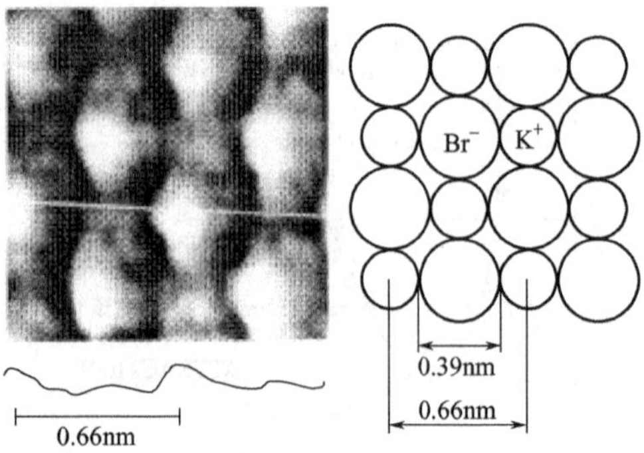

Fig.5.9 Atomic resolution of KBr surface using an AFM in ultra high vacuum [18]

5.2.3 Application of AFM

Fig.5.9 shows the contact mode and provides images of atomically resolved surfaces such as the potassium bromide surface (001). Here, the small and large protrusions are attributed to K^+ and Br^- ions. The tip exerts a large normal and lateral force in the repulsive regime when the tip is scanning the surface of the potassium bromide layer.

The atomic force microscope can be adapted to perform dip pen nanolithography (DPN) by adding a dip pen nanoprobe to AFM. Fig.5.10 shows the schematic diagram that allows nanofabrication of monolayer. The tip writes about 30nm lines on Au surface using 1-othodecanethiol as ink. In the dip pen mode, the AFM is operated in the ambient condition so that the monolayer film of self-assembled molecules is deposited from tip to substrate in a controlled manner. It seems logical that reservoirs of self-assembled molecules are provided in order to speed up the process of nanofabrication.

The dip pen nanolithography process can be also adapted to deposit carbon nanotubes, colloid particles, in addition to self-assembled monolayer. The process is likely to be developed into a nanomanufacturing process with the addition of many contacting probes containing reservoirs of colloidal fluids. One of the major uses of this technique is the deposition and positioning of carbon nanotubes and other structures of carbon in order to produce electronic components and other functional and structural products.

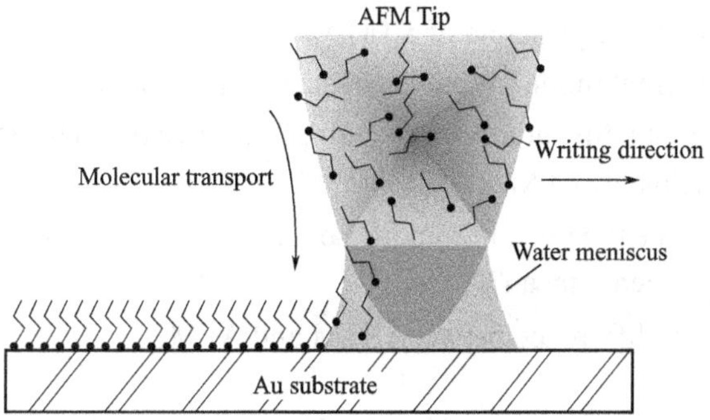

Fig.5.10 Schematic of the DPN process. The tip-applied ink molecules diffuse through the water meniscus and deposit on the surface to form nanostructures [19,20]

References

[1] G. Binnig, H. Rohrer, Helv. Phys. Acta, 55 (1982) 726.
[2] G. Binnig, H. Rohrer, C. Gerber, E. Weibel, Phys. Rev. Lett., 49 (1982) 57.
[3] G. Binnig, H. Rohrer, C. Gerber, E. Weibel, Phys. Rev. Lett., 50 (1983) 120.
[4] G. Binnig, C.F. Quate, C. Gerber, Phys. Rev. Lett., 56 (1986) 930.
[5] F.J. Giessible, Advances in atomic force microscopy, available at xxx. lanl. gov/ arXiv: cond-mat/030 5119.
[6] P.K. Hansma, J. Tersoff, J. Appl. Phys., 61(1987) R1.
[7] T. Pradeep, Nano: the essentials, (2007) McGraw-Hill Companies, Inc. Publisher, ISBN:0-07-154829-7.
[8] M.F. Crommie, C.P. Lutz, D.M. Eigler, Science, 262 (1993) 218-220.
[9] D.M. Eigler, E.K. Schweizer, Nature, 344 (1990) 524.
[10] K.M .Shakesheff, M.C. Davies, C.J. Roberts, S.J.B. Tendler, P.M. Williams, Crit.Rev.Therap. Drug Carrier Syst., 13 (1996) 225.
[11] S.O. Vansteenkiste, M.C. Davies, C.J. Roberts, S.J.B. Tendler, P.M. Williams, Prog. Surf. Sci., 57(2) (1998) 95-136.
[12] D.J. Johnson, N.J. Miles, N. Hilal, Quantification of particle-bubble interactions using atomic force microscopy: A review, Adv. Colloid Interface Sci., 127 (2006) 67-81.
[13] A.DeStefanis, A.A.G. Tomlinson, Scanning Probe Microscopies: From Surface Structure to Nanoscale

Engineering, (2001)Trans Tech, Enfield, NH.
[14] B.L. Gabriel, SEM: A User's Manual for Materials Science, (1985)American Society for Metals, Metals Park, OH.
[15] H. Hansma, R. Sinsheimer, J. Groppe, T. Bruice, V. Elings, M. Bezanilla, et al, Scanning, 15(5) (1993) 296-299.
[16] H.G. Hansma, J. P. Cleveland, M. Radmacher, D. A. Walters, P. E. Hillner, M. Bezanilla, et al, Appl. Phys. Lett., 64(13) (1994) 1738-1740.
[17] Q. Zhong, D. Inniss, K. Kjoller, V. Elings, Surf. Sci. Lett., (1993) L688-692.
[18] F.J. Giessibl, G. Binnig, True atomic resolution on KBr with a low temperature atomic force microscope in ultra high vacuum, Ultramicroscopy, 42-44 (1992) 281-286.
[19] R.D. Piner, J. Zhu, F. Xu, S.H. Hong, C.A. Mirkin, Science, 283 (1999) 661.
[20] M.J. Jackson, Micromanufacturing and nanomanufacturing, (2006) Taylor & Francis Group, LLC, ISBN-10: 0-8247-2431-3 (Hardback), ISBN-13: 978-0-8247-2431-3 (Electronic).

Review questions

1. Please elucidate the basic principle of STM.
2. What is constant current mode? What is constant height mode?
3. What does the manipulating atoms on the sample surface by STM depend on?
4. What is basic principle of AFM?
5. Please briefly describe the operation modes of AFM.
6. What does the “dip pen” mean?
7. What is the essential difference between STM and AFM?

Vocabulary

amplitude ['æmplɪtjuːd] n. 振幅；丰富，充足；广阔
bending ['bendɪŋ] n. 弯曲度；v. 弯曲
bias ['baɪəs] n. 偏见；偏爱；斜纹；乖离率；vt. 使存偏见；adj. 偏斜的；adv. 偏斜地
cantilever ['kæntɪliːvə] n. [建] 悬臂梁
corral [kə'rɑːl] n. 畜栏；围栏；环形车阵；vt. 把…关进畜栏；捕捉；把…布成车阵
deflection [dɪ'flekʃ(ə)n] n. 偏向；挠曲；偏差，偏斜，变形，弯沉
feedback ['fiːdbæk] n. 反馈；成果，资料；回复
ground [graʊnd] n. 地面；范围；vt. 使接触地面；打基础；vi. 着陆；adj. 土地的；磨碎的；磨过的
iridium [ɪ'rɪdɪəm; aɪ-] n. [化] 铱（Ir）
junction ['dʒʌŋ(k)ʃ(ə)n] n. 连接，接合；交叉点；接合点
laser diode 激光二极管
lateral force 侧向力，横向力
normal force 正交力，法向力，纵向力
orbital ['ɔːbɪt(ə)l] adj. 轨道的
oscillate ['ɒsɪleɪt] vt. 使振荡；使振动；使动摇；vi. 振荡；摆动；犹豫
1-othodecanethiol 1-奥索癸烷硫醇
photodetector ['fəʊtəʊdɪˌtektə] n. [电子] 光电探测器
piezoelectric [piːˌeɪzəʊɪ'lektrɪk] adj. [物] 压电的
piezoelectric actuator 压电传动装置
piezoscanner 压电扫描仪
platinum-iridium 铂-铱
potassium bromide 溴化钾
profile ['prəʊfaɪl] n. 轮廓；侧面；外形；剖面
protrusion [prə'truːʒn] n. 突出；突出物
pyramid ['pɪrəmɪd] n. 角锥体金字塔；vi. 渐增；上涨；成金字塔状；vt. 使…渐增
quadrant ['kwɒdr(ə)nt] n. 象限；象限仪；四分之一圆
repulsive [rɪ'pʌlsɪv] adj. 排斥的；令人厌恶的；击退的；冷淡的
scanner ['skænə] n. 扫描仪；扫描器；光电子扫描装置
sharp [ʃɑːp] adj. 急剧的；锋利的；刺耳的；敏捷的；强烈的；adv. 锐利地；急剧地
topography [tə'pɒgrəfɪ] n. 地势；地形学；形貌
torsion ['tɔːʃ(ə)n] n. 扭转，扭曲；转矩，扭力
wavefunctions 波函数

5. 扫描隧道显微镜和原子力显微镜

Binnig 和 Rohrer 在 1982 年发明了扫描隧道显微镜（STM），为此，他们获得了 1986 年的诺贝尔奖 [1~3]。后来，基于扫描隧道显微镜原理，为了解决绝缘体和导体材料的表面结构问题，开发了原子力显微镜（AFM）[4]。STM 和 AFM 是获取材料表面原子形貌信息的强有力的工具。此外，通过纳米操纵，人们可以采用扫描隧道显微镜和原子力显微镜制造纳米尺寸的物件或器件。

5.1 扫描隧道显微镜（STM）

5.1.1 STM 的基本原理

STM 利用电子的隧穿获取样品的表面形貌的图像。当探针和样品表面非常接近时，它们的波函数能够重叠在一起。当给样品和探针间施加一个小的偏置电压 V 时，重叠的电子波函数出现量子力学隧穿，并且产生隧穿电流。

扫描隧道显微镜有一个连接在悬臂上的非常尖锐的针尖，它与具有导电性能的原子组成的材料表面相互接触。针尖是 W、Ir 或 Pt-Ir 合金的材料，通过化学抛光或磨制而成。在超高真空的条件下，尖锐的金属针尖与导体表面非常靠近（小于 1nm），如图 5.1 所示[5]。同时，在探针和样品的连接处施加一个偏压。

电子能够量子遂穿而穿过空气间隙且产生隧道电流。当探针和样品之间的间隙每增加 0.1nm 时，隧道电流（大约 1nA）将降低到初始值的 1/10。将这种隧道电流与参比电流相比较，产生的误差信号传递给间隙控制的 z-轴压电扫描器，该扫描器向上或向下移动以保持一个恒定的隧道电流（等于参比电流值），这时探针在样品上扫描并记录图像。

5.1.2 操作模式

可以假定电子态（轨道）被局限在每个原子的位置。在样品上扫描的探针响应，能够给出样品表面原子结构的图像。

为了产生图像，扫描隧道显微镜有两种操作模式，即恒流模式和恒高模式，如图 5.2 所示 [6]。恒流模式是探针扫描以记录探针在垂直方向上下移动时的反馈控制信息，而在每个 x-y 位置上保持恒定的隧道电流。

另一种成像模式是恒高模式，该模式是探针保持恒定高度和恒定的偏压。当探针在样品表面扫描时，由于样品表面的形貌结构变化，改变着样品-针尖的距离，相应出现隧道电流的变化 [6]。

5.1.3 STM 的应用

使用 STM 的纳米制造是一种极有前景的生产纳米基电子元件的方法。在一个常规的 STM 成像过程中，探针-样品之间相互作用力很小以保证无损分析。然而，如果它们之间相互作用力很大，探针就会在基质表面上移动原子。使用 STM 操纵原子就成为可能。

操纵原子的过程在技术上是简单的,如图 5.3 所示[7]。针尖位于样品原子的上部，然后,针尖下降以增加针尖与原子之间的相互作用。这一过程是通过增加隧道电流来实现的（共计约

30nA）。值得注意的是，用于成像的典型的隧道电流为1nA。增强针尖与原子之间的相互作用，其结果是针尖可以移动到所需的位置。探针和原子间的相互作用的势能大到足以使它能够克服能垒（原子与其表面之间的能垒）以致原子能在样品表面上滑动。当移动原子到所需的位置之后，通过降低隧道电流，探针被返回原位。这个过程可以重复操纵，以获得理想的一个一个的原子结构。

通过在铜的（111）晶面上排列48个Fe原子所构建的量子围栏，如图5.4所示[8]，在量子围栏内，可以看见电子干涉图案。1990年，Eigler报道了另一个研究结果，在低温和超真空条件下，他使用STM在Ni（110）晶面上操纵了Xe原子[9]。

除此之外，隧道电流可用来破坏化学键，对有机分子可进行操纵，切开的碎片可以移动，然后再以理想的状态重新结合。

5.2 原子力显微镜（AFM）

5.2.1 AFM的基本原理

原子力显微镜（AFM）（也称为扫描力显微镜）通常可在常温常压条件下及超高真空条件下使用。样品可以是导体或绝缘体。AFM由三个主要的部分所组成：悬臂、样品台及由一个激光二极管和光电探测器组成的光学偏转系统，如图5.5和图5.6所示。激光束被反射到悬臂的背面。依据相应的纵向力和横向力，一个四象限光电探测器测量悬臂的纵向弯曲和转矩的大小，悬臂通常由特定尺寸的100～300μm长、10～30μm宽、0.5～3μm厚的硅或氮化硅通过微制造方法获得。探针通常由具有锋利针尖的氮化硅棱锥（针尖的半径约为10～20nm）所组成。探针安装在弹性的硅悬臂的端点。

有两种设计方法。一种是样品的位置由压电陶瓷所控制，相对于悬臂而言，压电陶瓷在三维方向上可移动样品，如图5.5所示。另一种是将悬臂安装在压电驱动器上，这样就可以扫描针尖，代替扫描样品，如图5.6所示。操纵与悬臂梁相连的针尖使其与样品表面接触，当样品表面经扫描后，悬臂的纵向弯曲和转矩被悬臂上方的激光束和四象限光电探测器所测量。

悬臂和探针遵循着一定的物理定律，因为悬臂充当一个弹性系数（0.01～100N/m）已知的弹簧，并且服从虎克定律（$F=-kz$），式中的F代表的是作用在悬臂上的力；k是悬臂弹性系数；z是悬臂端的垂直位移。因此，悬臂的弯曲就是针尖-样品相互作用的一种量度，悬臂的弯曲由一个光学检测系统所测量（AFM可测量10^{-7}～10^{-15} N的相互作用力）。

5.2.2 AFM的操作模式

对于原子力显微镜而言，有很多成像模式以提供样品表面的信息。三种主要成像模式如图5.7所示。当探针靠近并接触样品表面时，图5.7描述了探针与表面相互作用力的示意图，通常斥力为正值，而引力为负值[12]。

（1）接触模式

接触模式意味着探针总是与样品保持接触，以至于探针与样品之间的作用力表现为斥力，如图5.7和图5.8（a）所示。这是原子力显微镜的最简单的操作模式。

该技术又分为两种模式:恒力模式和变力模式。恒力模式是利用反馈回路机制保持悬臂变形，而施加给悬臂的力保持恒定。由于悬臂变形，z-轴高度发生改变，造成悬臂欲返回到原变形点或“起始点”，检测z-轴位置的改变，这种信息就形成了样品表面的形貌图像。在变力成像模式中，反馈回路机制被关掉，z-轴高度保持恒定，检测到的弯曲信息产生一种形貌图像[11,12]。

取决于大多数针尖的半径 [13,14]，接触模式的垂直分辨率在 2～10nm 之间，而横向分辨率可能小于 1nm。接触模式的缺点在于它可能会损害样品和探针。在软性材料的表面上，探针会留下划痕，在坚硬粗糙的表面上，针头就会遭到破坏。因此，针对所研究的表面，扫描速度就会保持很低（通常小于 1Hz）。

（2）间歇接触（轻敲）模式

如前所述，为了克服接触模式成像时的缺陷，一种间歇或称为轻敲成像模式得到了发展，如图 5.8（b）所示[15~17]。该模式允许悬臂以 100～500kHz 的频率、约 20nm 的振幅进行振荡。在每一个循环中，探针接触样品表面的时间非常短。当探针遇到样品的不同形貌时，悬臂的摆动振幅发生改变。通过使用反馈回路机制改变 z-轴（压电陶瓷）的高度，以保持恒定的振幅，这样得到的形貌图像与接触模式成像相似。轻敲模式的分辨率与接触模式相同，然而，与接触模式相比，该模式扫描速度更快并对样品的损伤更小。

（3）非接触模式

如果设定值选择的是距离样品表面为 5～10nm 时，由于范德华力、静电引力的长程相互作用出现在探针与样品的原子之间，探针就会被吸引到样品的表面，从而导致悬臂弯向物体表面。经由一个反馈系统使悬臂弯曲保持恒定值，样品表面轮廓就会被记录下来。这称为非接触式模式。

非接触模式AFM的分辨率比接触模式的分辨率低得多。然而，非接触模式的扫描速度更快，该方法适应于研究大而粗糙表面的各种性质。

5.2.3 AFM 的应用

图 5.9 给出的是接触模式，并提供了原子分辨率的表面图像，如溴化钾（001）面的表面图像。图中小的突出部分和大的突出部分别是 K^+和 Br^-。当探针扫描溴化钾层状表面时，在斥力范围内，针尖向样品表面施加一个大的垂直的和横向的力。

通过添加一个“浸蘸笔”纳米探针到 AFM 上，原子力显微镜就能适用于进行“浸蘸笔”纳米光刻，图 5.10 是单层纳米制造的示意图。选用 1-奥索癸烷硫醇作为墨水，使用针尖在 Au 的表面写出大于 30nm 的线条。在“浸蘸笔”模式下，AFM 在常温常压条件下操作，在可控的模式下，自组装分子的单层膜从针尖沉积到基质上。这似乎是合乎逻辑的，只要源源不断地提供自组装分子，就能加速纳米制造过程。

“浸蘸笔”除了能自组装单分子层外，其纳米光刻工艺也能用于沉积碳纳米管和胶体颗粒。通过外加含有溶胶源的接触性探针，该工艺就可能发展成为纳米制造工艺。这种技术的主要用途之一是沉积和组装碳纳米管以及其它的碳结构，以制造出电子元件以及其它功能和结构产品。

复 习 题

1. 请阐述 STM 的基本原理。
2. 什么是横流模式？什么是横高模式？
3. 采用 STM 在样品表面上操纵原子的依据是什么？
4. AFM 的基本原理是什么？
5. 请简要描述 AFM 的操作模式。
6. “浸蘸笔”的含义是什么？
7. STM 和 AFM 的本质区别是什么？

6. Synthesis of Carbon Nanomaterials

6.1 Carbon family

Carbon is an unusual element. The isolated carbon atom has filled 1s and 2s states and two electrons in the 2p state for a configuration of ($1s^2 2s^2 2p^2$).

6.1.1 Graphite and diamond

Generally speaking, carbon was thought to exist in only two allotropic forms, that is, carbon bonded only to other carbon atoms, as represented by graphite and diamond. Diamond, the sparkling and extremely hard crystalline structure, is chemically described by sp^3 hybridized carbon atoms bonded to four other carbon atoms in a perfect tetrahedral (C—C—C bond angles, 109.5°) 3D structure. Whereas, the structure of graphite (common pencil "lead") is quite different and is described as sp^2 hybridized orbitals (C—C—C bond angles, 120°) corresponding hexagonal crystal structure. The different properties of two materials are more clearly understood based on their "nanolevel architecture" . For example, the highly delocalized π bonding network in graphite accounts for the higher electrical conductivity of graphite than that of diamond. Whereas, the strong covalent interlocking network of sp^3 carbons in diamond is responsible for the characteristic hardness. Besides, diamond is an insulator due to its very large bandgap (5.5eV) while graphite is a conductor (bandgap being in vicinity of 0.25eV or low). Two examples immediately give us a clue that there is a possibility of millions of sp^3 and sp^2, giving an innumerable variety of carbon with varied bandgaps in the range of 0.25eV to 5.5eV.

The lowest-energy state of elemental carbon at ambient pressure and temperature is graphite which consists of individual graphene layers. Each layer is composed of interlinked hexagonal carbon rings tightly bonded to each other, loosely stacked into a three dimensional material. Within a single graphene layer, oriented in the *x-y* plane, each carbon atom is tightly bonded to three neighbors within a plane; these planes are then very weakly bonded in the form of dangling bonds to each other.

6.1.2 Allotrope of carbon

The nature of the chemical bonds in carbon structures leads to a classification scheme where each valence state corresponds to a certain form of a simple substance. Elemental carbon exists in four bonding states corresponding to sp^3, sp^2, sp and sp^n hybridization of the atomic orbitals. The corresponding four carbon allotropes are diamond, graphite, carbyne and fullerenes as shown in Fig.6.1 [1,2]. All other forms of carbon constitute "transitional" forms that can be divided into two large groups.

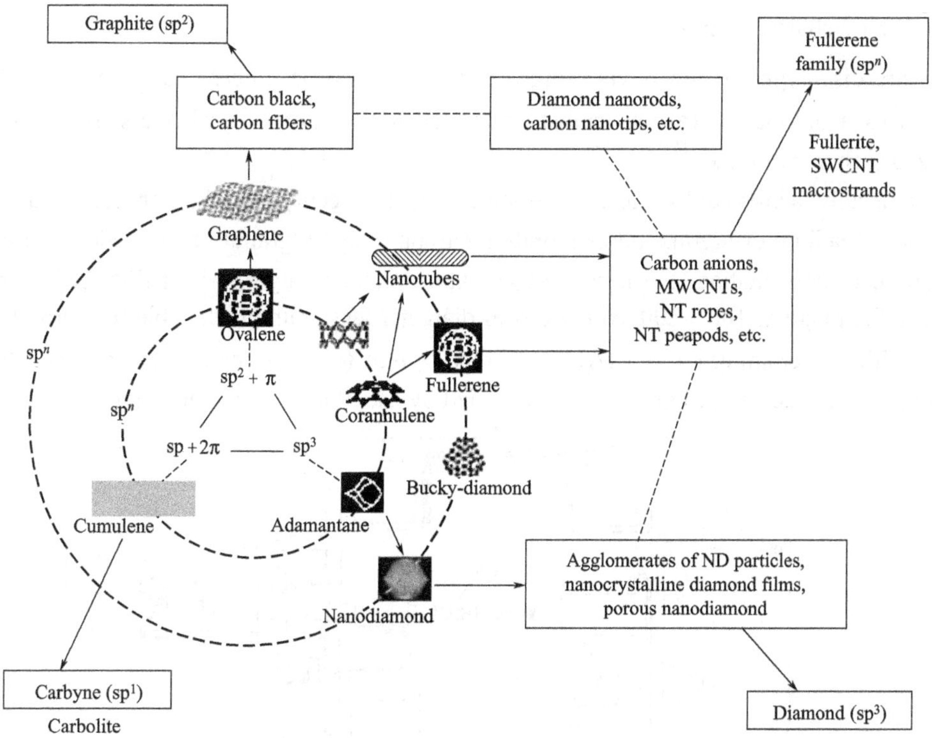

Fig.6.1 Family of carbon structures at the nanometer scale [1,2]

The first group is composed of mixed carbon forms with short-range order (e.g., diamond-like carbon, vitreous carbon, soot, carbon blacks), as well as numerous hypothetical structures such as graphynes and "superdiamond".

The second group includes intermediate carbon forms with a noninteger degree of carbon bond hybridization, sp^n. The subgroup with $1 < n < 2$ includes various monocyclic carbon structures. For the subgroup with $2 < n < 3$, the intermediate carbon forms comprise closed-shell carbon structures such as fullerenes, carbon anions, and CNTs.

Carbon nanostructures are defined as carbon materials, in which either their sizes or their structures are controlled at the nanometer scale (less than 100nm). The dimensionalities are represented within the extended carbon family: zero-dimensional structures (fullerenes, diamond clusters), one-dimensional structures (carbon nanotubes), two-dimensional structures (graphite sheets) and three-dimensional structures (amorphous carbon particles, diamond particles).

6.2 Fullerenes

Fullerenes were discovered by Kroto et al [3] in 1985 and were synthesized in large quantity by Kratschmer et al [4] in 1990. This allotrope of carbon has attracted intense interests because of its unique physical and chemical properties.

6.2.1 Synthesis of C_{60}

Fullerenes can be extracted from the soot by either combustion [5,6] or pyrolysis of aromatic hydrocarbons [7~12] or by an arc-discharge process [3].

(1) Arc-discharge method

The arc discharge method is via a high temperature-plasma process that approaches 3700℃ and produces a number of carbon nanostructures such as fullerenes, whiskers, soot and highly graphitized carbon nanotubes [13].

It is usually achieved by electric arc-discharge between double electrodes in a constant atmosphere of helium or argon, and evaporating carbon (usually graphite) electrodes (see Fig.6.2). In the arc-discharge method, the best inert gas atmosphere is high-purity helium with preferred pressure in the range of 100～200 Torr. Rods of diameter less 5mm are suitable for currents of 100 amperes. Either alternating current (AC) or direct current (DC) discharge may be available. The consumption of anode can be continuously provided by use of anode rod autoloaders.

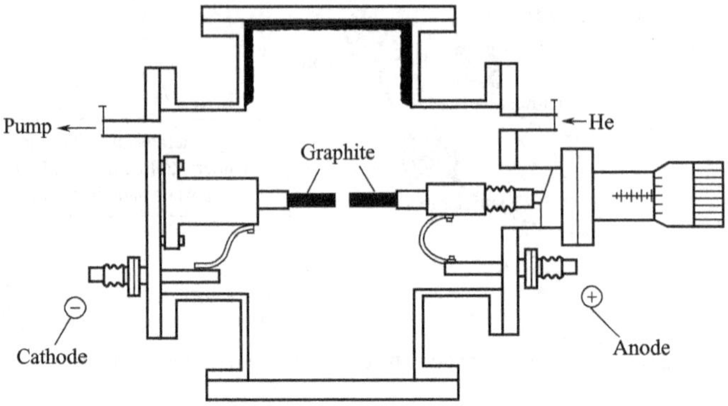

Fig.6.2 Generation of fullerenes by arc-discharge

(2) Laser method

The experimental apparatus used in the discovery of C_{60} is shown in Fig.6.3 [14,15]. A rotating graphite disk is evaporated by radiation of a laser. The evaporated carbon species are cooled by a helium stream and the carbon clusters are produced simultaneously. The cluster beam emanating from the nozzle is expanded into vacuum. The soot is detected by time of flight mass spectrometry. This set-up is now commercially available for the study of a variety of clusters.

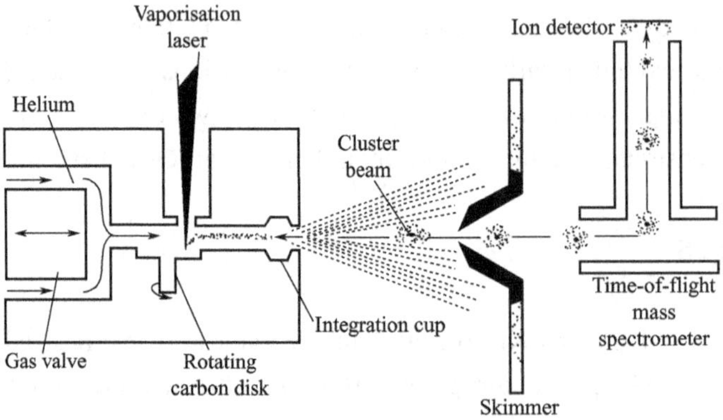

Fig.6.3 Experimental set-up used to discover C_{60} [14,15]

(3) Synthesis of C_{60} single-crystal

The vapor growth technique for synthesizing single- crystal fullerenes usually uses a horizontal quartz tube (usually about 50cm long by 1cm in diameter) placed in a furnace with a temperature gradient as shown in Fig.6.4 [16~19]. The initial C_{60} powder is filled in a gold boat and heated in the vacuum furnace at about 250℃ under dynamic vacuum for 6~72h in order to remove the solvent intercalated in the powder. Subsequently the Au boat is heated to 590~620℃ (above the C_{60} sublimation temperature), and high-purity, dry helium flow kept at a flow rate below 5ml/min is employed to carry the C_{60} vapor to the colder part of the quartz tube. The growth rate in this method is expected to be about 15mg per day. The vapor transport procedure is always kept for several days to yield larger C_{60} crystals. Several modified techniques based on the vapor transport growth method have been developed to grow larger and more perfect C_{60} crystals [20~26].

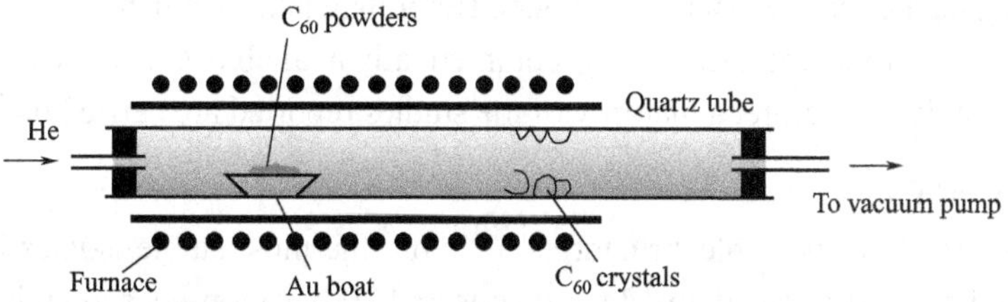

Fig.6.4 Schematic view of a horizontal quartz tube furnace apparatus for the vapor transport growth of C_{60} single crystals [16~19]

6.2.2 Purification of fullerenes

The soot generated is collected on water-cooled surfaces which could even be the inner walls of the vacuum chamber. The soot is collected and is extracted by soxhlet for 5~6h in toluene or benzene, resulting in a dark reddish-brown solution which is a mixture of fullerenes. The 20%~30% soot collected in this manner is soluble. This soluble material is subjected to chromatographic separation in Fig.6.5 [27,15].

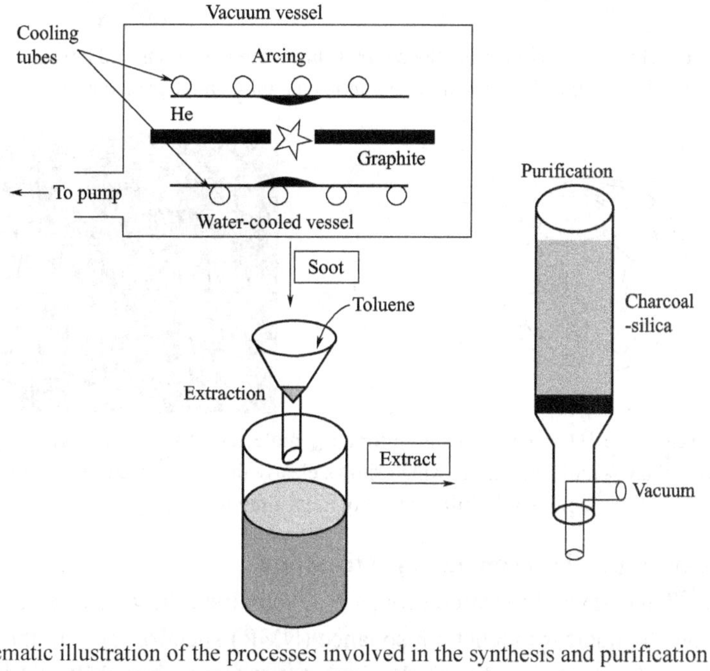

Fig.6.5 Schematic illustration of the processes involved in the synthesis and purification of fullerenes. Graphite rods are evaporated in an arc under He atmosphere. The soot collected is extracted with toluene and subjected to chromatograph [27,15]

Liquid column chromatography is the main technique used for fullerene purification. Here purification means the separation of the fullerenes to distinguish C_{60}, C_{70} and etc.

Over the years, several simple methods have been discovered including a filtration technique over an activated charcoal–silica gel column. About 80% of the soluble material is C_{60}, which can be collected by using toluene as the mobile phase. C_{70} can be separated by using toluene/*o*-dichlorobenzene mixture as eluant. Repeated chromatography may be necessary to get pure C_{70}. C_{60} solution is violet in colour while that of C_{70} is reddish-brown. Higher fullerenes, such as C_{76}, C_{78} and C_{82}, require high performance liquid chromatography (HPLC) for purification. The spectroscopic properties of several of these fullerenes are now known. Normally, the preparation of a gram of C_{60} from graphite requires about five hours. But it may take as much as 250hours to make 1mg in the case of the higher fullerenes. C_{60} is now commercially available. C_{60} costs about $25 a gram but C_{70} is not yet affordable. Therefore, a majority of the studies reported are carried out on C_{60}.

6.2.3 Structure of C_{60}

For C_{60} there are 1812 possible structures [28,29], but the most stable isomer is a cage structure which is composed of 32 surfaces with 12 pentagons and 20 hexagons as shown in Fig.6.6.

C_{60} is an aromatic molecule as shown in Fig.6.6 and Fig.6.7 [30]. All the pentagons in fullerene structure are isolated by hexagons to be stabilized against structures with adjacent pentagons [31,32]. The mean outer diameter of a C_{60} molecule is about 1nm [30,33].

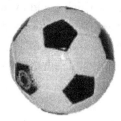

Fig.6.6 Buckminsterfullerene. Soccer ball shows the pentagons and hexagons (left) and C_{60} is a beautiful spherical hollow molecule with an outer diameter of ~1nm

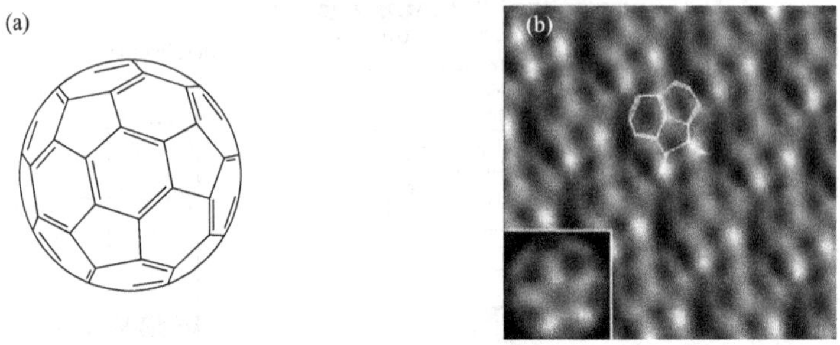

Fig.6.7 (a) Schematic 3D structural diagram for a C_{60} molecule; (b) Scanning tunneling microscopy images of a C_{60} molecular array which shows clearly the cage structure and double and single bonds. Inset is a theoretical simulation image [30]

6.2.4 ^{13}C nuclear magnetic resonance spectroscopy

Taylor et al. [34] and several research groups [35,36] obtained the structure data of C_{60} molecules in solution by using ^{13}C nuclear magnetic resonance (NMR) spectroscopy. In the spectrum, there is only one peak which shows extremely simple character at a chemical shift of 143.2ppm relative to tetramethyl silane (the standard reference molecule for ^{13}C NMR spectroscopy). This only one peak character confirms that each carbon atom in the C_{60} molecule is equivalent in the solution environment. In comparison with C_{60}, the C_{70} molecule has five in equivalent carbon sites corresponding to 130.9ppm, 145.4ppm, 147.4ppm, 148.1ppm, and 150.7ppm in a ratio of 10:20:10:20:10 as shown in Fig.6.8 [34].

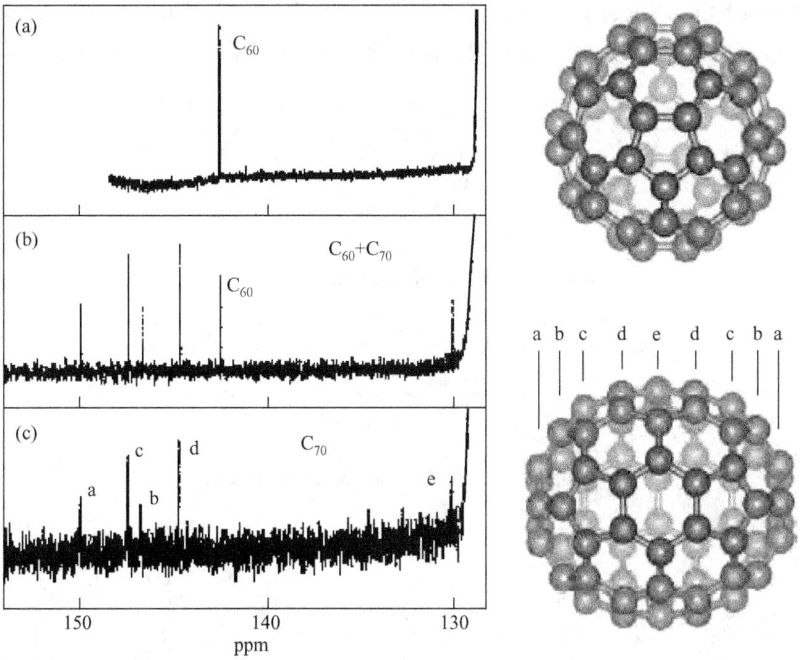

Fig.6.8 ^{13}C NMR spectra for C_{60}, C_{70}, and C_{70}/C_{60} mixture in benzene solution at 298K. (a) C_{60} sample; (b) C_{60}/C_{70} mixture sample and (c) C_{70} sample, peaks labeled a, b, c, d, and e are assigned to C_{70} and correspond to the distinct five carbon atoms in a C_{70} molecule [34]

6.2.5 Endofullerenes

Some important examples are the encapsulations of N or P atoms in carbon cage and these compounds are referred to as endofullerenes in Fig.6.9. The chemical formula is represented as $N@C_{60}$. Various noble gases (He, Ar, Xe) have also been encapsulated in carbon cages at elevated temperatures of 600～1000℃ and pressure of 40000psi.

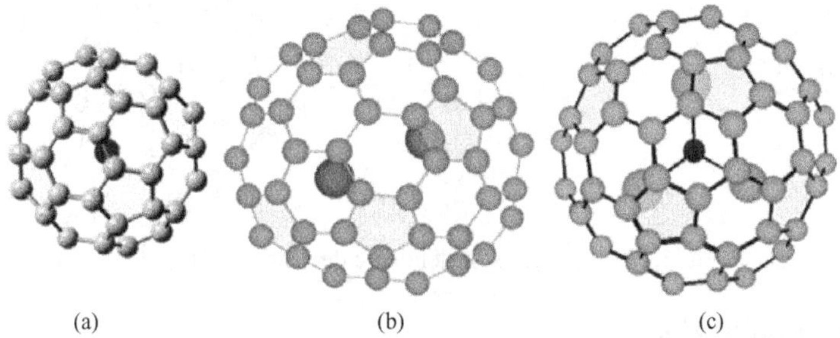

Fig.6.9 Three endofullerenes.(a) $N@C_{60}$; (b) $Sc_2@C_{82}$ and (c)$Lu_3N@C_{80}$

Representative metallofullerenes are shown in Fig.6.9, such as $Sc_2@C_{82}$ and $Lu_3N@C_{80}$. It should be noted that transition metallofullerenes (e.g., Co, Fe, and Ni) have not been commonly prepared, but these metals are common catalysts for preparing nanotubes in the Kratschmer-Huffman electric-arc approach. One of the important features of the cage with at least 80 carbon atoms is the internal size (about 0.8nm), and this cage is large enough to accommodate a molecular cluster composed of four atoms.

6.2.6 Nucleophilic addition reactions

The second important feature of fullerenes is their high reactivity at the carbon junction between two hexagon rings. Thus, one of the most common reactions of fullerenes is nucleophilic addition reaction of the Bingel-Hirsch type as shown in Fig.6.10. The corresponding reaction for C_{70} is shown in Fig.6.11. The synthesis of the dendritic derivative of C_{60} is shown in Fig.6.12 [37].

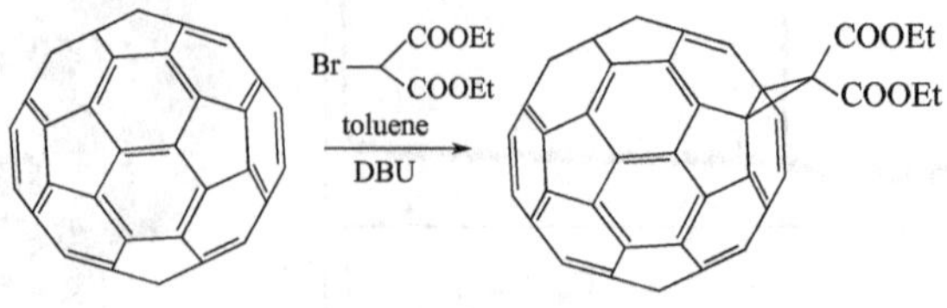

Fig.6.10 Bingel-Hirsch reaction for C_{60}

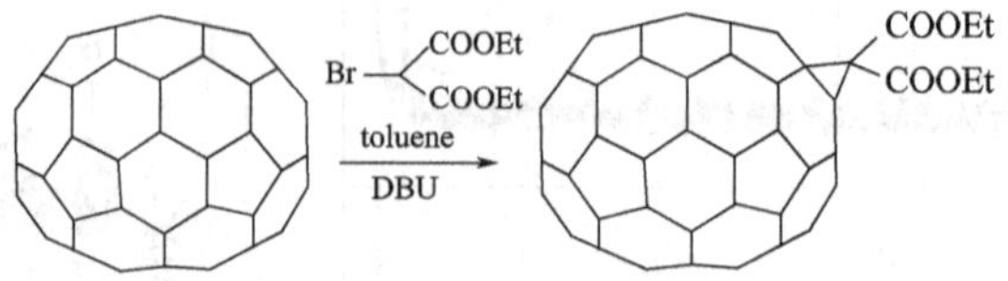

Fig.6.11 Bingel-Hirsch reaction for C_{70}

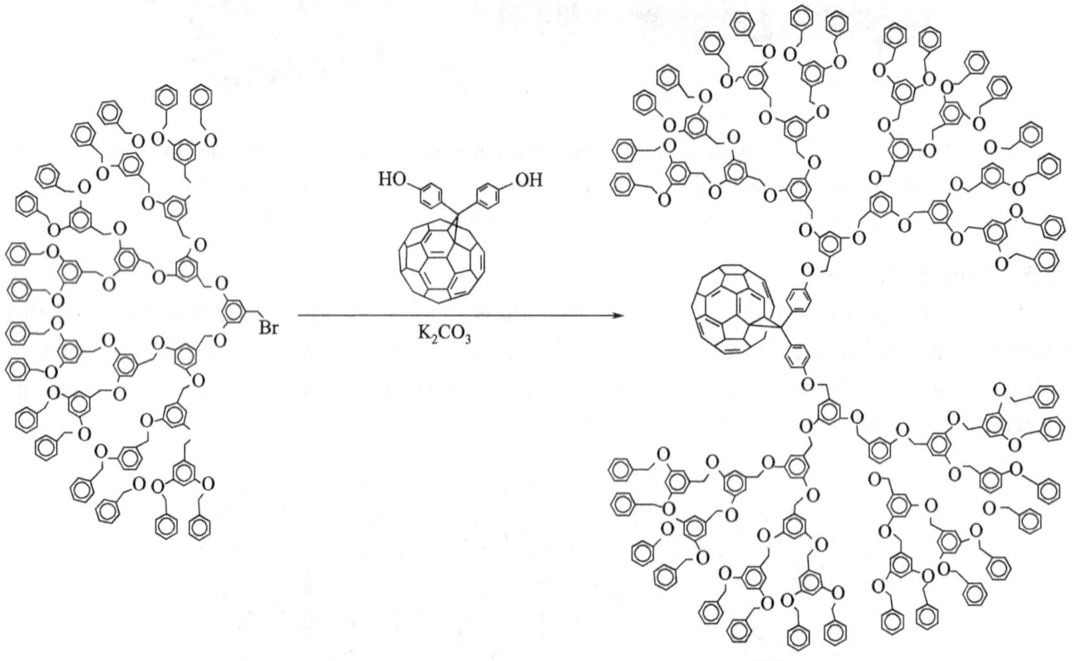

Fig.6.12 Synthesis of dendritic derivative of C_{60} [37]

6.2.7 Polymerization of C_{60}

Pressure-induced polymerization has been studied extensively under various temperatures and pressures in recent years [38].

Several polymerized fullerites have been synthesized by applying moderate pressures and elevated temperatures [39~41]. Three different phases possessing one-or two-dimensional C_{60}-C_{60} networks have been identified: one-dimensional orthorhombic phase, two-dimensional tetragonal phase and two-dimensional rhombohedral phase in Fig.6.13 [41,42].

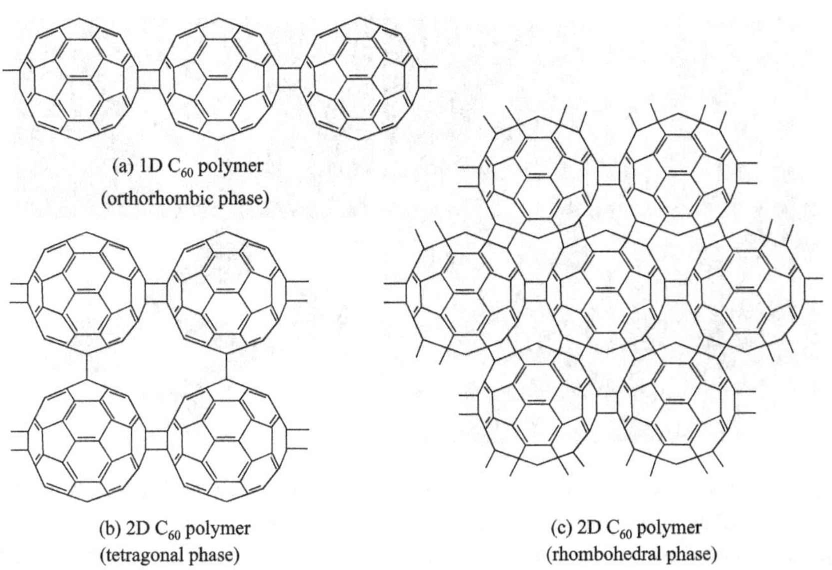

Fig.6.13 Geometric structures of C_{60} polymers. (a) 1D C_{60} polymer (orthorhombic phase); (b) 2D C_{60} polymer (tetragonal phase); (c) 2D C_{60} polymer (rhombohedral phase) [41,42]

6.2.8 Fabrication of nanocar

Fullerene (C_{60}) structures exhibit rolling characteristics on various surfaces because of their high symmetry. The design and synthesis of the nanocar molecule was directed to enable controllable surface transport by means of the rotation of wheel-like fullerenes (Fig.6.14) [43,44] and STM image exhibits the nanocar rolling on Au (111) surface in Fig.6.15.

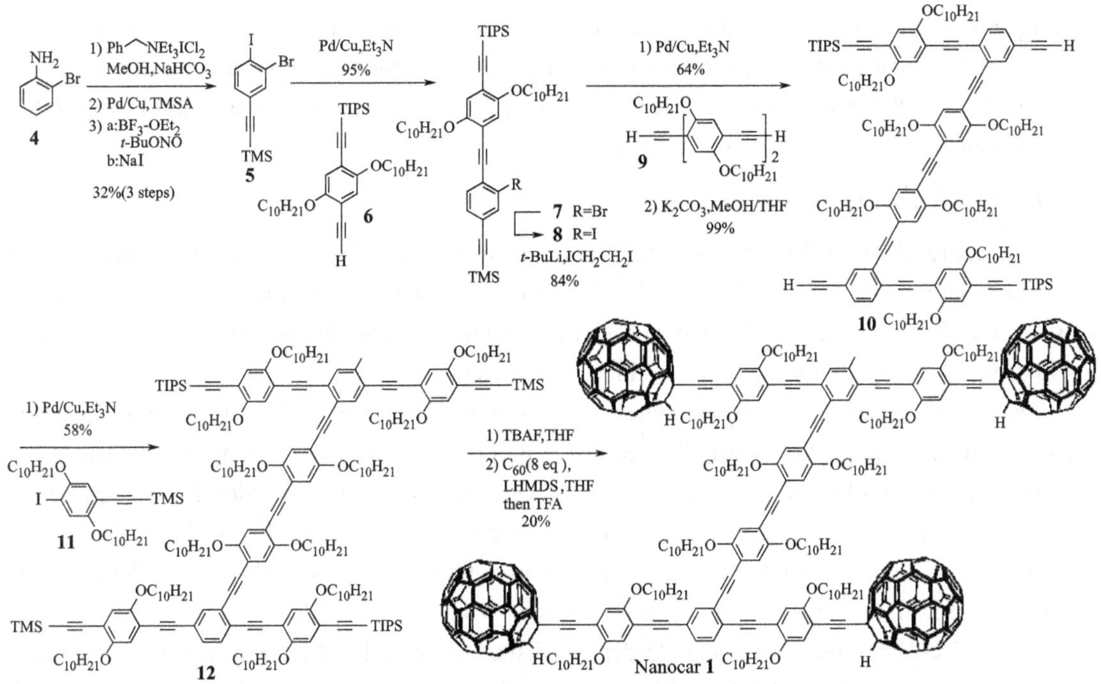

Fig.6.14 Synthesis and structure of nanocar 1. The synthesis of nanocar 1 involves extensive Pd-catalyzed coupling reactions, and four fullerene "wheels" were attached in one reaction step [43,44]

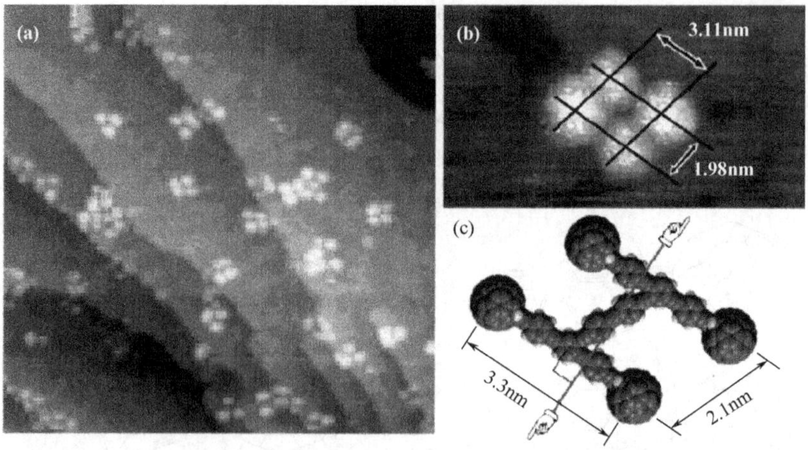

Fig.6.15 Nanocar **1** on Au (111) surface. (a) STM image of the nanocars **1** deposited on Au(111) by dosing valve; (b) High-resolution STM image; (c) Modeled distances at 3.3nm across the width and 2.1nm across the length, or the axle and chassis directions, respectively. The fingers indicate the expected direction of rolling perpendicular to the axles

6.3 Carbon nanotubes [45]

6.3.1 Synthesis of nanotubes

Iijima and co-workers firstly discover carbon nanotubes in the deposit on the cathode of carbon arc [46,47]. Since then, various approaches have been adopted to prepare carbon nanotubes. Nowadays, carbon nanotubes are produced by diverse techniques such as arc-discharge, pyrolysis of hydrocarbons over catalysts, laser evaporation of graphite and electrolysis of metal salts. All these techniques would be difficult to discuss in detail. Hence, only a few of them are highlighted here.

The carbon nanotube consists of one or multiple coaxial structures of rolled graphite sheets. A single wall nanotube (SWNT) has a diameter about 1nm, while the multiwalled nanotubes (MWNT) may have diameters in the range of 2～100nm.

(1) Arc-discharge method

To prepare MWNTs, DC arc discharge between two pure graphite rods has been carried out in a vacuum chamber, as shown in Fig.6.16 [48]. Of course, the alignment of the two graphite rods can be vertical or horizontal. After the chamber is evacuated to a vacuum higher than 10^{-5} Torr by using an oil diffusion pump, an inert gas is introduced into the chamber.

Then a DC arc voltage of approximately 25 V is applied between the two graphite rods. As the rods are brought closer, plasma was formed at the moment of a discharge. A hole is drilled in the carbon anode and is filled with a mixture of metal catalyst and graphite powder. In this case, most nanotubes are found in the soot deposited on the arc chamber wall. It has been found that vaporization of an anode containing Co, Co-Ni, Co-Y, Co-Fe, Ni, Ni-Y, Ni-Lu, and Ni-Fe gives deposits looking like a lace collar or a soft belt formed around the cathode.

HRTEM observation of the cathode deposit dispersed on a TEM microgrid has been carried out. There is a coaxial structure of rolled graphite sheets with some helical arrangement [49], as shown in Fig.6.17 [46].

The distance between each layer is the same as the [002] lattice spacing of graphite, namely, 0.34nm. A schematically atomic model of a four-wall MWNT is shown in Fig.6.18 [49], which has a helical arrangement and belongs to zigzag type.

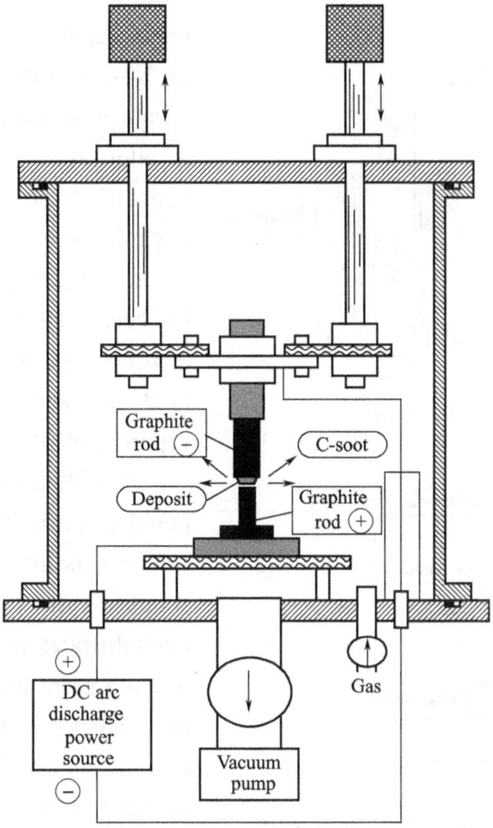

Fig.6.16 Schematic diagram of arc apparatus producing MWNTs [48]

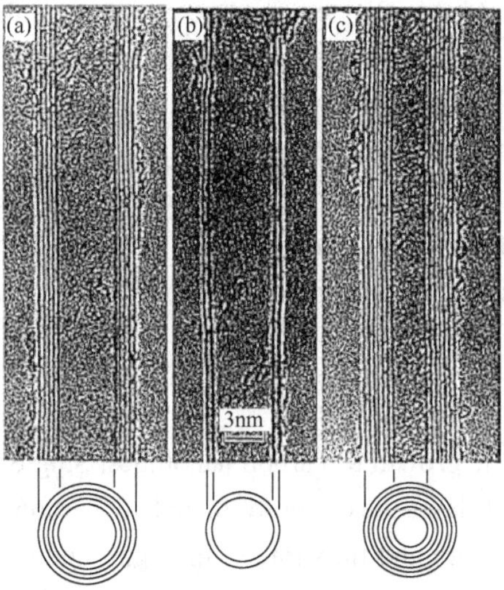

Fig.6.17 HRTEM micrographs of MWNTs
(a) five-wall, diameter 6.7nm;
(b) two-wall, diameter 5.5nm;
(c) seven-wall, diameter 6.5nm and the smallest hollow diameter 2.2nm [46]

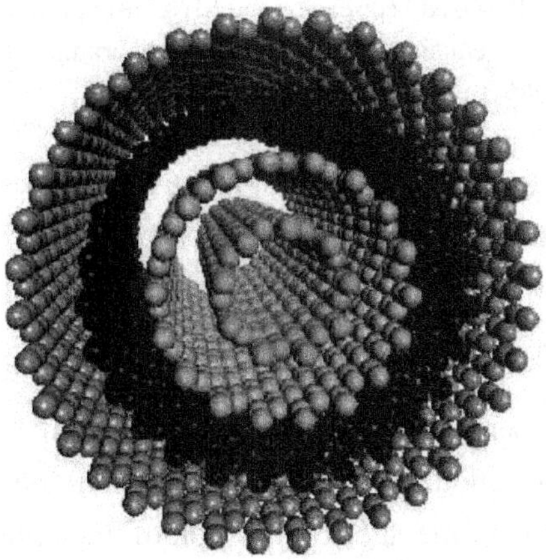

Fig.6.18 Atomic model of four-walled carbon nanotube [49]

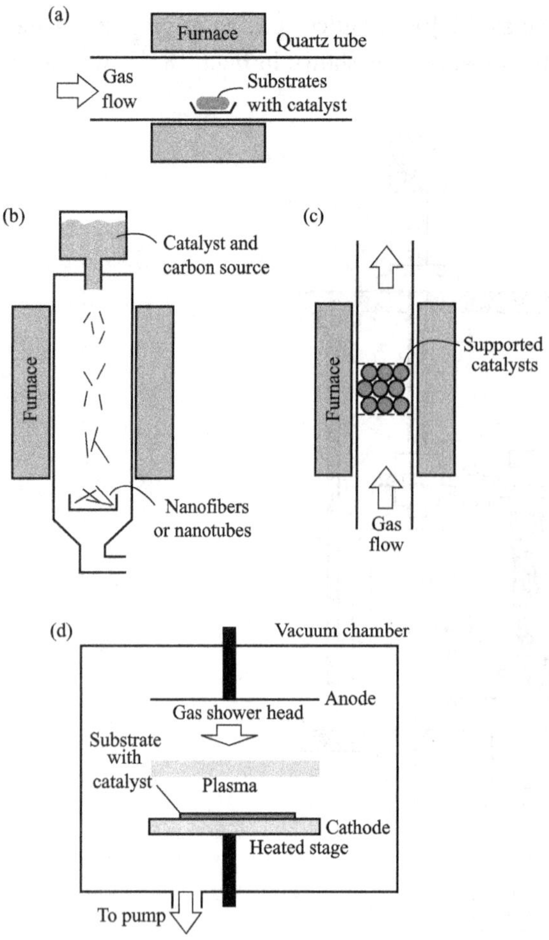

Fig.6.19 Types of chambers used for catalytic CVD of nanofibers and nanotubes. The most commonly used is the horizontal furnace (a); For mass production, the vertical furnace (b) has been employed; (c) The fluidized bed reactor and (d) a basic plasma enhanced CVD system based around a vacuum chamber [53]

(2) Catalytic-assisted pyrolysis method

Catalytic pyrolysis of hydrocarbons (chemical vapor deposition) was used to prepare carbon fibers even before the discovery of NTs and fullerenes [50]. In this method, chemical precursor along with inert carrier gas is allowed to pass through a quartz tube maintained at some specific temperature. Organic precursor decomposes and the products deposit on alumina plate. This method has been used to manufacture a large quantity of nanotubes. Besides, two furnaces could be used one for pyrolysis and the other for vaporization of the precursor [51,52].

The horizontal furnace is the most popular configuration for the production of carbon nanofibers and nanotubes. This is the simplest form to heat quartz tube in which the substrates and catalysts are placed. The reactant gases are flowed over the substrates and catalysts which sit in a removable ceramic boat/holder in the center of the quartztube in Fig.6.19 [53]. The horizontal furnace is advantageous because there is no (or small) temperature gradient within the heated zone. In most cases, the length of the nanotubes/nanofibers can be simply controlled by the deposition time.

Sharon and his group [54] have been trying to synthesize carbon nanotubes by pyrolysis of natural precursors. Their group has been able to synthesize various forms of carbon nanotubes from natural precursor like turpentine and camphor as shown in Fig.6.20 [54].

(3) Laser ablation

Laser technique has been successfully used for the growth of carbon nanotubes. Single walled carbon nanotube yield depends upon the power of laser beam used to ablate carbon target, irradiation time, the composition of the target, the pressure and flow rate of inert gas, and the target temperature. Wolfgang et al. reviewed in detail the production of carbon nanotubes [55]. Some standard conditions for synthesis of nanotubes (SWNT and MWNT) using laser are given in Table 6.1. The schematic diagrams of the laser evaporation technique are shown in Fig.6.21 [55] and Fig.6.22 [55], respectively.

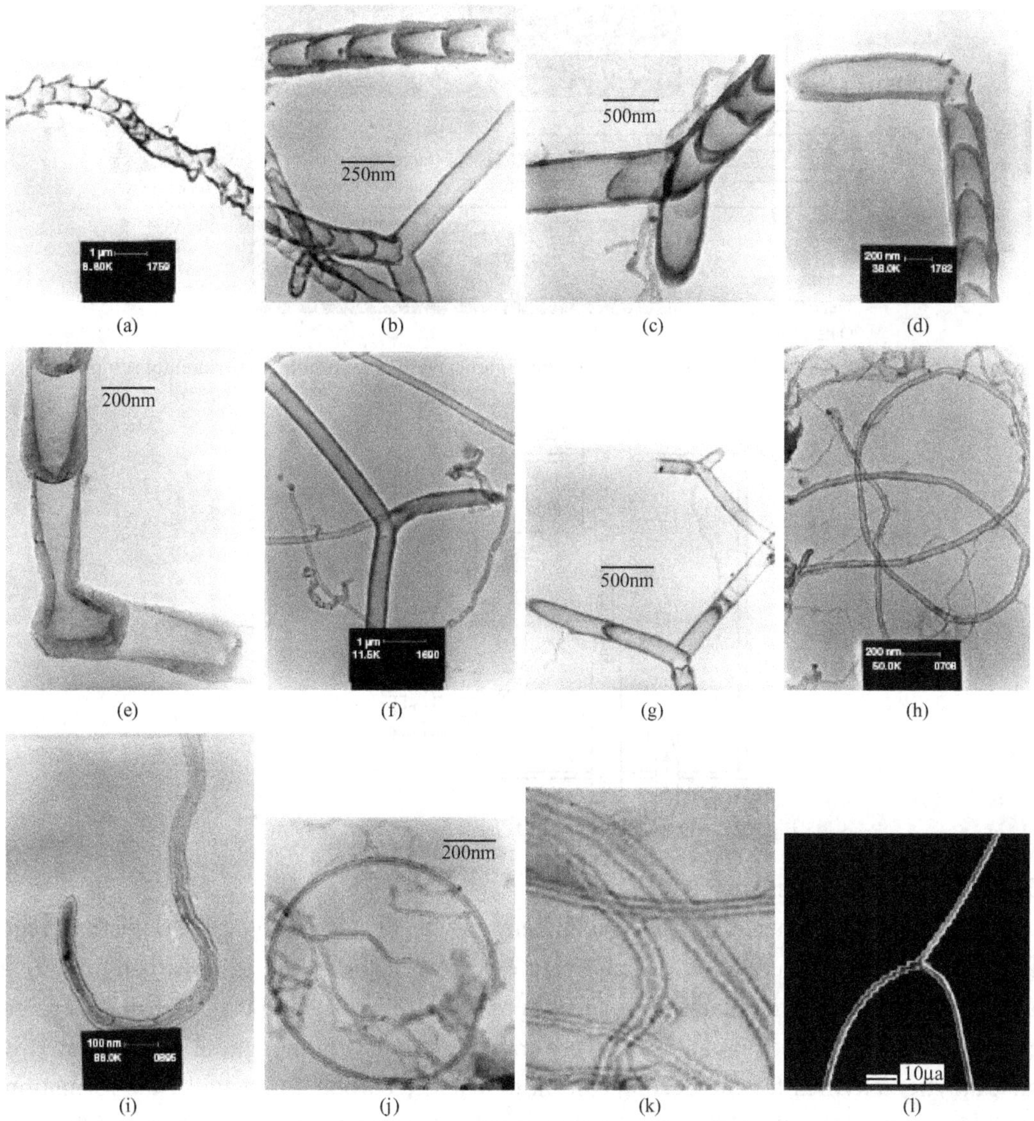

Fig.6.20 Various types of carbon nanomaterials synthesized from turpentine and camphor. Bamboo-shaped branched CNT [(a), (b), (c), (f)]; Y-junction CNT [(d), (e)] coiled double-walled CNT[(g), (i), (j)]; enlarged portion of (g) to show double-walled CNT (h); formation of branching in CNT (k); Yofork type CNT (l) [54]

Table 6.1 Some favorable standard laser ablation conditions for synthesis of nanotubes

Source	Wavelength	Target	Furnace /℃	Gas	Results
Single pulsed Nd:YAG	532nm	Graphite(G)	1200	Ar	MWNTs
Double pulsed Nd:YAG	1064nm	G(98)/Ni(0.6) Co(0.6) at.%	1200	Ar	SWNTs 60vol%~90vol%,1g/day
Cw-CO_2	1064nm	G(95)/Ni(4)/Y(1) at.%	No furnace	Ar	SWNTs 80vol%, 130mg/h
Solar light	Sunlight	G(96)/Ni(2) Co(2) at.%	No furnace	Ar	MWNTs and few SWNTs

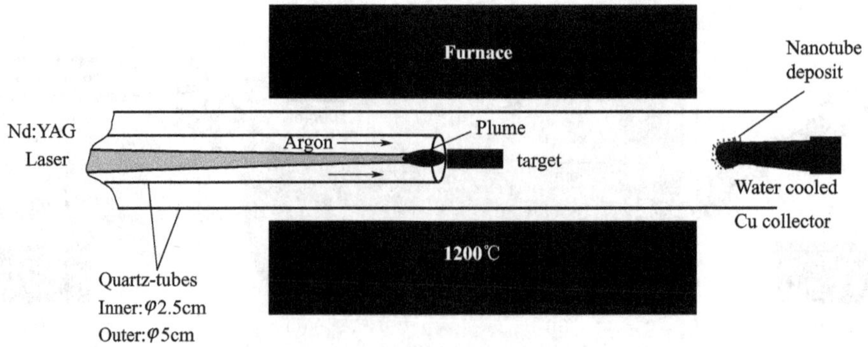

Fig.6.21 Sketch of Nd:YAG (yttrium aluminum garnet) laser evaporation experimental setup [55]

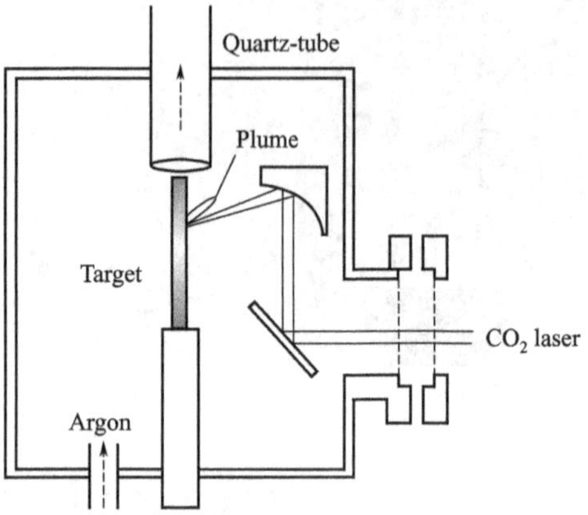

Fig.6.22 Drawing of cw-CO_2 laser evaporation system [55]

6.3.2 Growing mechanisms

In general, carbon nanotubes and nanofibers growth by the catalytic CVD method require catalyst nanoparticles (usually Fe, Co, or Ni), a carbon feedstock (e.g., hydrocarbon or CO), and heating. The diameter of the filament produced is often closely related to the physical dimension of the metal catalyst. The peculiar ability of these transition metals is thought to be associated with their catalytic activities for the decomposition of volatile carbon compounds, the formation of metastable carbides, and the diffusion of carbon through the metal particles [56]. Some of the growth models have been proposed both for nanotubes and nanofibers which are widely accepted by the research community [57~61].

The most commonly accepted mechanism was postulated by Baker et al [57] in the early 1970s, who explained the growth of carbon filaments by catalytic decomposition of the carbon feedstock. According to this mechanism (Fig.6.23), the hydrocarbon gas decomposes on the front-exposed surfaces of the metal catalysts to release hydrogen and carbon, which dissolve in the particle. The dissolved carbon diffuses through the particle and is precipitated at the trailing end to form the body of the carbon filament. Due to the exothermic decomposition of hydrocarbons, it is believed that a temperature gradient exists across the catalyst particles. Since the solubility of carbon in a metal is dependent on temperature, precipitation of excess carbon will occur at the colder zone behind the particle, thus allowing the solid filament to grow with the same diameter as the catalyst particles. Such a process will continue until the tip of the catalyst particle is “poisoned” or deactivated.

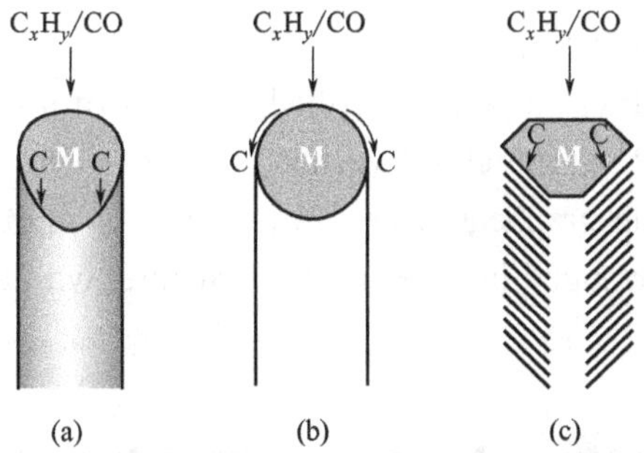

Fig.6.23 The growth of carbon nanotubes and nanofibers involves the catalytic decomposition of a carbon feedstock (hydrocarbon or CO), carbon diffusion, and its precipitation. In (a), the carbon diffuses through the bulk of the metal catalyst "M" as proposed by the Baker model [57,58]; In (b), the carbon diffuses over the surface of the metal catalyst and forms a tubular structure from the circumference of the catalyst as proposed by the Oberlin model [59]; In (c), angled graphene layers are precipitated from a faceted catalyst particle to form a nanofiber as proposed by Rodriguezand Terrones [60,61]

Two different growth modes [Fig.6.24(a) and (b)] can proposed by Baker [58] and Rodriguez [62] based on the interaction of the catalyst with its support. The interaction of the catalyst nanoparticles with the support can be characterized by its contact angle at the growth temperature, analogous to "hydrophobic" (weak interaction) and "hydrophilic" (strong interaction) surfaces. For example, Ni on silica (SiO_2) has a large contact angle (i.e., weak interaction) at 700℃ and thus tip growth is favored in this system [63] as shown schematically in Fig.6.24(a). On the other hand, it is reported that Co or Fe on silicon [64,65] favors the growth from bottom [Fig.6.24(b)], indicating that a strong interaction exists between Co or Fe and Si. Thus, the interaction between the catalyst and its support is an important factor which dictates the growth mode.

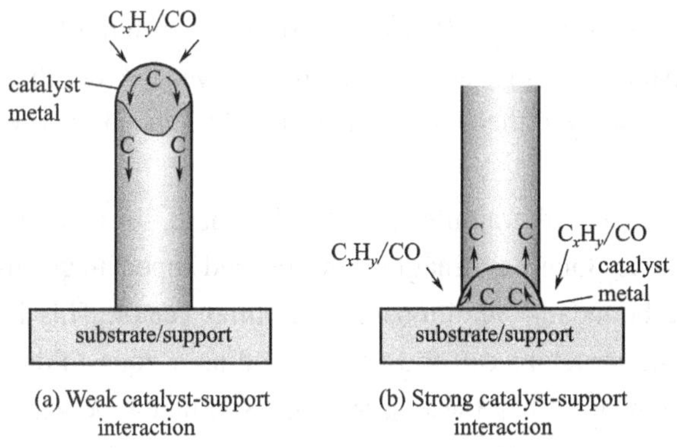

Fig.6.24 Two types of growth, namely tip or base growth, resulting from different catalyst–support interactions [58]

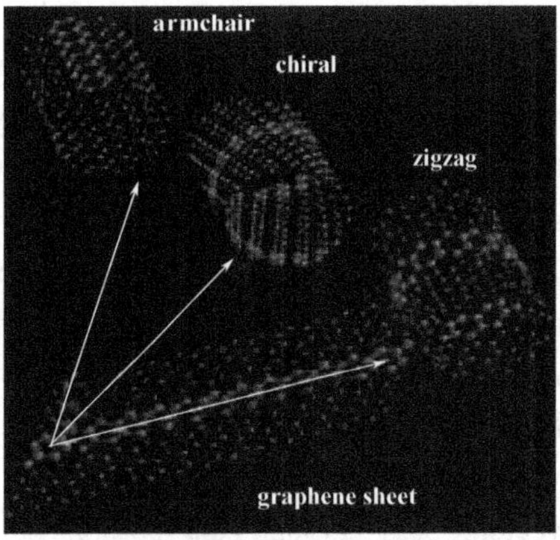

Fig.6.25 The different ways in which the graphitic wall of an individual carbon nanotube shell can be wrapped are generally presented as follows: the armchair, the zigzag, and the chiral nanotube [66]

6.3.3 Geometry of carbon nanotubes

The electronic properties in particular of a carbon nanotube are dependent on the geometry of the nanotube. Carbon nanotube structures are obtained by rolling up a sheet of graphite in one of the symmetry axes or along any other directions to produce a zigzag, armchair or chiral tubes, respectively, as shown in Fig.6.25 [66].

The structure of a cylindrical tube is described in terms of a tubule diameter d and a chiral angle θ as shown in Fig.6.26[67]. The chiral vector, $\boldsymbol{C} = m\boldsymbol{a}_1 + n\boldsymbol{a}_2$ represents the tube. The unit vectors $\boldsymbol{a}_1$ and $\boldsymbol{a}_2$ represent the graphene sheet. In a planar of graphene sheet (a single sheet of graphite), carbon atoms are arranged in a hexagonal structure, with each atom being connected to three neighbors.

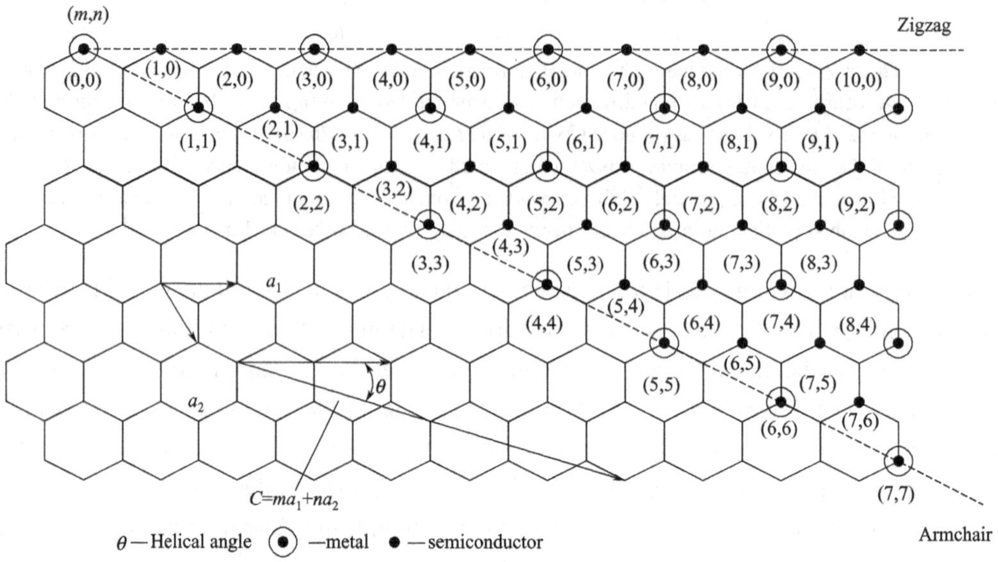

Fig.6.26 Geometry of carbon nanotubes [67]

There are numerous ways in which the tubes can be rolled. The tubes are called "zigzag tubes" where $\theta = 0°$. The tubes are called "armchair tubes" where $\theta = 30°$. These two types of tubes have high symmetry and a plane of symmetry is perpendicular to the tube axis, respectively. Any other tube is a chiral tube, where $0° < \theta < 30°$ which can be either left-handed or right-handed. These tubes will be optically active to polarized light.

A nanotube is in the form of a wrapped sheet of graphite and its electronic structure is analogous to the electronic structure of a graphene. One can imagine an unrolled, open form of nanotube, which is graphene subject to periodic boundary conditions on the chiral vector. This is known as 'zone folding'. A nanotube's electronic structure can thus be viewed as a zone-folded version of the electronic band structure of the graphene. When these parallel lines of nanotube wave vectors pass through the corners, the nanotube is metallic. Otherwise, the nanotube is a semiconductor with a gap of about 1eV, which reduces as the diameter of the tube increases. Within this simple approach, (n, m) nanotubes are metallic if $n\text{-}m = 3\times$ an integer. Consequently, all armchair tubes are metallic. Thus, the electronic structures of nanotubes are determined by their chirality and diameter, i.e. simply by their chiral vectors C.

Carbon nanotubes have attracted great interest as they are light, flexible, stiff and they have very large aspect ratios. The tubular shape and the network of carbon atoms in nanotubes with very strong covalent bonds make them very special materials. Some calculations and measurements showed that the tensile strength of carbon nanotubes can be as high as 300 times that of steel and a nanotube can be as stiff as diamond.

Current trends in microelectronics are to produce smaller and faster devices. Owing to their novel and unusual mechanical and electronic properties, carbon nanotubes appear to be potential candidates for meeting the demands of nanotechnologies. One of the most promising applications of carbon nanotubes is employing them as electron emitters in field emission devices. Recent experiments have shown that nanotubes have excellent field emission properties with high current density at low electric fields. Their emission characteristics and durability are found to be far better than other electron emitters. Nanotubes offer promising device applications such as flat panel displays and microwave power amplifiers.

6.4 Graphene

The current popularity of graphene (GR) in scientific research can be traced to the 2004 paper by the group of Nobel Laureates Andre Geim and Konstantin Novoselov [68]. The graphene [69~71] as an allotrope of carbon, is a plane of carbon atoms in 2-dimensions packing into a honeycomb network. When graphene is stacked in three dimensions forming the material, it called graphite which can be exfoliated layer by layer. When the graphene along a given direction is wrapped up into cylindrical form or wrapped into a sphere (i.e. a soccer ball of hexagons and pentagons), the allotropes are termed carbon nanotubes and buckyballs, respectively as shown in Fig.6.27 [69,72~75].

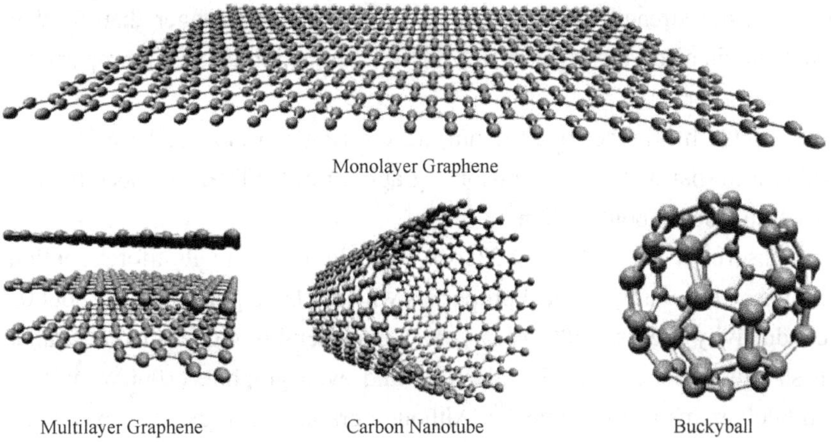

Fig.6.27 The structure of monolayer and multilayer graphene, carbon nanotubes and buckyballs [70]

6.4.1 Properties of graphene

Graphite is a highly anisotropic material with one carbon atom connected to 3 adjacent carbon atoms on the two dimensional plane by sp^2 bonds in a honeycomb lattice network with an interatomic distance of 0.142nm. Along the z-axis of graphite crystal, the p-orbitals of carbon in the graphene sheets are only held by weak van der Waals forces with an interlayer distance of 0.335nm. p(pi)-electrons are located above and below the graphene sheet with their p(pi) orbital overlapping to enhance carbon to carbon bonding.

The commercially available graphene is known as graphene nanoplatelets from a crystalline or flake form of graphite with many graphene sheets stacked together (Fig.6.28) [76]. However, single and bilayer graphene have also been developed [77].

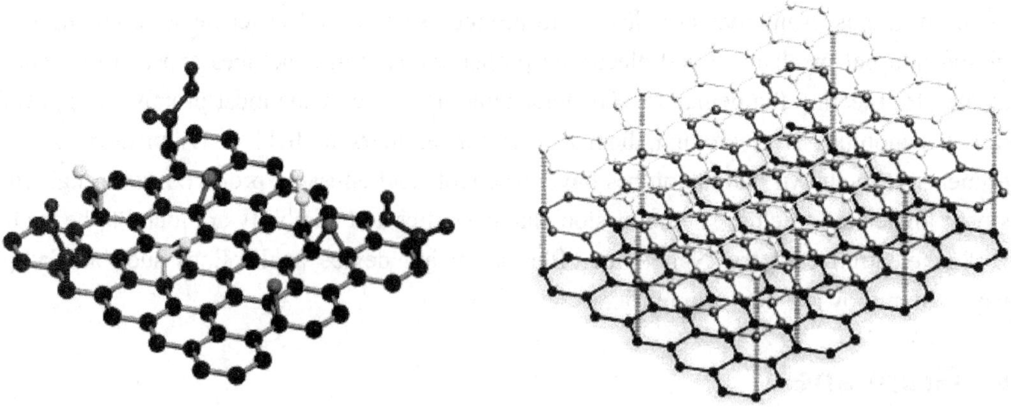

Fig.6.28 Single (left) and multi-layered (right) graphene. The yellow, red and blue groups in single-layered graphene are hydroxyl, ether and carboxyl groups, respectively [76]

The electronic mobility of graphene is >15000cm^2•V^{-1}•s^{-1} compared to a theoretical potential limit of 200000cm^2•V^{-1}•s^{-1} (limited by the scattering of graphene's acoustic photons) [71]. Graphene is a zero-overlap semimetal with holes and electrons as charge carriers. These charge carriers are able to travel sub-micrometer distances without scattering; a phenomenon known as ballistic transport. However, the quality of the graphene and the supporting substrate will be the limiting factors. With SiO_2 as the supporting substrate, mobility is potentially limited to 40000cm^2•V^{-1}•s^{-1}. Due to the strength of its 0.142nm long carbon bonds, graphene is the strongest material with a tensile strength of 130GPa, i.e., ~100 times stronger than steel with the same thickness. Graphene is also very light at 0.77mg m^{-2} and exhibits elastic properties, being able to retain its initial size after strain.

Graphene sheets (thicknesses of 2~8 nm) have spring constants of 1~5 N•m^{-1} and a Young's modulus (different to that of three-dimensional graphite) of 0.5TPa. The specific surface area of a single sheet of graphene is about 2630m^2•g^{-1} [78].

Graphene absorbs a rather large 2.3% of white light due to its afore-mentioned electronic properties. Graphene has unique optical properties with the band gap value of about 0~0.25 eV [79]. Its thermal conductivity is above 5000W•m^{-1}•K^{-1}, considerably higher than the value observed for other carbon structures such as CNTs, diamond and even graphite (1000W•m^{-1}•K^{-1}). Graphene with less than 6000 atoms is not stable [80]. Although graphite also exhibits excellent electrical and thermal conductivities, such conductivities are highly anisotropic, i.e., they differ along the different axes of the crystal.

Graphene is highly hydrophobic in nature although solvents such as dimethylformamide, *N*-methyl-2-pyrrolidone, dimethyl sulfoxide and γ-butyrolactone are useful in preventing the restacking between two graphene sheets [81]. A graphene oxide (GO) suspension is relatively stable in water due to the presence of oxygen-containing groups. As a result, the interlayer distance in GO increases from 0.335nm (graphene) to over 0.6 nm. The cohesion strength between GO layers become weaker compared to graphene and ultrasonication is effective in separating the layers [82].

6.4.2 Synthesis of gaphene

As graphene is a subunit of graphite, it makes sense that the earliest and simplest approach to its synthesis would be direct extraction from bulk graphite. At the start, it should be noted that not all graphite is created equal. There are two important varieties of graphite: natural and synthetic. The highest quality natural graphite possesses single crystalline domains with in-plane dimensions exceeding 1 mm and, consequently, single-layer graphene sheets obtained from natural sources are of exceptional crystal quality [83,84]. However, much of the work done on graphene has proceeded from large-area synthetic graphite, namely highly ordered pyrolytic graphite (HOPG) and Kish graphite.

HOPG is made from the thermal decomposition of hydrocarbons under pressure. This process yields graphite crystals which are much thicker than natural graphite. The larger dimensions make the material much easier to handle and exfoliate; however, the in-plane crystal domains tend to be much smaller than in natural graphite, on the order of 1μm. Kish graphite is produced by the fractional crystallization of carbon from molten steel, and its crystal properties are intermediate between HOPG and natural graphite [84]. The in-plane grain boundaries in these two graphenes impair electronic and phononic transports. Therefore, devices built from this material are typically easier to fabricate, with the tradeoff being that they are usually of lower quality than devices made from natural graphite. This observation that defects in the crystal domain directly degrade the superlative properties that theoreticians predict for graphene, is a major engineering challenge in the synthesis of graphene and graphene devices [85]. It is of importance not only in graphene produced from graphite, but also in graphene of non-graphitic origins. Much of the current research in graphene synthesis focuses on obtaining large-area single crystal graphene, or at least increasing crystal domain sizes and reducing the preponderance of grain boundaries, with the aim of producing graphene with material properties approaching the predictions of the theoreticians. Indeed, recent results show that CVD graphene growth has now surpassed natural graphite crystallinity with ~1 cm wide crystals [86].

(1) Micromechanical exfoliation

The micromechanical method itself is straightforward and can be performed without specialized equipment. A piece of adhesive tape is placed onto and then peeled off the surface of a sample of graphite. The flakes of graphite that adhere to the tape are cleaved preferentially along the plane of the crystal, leaving the exposed atomically flat surfaces [84]. To obtain few- and single-layer graphene, clean tape is pressed against the graphite flakes adhering to the first piece of tape. Peeling apart these two pieces of tape further cleaves the graphite into even thinner flakes. This process is repeated as many times as desired, with each iteration producing thinner sheets of graphite.

The top-down approach, known as exfoliation, is mechanical, chemical, or electrochemical to disrupt the van der Waals forces between the graphene layers of graphite (Fig.6.29). The graphite flakes are then pressed against a substrate such as SiO_2 on silicon, where further processing or device-building with the graphene sheets can be performed (Fig.6.30). The time consuming step of micromechanical cleavage is identifying single or few-layer sheets. This is typically achieved using optical microscopy, exfoliation onto a 300 nm thick SiO_2 film to enhance the optical contrast [68].

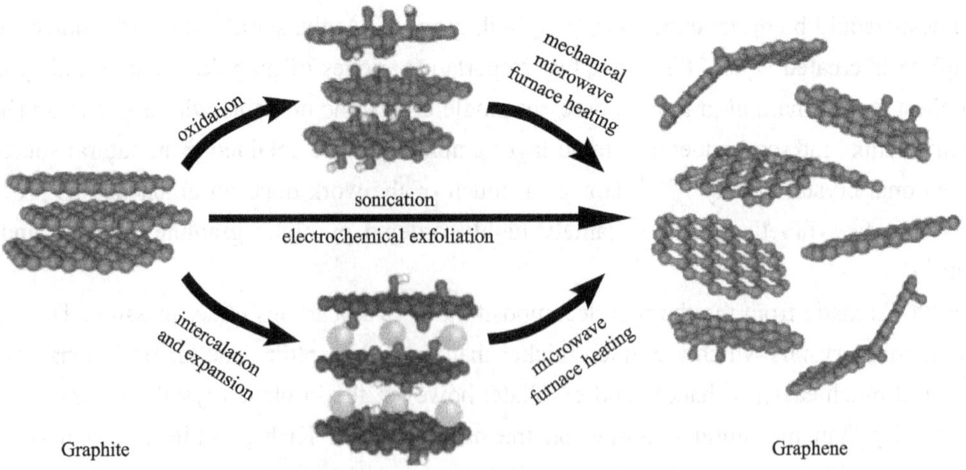

Fig.6.29 Exfoliation techniques to convert bulk amounts of graphite into graphene [87]

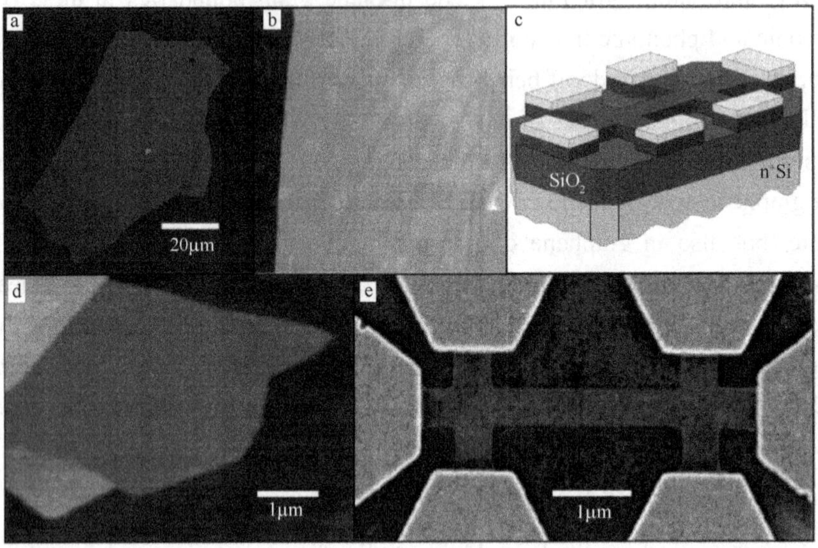

Fig.6.30 (a) Optical microscope image of multilayer graphene cleaved from bulk graphite using the "scotch tape method." (b) AFM image of an edge of the flake. (c) AFM image of few-layer graphene. (d) SEM of a device used by Geim and Novoselov for their studies of the electric field effect in graphene. (e) Schematic of device in d [68]

After the publication of the first graphene paper, many other groups began to use micromechanical cleavage to isolate graphene and to build devices out of it. As interest in and demand for graphene grew, and as the new material's potential as a game-changer in electronics became apparent, researchers realized that this simple production method was not sufficiently scalable to meet future demand. As a result, other methods of graphene production were sought out as described below. However, micromechanical cleavage remains a popular method for graphene production as well as for obtaining high quality samples whose surfaces are exceptionally clean.

The elegant simplicity of micromechanical cleavage has captured the public's imagination: in spite of the construction of massive multibillion dollar research facilities and the passing of global science policy initiatives, an experiment involving little more than adhesive tape and pencil lead can produce Nobel Prize-winning science [88]. However, it has been evident almost from the outset that the scotch tape method is far too inefficient to be useful in mass-producing graphene for practical applications.

(2) Liquid-phase mechanical exfoliation

Most forms of elemental carbon with the notable exception of the fullerenes are completely insoluble under ordinary laboratory conditions. Some liquid-phase approaches have been adopted by graphene researchers to exfoliate individual sheets of graphene from bulk graphite. In one study, scientists sonicated graphite powder in N-methylpyrrolidone and centrifuged the resultant to remove large unexfoliated graphite pieces. They produced suspensions of few-layer graphene with concentrations of up to 1 wt.% that showed remarkable stability toward aggregation over a period of weeks to months [89]. Surfactant-assisted graphene exfoliation is another well established technique, with many surfactants being employed [90,91]. The method is presented to produce graphene dispersions, stabilized in water by the surfactant sodium cholate, at concentrations up to 0.3 mg/mL. The process uses low power sonication for long times (up to 400h) followed by centrifugation to yield stable dispersions. The dispersed concentration increases with sonication time while the best quality dispersions are obtained for centrifugation rates between 500 and 2000 rpm. The method extends the scope for scalable liquid-phase processing of graphene for a wide range of applications.

The solvent-aided and surfactant-aided sonication methods have been explored as ways to efficiently generate ultrathin carbon films with an eye toward device fabrication. But even using liquid-phase exfoliation, the highest concentration of few-layer graphene that one can reasonably achieve is less than 0.1 mg/mL of solvent.

(3) Chemical vapor deposition (CVD)

CVD graphene growth is most commonly performed on copper and nickel surfaces via a thermal method. Thermal growth procedures are quite similar on both metals. High-quality sheets of few-layer graphene have been synthesized via CVD on thin nickel films using multiple techniques. Typically, the thin nickel film (or copper foil) is exposed to Argon gas at 900～1000℃ under a continuous flow of a hydrogen-argon mixture in a 1:10 volumetric ratio, in which a small amount of hydrogen maintain a reducing environment and argon can be used as a carrier gas in order to fine-tune the pressure of the growth experiments [92]. Methane is mixed into the gas, and the carbon source from methane is absorbed into the substrate (nickel or copper foil). The nickel–carbon solution is then cooled down in argon gas and the carbon diffuses out of the nickel to form graphene films during the cooling process [93].With a copper foil as the substrate at very low pressure, the growth of graphene automatically stops after the formation of a single layer in Fig.6.31 [94,95]. The single layer growth is also due to the low concentration of carbon in methane, whereas hydrocarbons such as ethane and propane produce bilayer coatings [96]. Atmospheric pressure CVD growth produces multilayered graphene on Cu, similar to Ni [97].

(4) Chemical synthesis of graphene nanoribbons

One of the most intriguing new methods of graphene synthesis involves using chemical techniques to polymerize and cyclo-dehydrogenate from simple aromatic precursors and thereby obtain atomically precise graphene nanostructures (Fig.6.32). This method was pioneered by the groups of Müllen and Fasel, who originally exposed anthracene and triphenylene derivatives to a hot gold surface to assemble graphene nanoribbons with rationally designed geometry [99]. Recent developments include building working devices from these nanoribbons, as well as synthesizing soluble and heteroatom-substituted nanoribbons. This interdisciplinary effort between synthetic chemists and materials scientists holds a great deal of promise, in that the technique of bottom-up synthesis of nanomaterials allows for atom by atom control of the material. The major drawback to these synthetic techniques right now is that they are still in their infancy, but exciting progress in this direction is being made quickly.

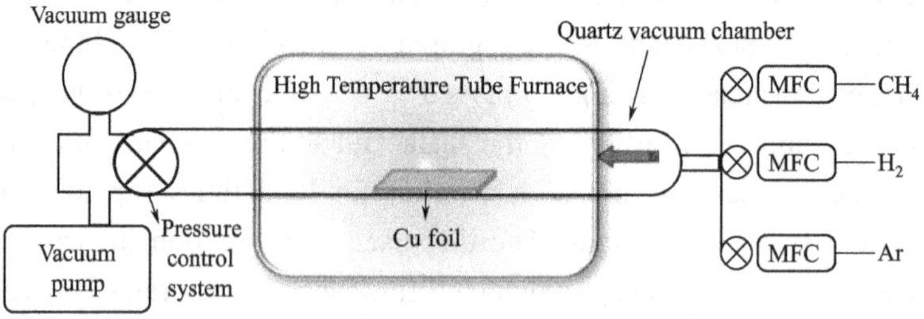

Fig.6.31 CVD for the synthesis of graphene from methane (MFC: mass flow controller) [98]

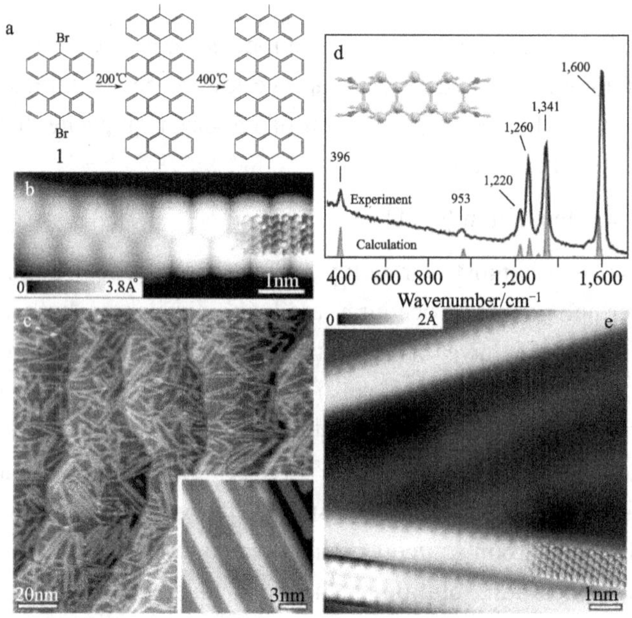

Fig.6.32 (a) Scheme of bottom-up synthesis of graphene nanoribbons from simple aromatic precursors. (b) STM image of polymeric anthracene before full cyclodehydrogenation. (c and e) STM image of fully aromatic graphene nanoribbons after cyclodehydrogenation. (d) Raman spectrum of graphene nanoribbons, showing the radial breathing mode at 396 cm^{-1} [99]

(5) Reduced graphite oxide

In 1859, in an attempt to measure the atomic weight of carbon, Benjamin Brodie performed some of the first experiments on the chemical reactivity of graphite [100]. He noticed that graphite could be extensively oxidized by repeatedly exposing it to a mixture of nitric acid and potassium chlorate for several days. At the end of his experiments, he obtained "···a substance of a light yellow colour, consisting of minute transparent and brilliant plates." He went on to evaluate many properties of what he called "graphic acid," which we today call graphite oxide (GO). Nearly a century later, Hummers and Offeman pioneered a significantly safer synthesis of graphite oxide using a mixture of graphite with sodium nitrate, sulfuric acid, and potassium permanganate [101]. The Hummers method is still the favored method for the production of graphite oxide in the lab.

GO can be prepared from chemical exfoliation of graphite using a strong oxidizing agent, e.g., fuming nitric acid [102]. However, highly oxidized GO should be prepared using a mixture of concentrated sulfuric acid and fuming nitric acid [103]. An alternative method is based on $KMnO_4$ and $NaNO_3$ in concentrated H_2SO_4 [104]. In general, GO is subject to structural damages and defects with different oxygen groups during oxidation as shown in Fig.6.33.

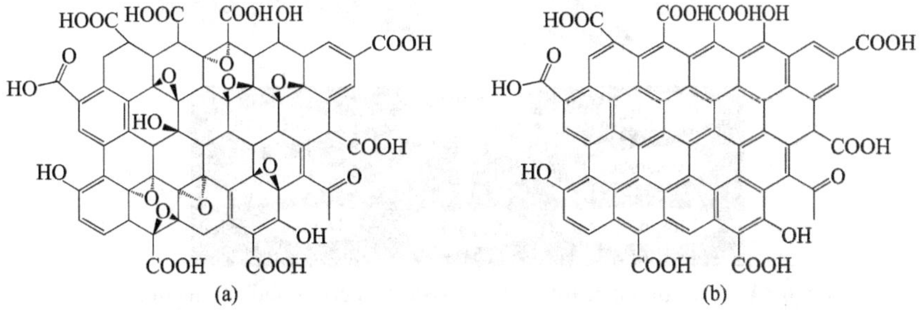

Fig.6.33 Idealized structure proposed for graphene oxide (GO), adapted from C. E. Hamilton,PhD thesis, Rice University, 2009(a). Idealized structure proposed for reduced graphene oxide (rGO)(b)

Reduced GO (rGO) can be easily prepared by reducing GO with a reducing agent, e.g., hydrazine. Soluble fragments of graphene are prepared by versatile wet chemistry by treating microcrystalline graphite with a sulfuric–nitric acid mixture. A series of oxidation and exfoliation steps produces small graphene plates with carboxyl groups at their edges [105] as shown in Fig.6.33.

Functional groups such as carboxyl, hydroxyl, sulfonic acid and amine are introduced for subsequent bioconjugation [106~108]. On account of its favorable characteristics, graphene has been utilized in diverse areas, such as nanoelectronics, optoelectronics, chemical and biochemical sensing, polymer composites, H_2 production and storage, intercalation materials, drug delivery, supercapacitors, catalysis and photovoltaics [109~114].

Today, nanocarbon materials including fullerene, carbon nanotube and nano-porous carbon as a key material in the 21st century have being widely applied for the industry, environment, energy, IT and other fields due to their unique morphology and nano-sized scale as shown in Fig.6.34 [115]. The inherent advantage of carbon materials, originating from the extraordinary ability of the chemical element carbon to combine itself and other chemical elements in different ways, is blooming in the 21st century.

The facing challenge is how to utilize their intrinsic properties of mechanical, thermal and electrical properties. Tailoring of carbon structure, that is, controlling the physicochemical properties of carbon materials on the nanometer scale by making use of sp^3, sp^2 and sp or hexagonal carbon structure, will be core technology for obtaining novel carbons with new and extraordinary functions. As some scientists ever reported "Carbon is hot now". At present, many scientists are being "hot" on carbon materials.

It can be said "21st century is the carbon age".

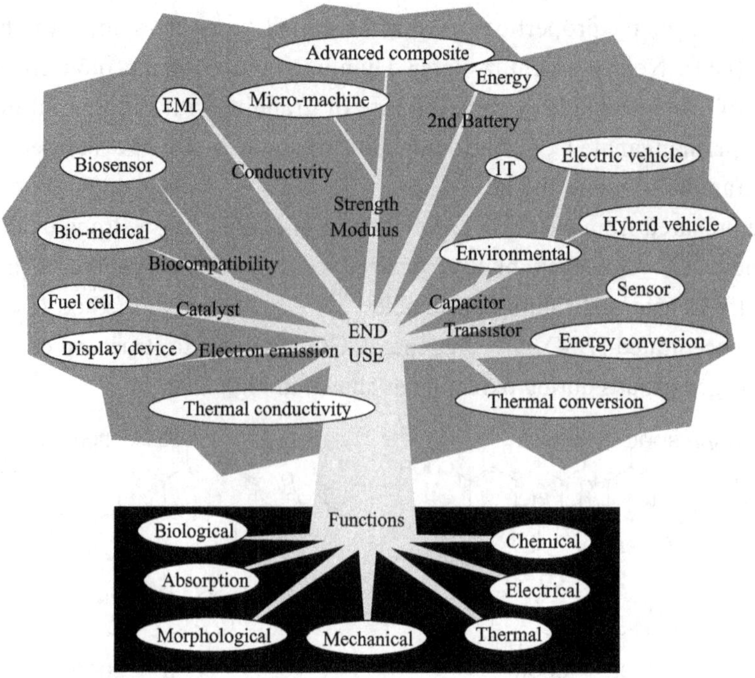

Fig.6.34 Tree of nanocarbons with basic properties and applications [115]

References

[1] W.A. Goddard, D.W. Brenner, S.E. Lyshevski, G.J. Lafrate, Handbook of nanoscience, engineering, and technology, Second edition, (2007) Taylor & Francis Group, LLC, ISBN: 0-8493-7563-0, P27-34.

[2] O. Shenderova, Z. Hu, et al., Carbon family at the nanoscale, in Synthesis, Properties and Applications of Ultrananocrystalline Diamond, D. Gruen, O. Shenderova, A. Vul, (2005) Springer, Amsterdam, P1–14.

[3] H.W. Kroto, J.R. Heath, S.C. O'Brien, R.F. Curl, R.E. Smalley, Nature, 318 (1985) 162.

[4] W. Kratschmer, L.D. Lamb, K. Fostiropoulos, D.R. Huffman, Nature, 347 (1990) 354.

[5] J.B. Howard, J.T. McKinnon, Y. Makarovsky, A.L. Lafleur, M.E. Johnson, Nature,352 (1991) 139.

[6] D.S. Bethune, C.H. Kiang, M.S. Devries, Nature, 363 (1993) 605.

[7] R.F. Curl, Carbon, 30 (1992) 1149.

[8] R. Taylor, G.J. Langley, H.W. Kroto, D.R. M.Walton, Nature, 366 (1993) 728.

[9] M.C. Zumwalt, D.R. Huffman, in Proc. Eletrochem. Soc., 360 (1994) 94–124.

[10] C. Crowley, R. Taylor, H.W. Kroto, D.R.M. Walton, P.-C. Cheng, L.T. Scott, Synth. Met., 77 (1996) 17.

[11] J. Osterodt, A. Zett, F. Vogtle, Tetrahedron, 52 (1996) 4949.

[12] N.R. Conley, J.J. Lagowski, Carbon, 40 (2002) 949.

[13] T. W. Ebbesen, J. Tabuchi, K. Tanigaki, The mechanistics of fullerene formation, Chem. Phys. Lett., 191(1992) 336.

[14] P.J. Brucat, L.S. Zheng, C.L. Pettiette, S. Yang, R.E. Smalley, J. Chem. Phys., 84 (1986) 3078.

[15] T. Pradeep, Nano: the essentials, (2007)McGraw-Hill Companies, Inc. Publisher, ISBN: 0-07-154829-7.

[16] H.S. Nalwa, Encyclopedia of Nanoscience and Nanotechnology, 1(1) 2004 American Scientific Publishers, P428.

[17] J.Z. Liu, J.W. Dykes, M.D. Lan, P. Klavins, R.N. Shelton, M.M. Olmstead, Appl. Phys. Lett., 62 (1992) 531.

[18] J.-B. Shi, W.-Y. Chang, S.-R. Su, M.-W. Lee, Jpn. J. Appl. Phys.,35 (1996) L45.

[19] X.D. Xiang, J.G. Hou, G. Briceno, W.A. Vareka, R. Mostovoy, A. Zettl, V.H. Crespi, M.L. Cohen, Science, 256 (1992) 1190.

[20] J. Li, S. Komiya, T. Tamura, C. Nagasaki, J. Kihara, K. Kishio, K. Kitazawa, Physica C ,195 (1992) 205.

[21] M. Haluska, H. Kuzmany, M. Vybornov, P. Rogl, P. Fejdi, Appl. Phys. A, 56 (1993) 161.
[22] M. Tan, B. Xu, H. Li, Z. Qi, Y. Xu, J. Crystal Growth ,182 (1992) 375.
[23] K.-C. Chiu, J.-S. Wang, C.-Y. Lin, T.-Y. Lin, C.-S. Ro, Mater. Res. Bull., 30 (1995) 883.
[24] K. Matsumoto, E. Schonherr, M. Wojnowski, J. Crystal Growth, 135 (1994) 154.
[25] K. Murakami, K. Matsumoto, H. Nagatomo, E. Schonherr, M. Wojnowski, Ultramicroscopy, 73 (1998) 191.
[26] H. Li, Y. Xu, J. Zhang, P. He, H. Li, T. Wu, S. Bao, Progr. Natural Sci., 11 (2001) 427.
[27] J.L. Atwood, G.A. Koutsantonis , C.L. Raston, Nature, 368 (1994)229.
[28] J .G. Hou, A.D. Zhao, T. Huang, S. Lu, Encyclopedia of Nanoscience and Nanotechnology, 1 (1) (2004) American Scientific Publishers, P409-474.
[29] D.E. Manolopolous, Chem. Phys. Lett. ,192 (1992) 330.
[30] J.G. Hou, J. Yang, H. Wang, Q. Li, C. Zeng, L. Yuan, B. Wang, D.M. Chen, Q. Zhu, Nature ,409 (2001) 304.
[31] H.W. Kroto, Nature, 329 (1987) 529.
[32] S.T.G. Chmalz, W.A. Seitz, D.J. Klein, G.E. Hite, J. Am. Chem. Soc. ,110 (1988) 1113.
[33] R.D. Johnson, D.S. Bethune, C.S. Yannoni, Acc. Chem. Res., 25 (1992) 169.
[34] R. Taylor, J.P. Hare, A.K. Abdul-Sada, H. W. Kroto, Chem. Commun. ,1423 (1990).
[35] R.D. Johnson, G. Meijer, J.R. Salem, D.S. Bethune, J. Am. Chem. Soc.,113 (1991) 3619.
[36] R.D. Johnson, D.S. Bethune, C.S. Yannoni, Acc. Chem. Res.,25 (1992) 169.
[37] N. Matsuzawa, D. Dixon, T. Fukunaga, J. Phys. Chem., 96 (1992) 7594.
[38] B. Sundqvist, Adv. Phys. ,48 (1999) 1.
[39] Y. Iwasa, T. Arima, R.M. Fleming, T. Siegrist, O. Zhou, R.C. Haddon, L.J. Rothberg, K.B. Lyons, H.L. Jr. Carter, A.F. Hebard, R. Tycko, G. Dabbagh, J.J. Krajewski, G.A. Thomas, T. Yagi, Science ,264 (1994) 1570.
[40] G. Oszlanyi, L. Forro, Solid State Commun. ,93 (1995) 265.
[41] M. Nunez-Regueiro, L. Marques, J.-L. Hodeau, O. Bethoux, M. Perroux, Phys. Rev. Lett. ,74 (1995) 278.
[42] H.S. Nalwa, Encyclopedia of Nanoscience and Nanotechnology, 1(1) 2004 American Scientific Publishers, P446.
[43] Y. Shirai, A.J. Osgood, Y. Zhao, K.F. Kelly, J.M. Tour, Nano Lett., 5(11) 2005 2330-2334.
[44] Y. Shirai, Y. Zhao, L. Cheng, J.M. Tour, Org. Lett., 6 (2004) 2129-2132.
[45] M. Sharon, Encyclopedia of Nanoscience and Nanotechnology, 1(1) 2004 American Scientific Publishers, P517-546.
[46] S. Iijima, Nature, 354 (1991) 56.
[47] S. Iijima, T. Ichihashi, Y. Ando, Appl. Phys., 32 (1993) L107.
[48] X. Zhao, M. Wang, M. Ohkohchi, Y. Ando, Bull. Res. Inst. Meijo Univ., 1 (1996) 7.
[49] R. Saito, G. Dresselhaus, M.S. Dresselhaus, Physical Properties of Carbon Nanotubes, (1998) Imperial College Press.
[50] C.J. Lee, S.C. Lyu, H.-W. Kim, C.-Y. Park, C.-W. Yang, Chem. Phys. Lett., 359 (2002) 109.
[51] K. Mukhopadhyay, M. Sharon, Mater. Chem. Phys., 49 (1997) 105.
[52] D. Pradhan, M. Sharon, Mater. Sci. Eng. B ,96 (1996) 24.
[53] H.S. Nalwa, Encyclopedia of Nanoscience and Nanotechnology,1(1) 2004 American Scientific Publishers, P675.
[54] H.S. Nalwa, Encyclopedia of Nanoscience and Nanotechnology,1(1) 2004 American Scientific Publishers, P535.
[55] K. Wolfgang, M. Ana, M. Benito, T. Martinez, Carbon ,40 (2002) 1685.
[56] S.B. Sinnott, R. Andrews, D. Qian, A.M. Rao, Z. Mao, E.C. Dickey, F. Derbyshire, Chem. Phys. Lett. ,315 (1999) 25.
[57] R.T.K. Baker, M.A. Barber, P.S. Harris, F.S. Feates, R.J. Waite, J. Catal., 26 (1972) 51.
[58] R.T.K. Baker, Carbon ,27 (1989) 315.

[59] A. Oberlin, M. Endo, T. Koyama, J. Crystal Growth, 32 (1976) 335.
[60] N.M. Rodriguez, A. Chambers, R.T.K. Baker, Langmuir, 11 (1995) 3862.
[61] H. Terrones, T. Hayashi, M. Munoz-Navia, M. Terrones, Y.A. Kim, N. Grobert, R. Kamalakaran, J. Dorantes-Davila, R. Escudero, M.S. Dresselhaus, M. Endo, Chem. Phys. Lett., 343 (2001) 241.
[62] N.M. Rodriguez, J. Mater. Res., 8 (1993) 3233.
[63] M. Chhowalla, K.B.K. Teo, C. Ducati, N.L. Rupesinghe, G.A.J. Amaratunga, A.C. Ferrari, D. Roy, J. Robertson, W.I. Milne, J. Appl. Phys., 90 (2001) 5308.
[64] C. Bower, O. Zhou, W. Zhu, D.J. Werder, S.H. Jin, Appl. Phys. Lett., 77 (2000) 2767.
[65] J.I. Sohn, S. Lee, Y.H. Song, S.Y. Choi, K.I. Cho, K.S. Nam, Appl. Phys. Lett., 78 (2001) 901.
[66] N.P. Mahalik, Micromanufacturing and Nanotechnology, (2006) Springer-Verlag Berlin Heidelberg, ISBN-10: 3-540-25377-7, ISBN-13: 978-3-540-25377-8.
[67] M.J. Bonard, H. Kind, T. Stockli, O. L. Nilsson, Field emission from carbon nanotubes:The first five years, Solid-State Electron. ,45 (2001) 893-896.
[68] K.S. Novoselov, A.K. Geim, S.V.Morozov, D. Jiang, Y. Zhang, S.V. Dubonos, et al., Electric field effect in atomically thin carbon films, Science 306 (5696) (2004) 666-669.
[69] S. K. Vashist, J.H.T. Luong, Recent advances in electrochemical biosensing schemes using graphene and graphene-based nanocomposites, Carbon 84 (2015) 519-550.
[70] A.K. Geim, P. Kim, Scientif ic American, p. 90 (2008).
[71] A. K. Geim and K. S. Novoselov, The rise of graphee, Nat. Mater., 6 (2007) 183-191.
[72] A. K. Geim, Science, **324 (**2009) 1530-1534.
[73] D. Li and R. B. Kaner, Science, **320 (**2008) 1170-1171.
[74] S. Bai and X. Shen, *RSC* Adv., **2** (2012) 64-98.
[75] V. Singh, D. Joung, L. Zhai, S. Das, S. I. Khondaker and S. Seal, Prog. Mater. Sci., 56 (2011) 1178-1271.
[76] S.K.Vashist, Advances in graphene-based sensors and devices, J. Nanomed. Nanotech 04(01) (2012) 127.
[77] A.T. Valota, I.A. Kinloch, K.S. Novoselov, C. Casiraghi, A. Eckmann, E.W. Hill, et al. Electrochemical behavior of monolayer and bilayer graphene, ACS Nano. 5(11) (2011) 8809.
[78] M.D. Stoller, S. Park, Y. Zhu, J. An, R.S. Ruoff, Graphene-based ultracapacitors, Nano. Lett. 8(10) (2008) 3498-502.
[79] Y. Zhang, T.T. Tang, C. Girit, Z. Hao, M.C. Martin, A. Zettl, et al. Direct observation of a widely tunable bandgap in bilayer grapheme, Nature 459 (2009) 820-823.
[80] O. Shenderova, V. Zhirnov, D. Brenner, Carbon nanostructures, Crit. Rev. Solid State 27(3-4) (2002) 227-356.
[81] C.J. Shih, S. Lin, M.S. Strano, D.Blankschtein, Understanding the stabilization of liquid-phase-exfoliated graphene in polar solvents: molecular dynamics simulations and kinetic theory of colloid aggregation, J. Am. Chem. Soc. 132(41) (2010) 14638-14648.
[82] D.R. Dreyer, R.S. Ruoff, C.W. Bielawski, From conception to realization: an historial account of graphene and some perspectives for its future, Angew. Chem. Int. Ed. Engl. 49(49) (2010) 9336-9344.
[83] K.E.Whitener Jr., P.E. Sheehan, Graphene synthesis, Diamond Related Mate. 46 (2014) 25-34.
[84] M.S. Dresselhaus, G. Dresselhaus, Intercalation compounds of graphite, Adv. Phys. 51 (1) (2002) 1-186.
[85] F. Banhart, J. Kotakoski, A.V. Krasheninnikov, Structural defects in graphene, ACS Nano 5 (1) (2010) 26-41.
[86] Y. Hao, M.S. Bharathi, L. Wang, Y. Liu, H. Chen, S. Nie, et al., The role of surface oxygen in the growth of large single-crystal graphene on copper, Science 342 (2013) 720-723.
[87] M. Cai, D. Thorpe, D.H. Adamson, H.C.Schniepp, Methods of graphite exfoliation. J. Mater. Chem. 22(48) (2012) 24992-25002.
[88] D. Overbye, Physics Nobel Honors Work on Ultra-Thin Carbon, New York Times, 2010.
[89] Y. Hernandez, V. Nicolosi, M. Lotya, F.M. Blighe, Z. Sun, S. De, et al., High-yield production of graphene by liquid-phase exfoliation of graphite, Nat. Nanotechnol. 3 (9) (2008) 563-568.
[90] M. Lotya, Y. Hernandez, P.J. King, R.J. Smith, V. Nicolosi, L.S. Karlsson, et al., Liquid phase production of

graphene by exfoliation of graphite in surfactant/water solutions, J. Am. Chem. Soc. 131 (10) (2009) 3611-3620.
[91] M. Lotya, P.J. King, U. Khan, S. De, J.N. Coleman, High-concentration, surfactantstabilized graphene dispersions, ACS Nano. 4 (6) (2010) 3155-3162.
[92] Y. Zhang, L. Zhang, C. Zhou, Review of chemical vapor deposition of graphene and related applications, Acc. Chem. Res. 46 (10) (2013) 2329-2339.
[93] J. Rafiee, X. Mi, H. Gullapalli, A.V. Thomas, F. Yavari, Y. Shi, et al. Wetting transparency of grapheme, Nat, Mater, 11(3) (2012) 217-222.
[94] X. Li, W. Cai, J. An, S. Kim, J. Nah, D. Yang, et al. Large-area synthesis of high-quality and uniform graphene films on copper foils, Science 324(5932) (2009) 1312-1314.
[95] S. Bae, H. Kim, Y. Lee, X. Xu, J.S. Park, Y. Zheng, et al. Roll-to-roll production of 30-inch graphene films for transparent electrodes, Nat. Nanotechnol. 5(8) (2010) 574-578.
[96] J.K. Wassei, M. Mecklenburg, J.A. Torres, J.D. Fowler, B.C. Regan, R.B. Kaner, et al. Chemical vapor deposition of graphene on copper from methane, ethane and propane: evidence for bilayer selectivity, Small 8(9) (2012)1415-1422.
[97] D.R. Lenski, M.S. Fuhrer, Raman and optical characterization of multilayer turbostratic graphene grown via chemical vapor deposition, J. Appl. Phys. 110(1) (2011) 013720.
[98] A. Kumar, C.H. Lee, In: Aliofkhazraei DM, editor. Advances in graphene science. InTech; 2013.
[99] J. Cai, P. Ruffieux, R. Jaafar, M. Bieri, T. Braun, S. Blankenburg, et al., Atomically precise bottom-up fabrication of graphene nanoribbons, Nature 466 (7305) (2010) 470-473.
[100] B.C. Brodie, On the atomic weight of graphite, Phil. Trans. R. Soc. London 149 (1859) 249-259.
[101] W.S. Hummers, R.E. Offeman, Preparation of graphitic oxide, J. Am. Chem. Soc. 80 (6) (1958) 1339.
[102] B.C. Brodie, On the atomic weight of graphite, Philos. Trans. R. Soc. (1859) 249-59.
[103] L. Staudenmaier, Verfahren zur darstellung der graphitsa¨ure. Berichte der deutschen chemischen, Gesellschaft 31(2) (1898) 1481-1487.
[104] W.S. Hummers Jr, R.E. Offeman, Preparation of graphitic oxide, J. Am. Chem. Soc. 80(6) (1958) 1339.
[105] S. Eigler, M. Enzelberger-Heim, S. Grimm, P. Hofmann, W. Kroener, A. Geworski, et al., Wet chemical synthesis of grapheme, Adv. Mater. 25(26) (2013) 3583-3587.
[106] K. Hu, D.D. Kulkarni, I. Choi, V.V. Tsukruk, Graphene-polymer nanocomposites for structural and functional applications, Prog. Polym. Sci. 39(11) (2014) 1934-1972.
[107] T. Premkumar, K.E. Geckeler, Graphene-DNA hybrid materials: assembly, applications, and prospects, Prog. Polym. Sci. 37(4) (2012) 515-529.
[108] A. Ambrosi, C.K. Chua, A. Bonanni, M. Pumera, Electrochemistry of graphene and related materials, Chem. Rev. 114(14) (2014) 7150-7188.
[109] Y. Zhang, Z.-R. Tang, X. Fu, Y.-J. Xu, ACS Nano. 4 (2010) 7303-7314.
[110] X. An, J.C. Yu, RSC Adv. 1 (2011) 1426-1434.
[111] M.J. Allen, V.C. Tung, R.B. Kaner, Chem. Rev. 110 (2010) 132-145.
[112] Y. Zhu, S. Murali, W. Cai, X. Li, J.W. Suk, J.R. Potts, R.S. Ruoff, Adv. Mater. 22 (2010) 3906-3924.
[113] B. Seger, P.V. Kamat, J. Phys. Chem. C 113 (2009) 7990-7995.
[114] X. Li, X. Wang, L. Zhang, S. Lee, H. Dai, Science 319 (2008) 1229-1232.
[115] Z.K. Tang，P. Sheng, Nanoscale Phenomena Basic Science to Device Applications, (2008) Springer Science & Business Media, LLC, ISBN-13: 978-0-387-73047-9 (Hardback), ISBN-13: 978-0-387-73048-6 (Electronic).

Review questions

1. Please indicate carbon allotrope and their spatial dimensions.
2. What are the fullerenes and football molecule?

3. Please tell us the cage structure of C_{60}.

4. How to use the laser ablation method to preparation of C_{60}?

5. Please briefly describe the working principle making use of electric arc discharge method to fabricate C_{60}.

6. Please illustrate what is endofullerenes.

7. What kind of material can be used to manufacture the wheels of nanocar?

8. What are single walled carbon nanotube and multi-walled carbon nanotubes?

9. How to use the catalytic-assisted pyrolysis method to synthesize carbon nanotubes in a horizontal furnace?

10. Briefly describe the growing mechanism of carbon nanotubes.

11. Please dictate geometric shapes of the three kinds of carbon nanotubes by rolling up a sheet of graphite.

12. Why does somebody say "21st century is the carbon age"?

Vocabulary

ablate [æb'leɪt] vt. 烧蚀，消融；切除；使脱落；vi. 脱落

acetylene [ə'setɪliːn] n. 乙炔，电石气

alignment [ə'laɪnm(ə)nt] n. 结盟；队列，成直线；校准

allotrope ['ælətrəʊp] n. [化] 同素异形体

allotropic [ˌæləʊ'trɒpɪk] adj. [化] 同素异形的

alumina [ə'luːmɪnə] n. 氧化铝；矾土

ambient ['æmbɪənt] adj. 周围的；外界的；环绕的；n. 周围环境

ampere ['æmpeə(r)] n. 安培

anode ['ænəʊd] n. [电] 阳极，正极

argon ['ɑːgɒn] n. [化] 氩

aromatic [ærə'mætɪk] adj. 芳香的，芬芳的；芳香族的；n. 芳香剂；芳香植物

autoloader ['ɔːtəu,ləudə] n. 自动装货车，自动装卸机

benzene ['benziːn] n. [化] 苯

bioconjugation 生物耦合

camphor ['kæmfə] n. [化] 莰酮，樟脑

carbyne ['kɑːbain] 碳炔，卡宾

cathode ['kæθəʊd] n. 阴极

chamber ['tʃeɪmbə] n.（身体或器官内的）室，膛；房间；会所；vt. 把…关在室内；装填

chiral ['tʃɪrəl] adj. [化] 手性的；对掌性的

chromatograph [krəʊ'mætəgrɑːf] vt. 用色谱法分析；n. 套色版；色层

circulate ['sɜːkjʊleɪt] vi. 流通；循环；vt. 使循环；使流通；使传播

coaxial [kəʊ'æksɪəl] adj. 同轴的

column chromatography 柱层析

combustion [kəm'bʌstʃ(ə)n] n. 燃烧，氧化；骚动

commercially available apparatus 商品仪器

concentric [kən'sentrɪk] a. 同中心的

configuration [kənˌfɪgə'reɪʃ(ə)n; -gjʊ-] n. 配置；结构；外形

covalent [kəʊ'veɪl(ə)nt] a. 共有原子价，共价

curvature ['kɜːvətʃə] n. 弯曲，曲率

cylindrical [sɪ'lɪndrɪkəl] adj. 圆柱形的；圆柱体的

dangle ['dæŋg(ə)l] v. 摇晃地悬挂着

deactivate [diː'æktɪveɪt] vt. 使无效；使不活动；遣散；复员

delocalize [diː'ləʊk(ə)laɪz] vt. 使…离开原位；使…不受位置限制

dendritic [den'drɪtɪk] adj. 树枝状的；树状的

derivative [dɪ'rɪvətɪv] n. 衍生物，派生物；adj. 派生的；引出的

dichlorobenzene [daiˌklɔːrəu'benziːn] n. 二氯苯；二氯代苯

dictate [dɪk'teɪt] vt. 命令；口述；使听写；vi. 口述；听写；n. 命令；指示

electrolysis [ˌɪlek'trɒlɪsɪs;,el-] n. 电解，电解作用；以电针除痣

eluant ['ɛljʊənt] n. [分化]洗提液，洗脱剂

emanate ['eməneɪt] vt. 放射;发散；vi. 发源；发出；散发

encapsulate [ɪn'kæpsjʊleɪt; en-] vt. 压缩;将…装入胶囊；vi. 形成胶囊

endofullerene 富勒烯包合物

equivalent [ɪ'kwɪv(ə)l(ə)nt] adj. 等价的，相等的；同意义的；n. 等价物，相等物

evacuate [ɪ'vækjʊeɪt] vt. 排泄；疏散，撤退；vi. 撤退；疏散；排泄

evaporate [ɪ'væpəreɪt] vt. 使……蒸发；使……脱水；使……消失；vi. 蒸发，挥发；
exfoliate [ɪks'fəʊlɪeɪt; eks-] vi. 片状剥落；鳞片样脱皮；vt. 使片状脱落
filament ['fɪləm(ə)nt] n. 灯丝；细丝；细线；单纤维
furnace ['fɜːnɪs] n. 火炉，熔炉
geometry [dʒɪ'ɒmɪtrɪ] n. 几何学
graphene ['græfiːn] n. 石墨烯，单层石墨
graphite ['græfaɪt] n. 石墨;黑铅；vt. 用石墨涂
hardness ['hɑːdnəs] n. 硬度；坚硬；困难；冷酷
helical ['helɪk(ə)l; 'hiː-] a. 螺旋状的
helium ['hiːlɪəm] n. 氦
hexagonal [hek'sægənəl] a. 六角形的，六边的，底面为六角的
high performance liquid chromatography (HPLC) 高效液相色谱
honeycomb ['hʌnɪkəʊm] n. 蜂巢，蜂巢状之物
hydrocarbon [ˌhaɪdrə(ʊ)'kɑːb(ə)n] n. 碳氢化合物，烃
hydrophilic [ˌhaɪdrə(ʊ)'fɪlɪk] n. 软性接触透镜吸水隐形眼镜片；adj. [化] 亲水的
hydrophobic [haɪdrə(ʊ)'fəʊbɪk] adj. 狂犬病的；恐水病的；疏水的
hypothetical [ˌhaɪpə'θetɪk(ə)l] adj. 假设的；爱猜想的
infancy ['ɪnf(ə)nsɪ] n. 初期；婴儿期；幼年
interlock [ɪntə'lɒk] v. 互锁；连锁
isomer ['aɪsəmə] n. [化] 同分异构物；[核] 同质异能素
lace [leɪs] n. 花边；鞋带；饰带；少量烈酒；vt. 饰以花边；结带子
monocyclic [ˌmɒnə(ʊ)'saɪklɪk; -'sɪk-] adj. 单环的；单轮的；单周期性的
multiple ['mʌltɪpl] adj. 多重的；许多的；n. 并联；倍数
noble ['nəʊb(ə)l] adj. 惰性的；高尚的；n. 贵族
nomenclature [nə(ʊ)'meŋklətʃə; 'nəʊmənˌkleɪtʃə] n. 命名法；术语
nozzle ['nɒz(ə)l] n. 喷嘴；管口；鼻
nucleophilic [ˌnjuːkliəu'filik] adj. 亲核的；亲质子的
optoelectronics [ˌɒptəʊɪlek'trɒnɪks; -el-] n. 光电学，[物][电子] 光电子学
orthorhombic [ˌɔːθə(ʊ)'rɒmbɪk] adj. 正交晶的；斜方晶系的
pentagon ['pentəg(ə)n] n. 五角形
perpendicular [ˌpɜːp(ə)n'dɪkjʊlə] adj.陡峭的，直立的；n. 垂线，垂直的位置
physicochemical [ˌfɪzɪkəʊ'kemɪkəl] adj. 物理化学的
planar ['pleɪnə] adj. 平面的；二维的；平坦的
plasma ['plæzmə] n. 等离子体
polarizability ['pəuləˌraizə'biləti] n. 极化性；极化度
polarized light 偏振光
postulate ['pɒstjʊleɪt] vt. 假定；要求；视…为理所当然；n. 基本条件；假定
precursor [prɪ'kɜːsə] n. 前体物，前驱体
pyrolysis [paɪ'rɒlɪsɪs] n. [化] 高温分解；[化] 热解
pyrolytic [ˌpaiərəu'litik] adj. 热解的
quartz [kwɔːts] n. [矿] 石英
radioactive [ˌreɪdɪəʊ'æktɪv] adj. 放射性的；有辐射的
reddish ['redɪʃ] adj. 微红的；略带红色的
rhombohedra [ˌrɑmbə'hidrən] n. 菱面体
seamless ['siːmlɪs] adj. 无缝的
silane ['sɪleɪn] n. [化] 硅烷
skim [skɪm] vt. 略读；撇去…的浮物；从…表面飞掠而过；去除；vi. 浏览；掠过；n. 撇；撇去的东西；表层物；瞒报所得的收入；adj. 脱脂的；撇去浮沫的；表层的
sonication [ˌsɒnɪ'keɪʃən] n. 声波降解法
soot [sʊt] n. 煤烟，烟灰；vt. 用煤烟熏黑；以煤烟弄脏
spectroscopic [ˌspɛktrə'skɑpɪk] adj. 光谱的；分光镜的
substrate ['sʌbstreɪt] n. 基质；酶作用物；基片；底层
subunit ['sʌbˌjʊnɪt] n. 亚组；副族
tetragonal [tɪ'træg(ə)n(ə)l] adj. 四角形的
tetrahedral [ˌtetrə'hiːdrəl] adj. 四面体的；有四面的
tetramethyl [ˌtɛtrə'mɛθɪl] adj. [化] 四甲基的
toluene ['tɒljʊiːn] n. [化] 甲苯
turpentine ['tɜːp(ə)ntaɪn] n. 松节油；松脂；vt. 涂松节油
ultrasonication [ˌʌltrəˌsɔni'keiʃən] n. 声波降解法；超声破碎法
ultraviolet [ʌltrə'vaɪələt] adj. 紫外的；紫外线的；n. 紫外线辐射，紫外光
vector ['vektə] n. 矢量；带菌者；航线；vt. 用无线电导航
versatile ['vɜːsətaɪl] adj. 多才多艺的；通用的，万能的；多面手的
vertex ['vɜːteks] n. 顶点，最高点，头顶
vertical ['vɜːtɪk(ə)l] adj. 垂直的，直立的；头顶的，顶点的；n. 垂直线，垂直面
vicinity [vɪ'sɪnɪtɪ] n. 邻近，附近；近处
violet ['vaɪələt] n.堇菜；紫罗兰；adj. 紫罗兰色的；紫色的
vitreous ['vɪtrɪəs] adj. 玻璃的；玻璃状的；玻璃似的
volatile ['vɒlətaɪl] adj. 挥发性的；不稳定的；爆炸性的；反复无常的；n. 挥发物；
zigzag ['zɪgzæg] n. 之字形；曲折的；锯齿形的；之字形的；adv. 曲折地；vt.使曲折行进

6. 碳纳米材料的合成

6.1 碳族

碳是一种异乎寻常的元素。孤立碳原子的 1s 和 2s 电子轨道为全满，2p 轨道填充两个电子，形成 $1s^22s^22p^2$ 的电子结构。

6.1.1 石墨和金刚石

一般而言，碳仅存在两种同素异形体，也就是说，碳原子仅仅与其它的碳原子相键合，具有代表性的是石墨和金刚石。闪闪发光且非常坚硬的晶体结构的金刚石是由四个碳原子分别以 sp^3 杂化形式而相互结合，形成三维的正四面体结构（C—C 键之间的键角为 109.5°）。石墨（普通的铅笔芯）的结构完全不同于金刚石，碳原子采取 sp^2 杂化（C—C 键之间的键角为 120°）形成相应的六方晶体结构。基于它们“纳米量级的体系结构”，这两种材料的性质差异是十分显著的。例如，石墨中高度离域的π键网络结构表明，石墨比金刚石具有更高的电导率。而金刚石 sp^3 碳原子有很强的共价键联锁网状结构，具有很高的硬度。加之，由于金刚石很宽的带隙（5.5eV），因而金刚石是一种绝缘体，而石墨是一种导体（带隙约为 0.25eV 或者更小）。这两个例子给了我们这样一条思路，数以百万计的 sp^3 和 sp^2 杂化的碳原子结合，可得到带隙在 0.25～5.5eV 之间变化的无数种形态的碳。

常温常压下，碳单质的最低能态是由石墨烯单层组成的石墨。每一层是由相互连接的六边形的碳环紧密键合在一起，层与层之间疏松地堆积成一个三维材料。在一个石墨烯片层内，在 x-y 二维平面上，每个碳原子与其它三个相邻的碳原子在一个平面上紧密键合，而层与层之间以很弱的悬空键形式键合。

6.1.2 碳的同素异形体

依据碳结构中化学键的属性对其进行分类，每一个价态对应于单一物质的一种形态。单质碳以四种键合状态存在，分别对应于 sp^3、sp^2、sp 和 sp^n 四种碳原子的杂化轨道。相应的四种同素异形体分别为金刚石、石墨、卡宾碳和富勒烯，如图 6.1 所示[1,2]，所有其它形式的碳由“过渡型”组成，可分为两大类。

第一类由短程有序的混合碳所组成（如：类金刚石的碳、玻璃体碳、烟灰、炭黑），还有大量的假想结构如石墨炔和“超金刚石”。

第二类包括碳碳键非整数杂化的 sp^n 形式组成的过渡态碳。该类中的次类 $1<n<2$ 时，就包括各种单环碳结构。该类中的次类 $2<n<3$ 时，过渡态的碳形式由满壳的碳结构所组成，如富勒烯、碳阴离子和碳纳米管。

碳纳米结构被定义为其尺寸或结构都被控制在纳米范围（<100nm）的碳材料。在延伸的碳族中，其维度可表示为：零维结构（富勒烯、金刚石簇），一维结构（碳纳米管），二维结构（石墨片），三维结构（无定形的碳粒子、金刚石粒子）。

6.2 富勒烯

Kroto 等人[3]在 1985 年发现了富勒烯，1990 年，Kratschmer 等人[4]对其进行了大量的合成。由于富勒烯独特的物理和化学性质，这种碳的同素异形体已经引起了人们极大的兴趣。

6.2.1 C_{60}的合成

富勒烯可以从芳香烃燃烧[5,6]或芳香烃热解[7~12]，或电弧放电过程[3]的烟灰中提取。

（1）电弧法

电弧放电法经由一个温度可达 3700℃的高温等离子体过程，可产生出大量的碳纳米结构、如富勒烯、晶须、烟灰以及高度石墨化的碳纳米管[13]。

在充有一定 He 或 Ar 气的系统中，通过两个碳电极间的电弧放电，使碳电极（一般用石墨）蒸发而制备富勒烯，如图 6.2 所示。在电弧-放电法中，最好的惰性气氛是在系统中充以 100～200Torr 的高纯 He。直径小于 5mm 的电极棒电流负荷为 100A。交流或直流放电都可使用。阳极的消耗可通过自动加载棒连续地供给阳极。

（2）激光法

图 6.3 为发现 C_{60}时使用的实验装置[14,15]。利用激光对一个旋转的石墨盘进行辐射，使石墨蒸发。用 He 气流冷却蒸发的碳物种以形成碳簇。从喷嘴喷出的簇束进入真空而膨胀。然后，通过时间飞行质谱仪对碳簇进行检测。该套装置在商业上可用于各种簇的研究。

（3）C_{60}单晶的合成

用于合成单晶富勒烯的气相生长法，通常使用一个水平的石英管（一般长约 50cm，直径 1cm），石英管放置在具有温度梯度的炉子中，如图 6.4 所示[16~19]。将 C_{60}粉末放入金舟中，并在 250℃的真空炉中加热 6～72h，以去除粉末中夹杂的溶剂。然后将 Au 舟加热到 590～620℃（大于 C_{60}的升华温度），同时以小于 5mL/min 的流速通入高纯的干燥氦气，以带走 C_{60}蒸气至石英管中较冷的部分。该方法的生长速率为每天 15mg。气相传输过程运行数天，以获得较大的 C_{60} 晶体。基于蒸发传输生长方法的一些改进技术，已经发展到生长更大的和更完美的 C_{60}晶体[20~26]。

6.2.2 富勒烯的提纯

在真空腔内壁的水冷表面上可收集到生成的烟灰。在甲苯或苯溶液中，采用索氏提取器萃取 5～6h 后，形成一种深红褐色含有富勒烯的混合物溶液。用这种方法收集的烟灰，20%～30%的是可溶的。对这些可溶的材料进行层析分离，如图 6.5 所示[27,15]。

液相层析是富勒烯提纯的主要技术，这里的提纯是指对富勒烯中的C_{60}和C_{70}等进行分离。

多年以来，已发现了一些简单的方法，包括通过活性炭-硅胶柱的过滤技术。以甲苯为流动相，收集到的溶液中 80%的可溶物是 C_{60}。通过使用甲苯/邻二氯苯的混合液作为洗脱液可分离 C_{70}，要获得纯度高的 C_{70}，反复的层析是必不可少的。C_{60}的溶液是紫色的，而 C_{70}溶液是红棕色的。更大的富勒烯如 C_{76}、C_{78}、C_{82}分离提纯时要求高效液相色谱（HPLC）。人们已经了解了这些富勒烯中的一些光谱性质。通常情况下，从石墨中制备 1g C_{60}需要约 5h。但要制备 1mg 分子量更大的富勒烯需 250h。现在，C_{60}市场上有售，1g C_{60}约为 25 美元，而 C_{70}还无法买到。因此，所报道的富勒烯主要是关于 C_{60}的研究。

6.2.3 C_{60}的结构

C_{60}存在 1812 种可能的结构[28,29]，但其中最稳定的异构体是由 12 个五边形和 20 个六边形组成的 32 面体的笼状结构,如图 6.6 所示。

C_{60}是一种芳香分子，如图 6.6 和图 6.7 所示[30]。富勒烯结构中，所有的五元环被六元环所隔离，形成五边形不相邻的稳定结构[31,32]。一个 C_{60} 分子的平均外径约为 1nm[30,33]。

6.2.4 ^{13}C 核磁共振谱

Taylor 等人[34]以及其它的研究小组[35,36]通过使用 ^{13}C 核磁共振（NMR）谱，在 C_{60} 分子的溶液中获得了它的结构数据。在图谱中仅有一个峰，相对于四甲基硅烷（^{13}C 核磁共振谱的标准参考分子），其化学位移为 143.2，表明 C_{60} 极其简单的结构特征。仅有一个峰的特征表明在溶液环境下，C_{60} 分子中的每一个碳原子都是等价的。与 C_{60} 分子相比，C_{70} 分子具有 5 种等价的碳原子，分别对应的化学位移为 130.9、145.4、147.4、148.1 和 150.7，比例关系为 10:20:10:20:10，如图 6.8 所示[34]。

6.2.5 富勒烯包合物

一些重要的例子是将 N 或 P 原子封装在富勒烯的碳笼中，这些化合物称为富勒烯包合物，如图 6.9 所示。化学式表示为 N@C_{60}。在高温（600～1000°C）和高压下[40000 psi（1psi=6894.76Pa,下同）]，可以把多种惰性气体（He, Ar, Xe）封装在碳笼中。

具有代表性的金属富勒烯如图 6.9 所示，例如，Sc_2@C_{82} 和 Lu_3N@C_{80}，应当提及的是还没有制备出过渡金属（如 Co，Fe 和 Ni）富勒烯包合物，而这些金属是 Kratschmer–Huffman 电弧法制备纳米管时常用的催化剂。由至少 80 个碳原子组成的笼的一个重要的特征是笼的内径约 0.8nm，可以容纳由四个原子组成的分子簇。

6.2.6 亲核加成反应

富勒烯第二重要的特点是在两个六元环连接处的碳原子具有高的反应活性。这样，富勒烯最常见的反应之一是 Bingel-Hirsch 型的亲核加成反应，如图 6.10 所示。图 6.11 是 C_{70} 相应的加成反应。图 6.12 是 C_{60} 树枝状衍生物的合成[37]。

6.2.7 C_{60} 的聚合反应

近年来，各种温度和压力条件下的压力诱导聚合反应已经得到了深入的研究[38]。

在施加适当的压力和高温条件下，合成了数种聚合态的富勒体[39~41]。并可将具有一维或二维的 C_{60}-C_{60} 网状结构分为不同的三相：一维正交晶相，二维四方相和二维的斜方六面体相，如图 6.13 所示[41,42]。

6.2.8 纳米车的制造

由于具有高度的对称性，富勒烯（C_{60}）结构显示出可以在各种表面上滚动的特性。通过轮状的富勒烯的转动，设计和合成的纳米车分子可直接在可控的表面上跑动，如图 6.14 所示[43,44]，图 6.15 是纳米车在 Au（111）晶面上滚动的扫描隧道显微镜图片。

6.3 碳纳米管[45]

6.3.1 碳纳米管的合成

Iijima 和他的同事首次在碳电弧的阴极上发现了沉积的碳纳米管[46,47]，从此之后，有多种方法用于碳纳米管的合成。当前，制备碳纳米管有多种工艺，如电弧-放电法、烃的催化热裂解法、激光蒸发石墨、金属盐的电解。所有这些工艺很难在此详细讨论。在此，仅列举几个。

碳纳米管是由一个或多个同心轴结构的卷曲的石墨片组成。一个单壁纳米管的直径约为 1nm，而多壁纳米管的直径在 2～100nm 之间。

（1）电弧-放电法

为了制备多壁纳米管，可在一个真空腔中，通过在两个纯的石墨电极棒之间通过直流放电来实现，如图 6.16 所示[48]。当然，两个石墨棒可以是垂直排列，也可以是水平排列。利用一个油扩散泵，当真空度大于 10^{-5} Torr 之后，将一种惰性气体通入腔体中。

然后，在两个石墨电极之间施加约 25V 的直流电弧电压。当两电极棒不断靠近时，放电发生的瞬间形成了等离子体。在碳的阳极上钻出一个小孔，并填充金属催化剂的混合物及石墨粉。在这种情况下，在烟灰中形成的大部分是纳米管，沉积在电弧腔体内壁上。包含有 Co、Co-Ni、Co-Y、Co-Fe、Ni、Ni-Y、Ni-Lu 和 Ni-Fe 催化剂的阳极蒸发，在阴极上沉积形成像花边领或软带一样的物质。

利用高分辨率的透射电镜观察分散在 TEM 微网格上的阴极沉积物，发现存在一种具有螺旋状排列的卷曲的石墨片同轴结构[49]，如图 6.17 所示[46]。

层间距与石墨的（002）晶面的点阵间距相同，均为 0.34nm。四壁碳纳米管的示意原子模型如图 6.18 所示，它具有螺旋排列和锯齿形的结构[49]。

（2）催化辅助热裂解法

在碳纳米管和富勒烯发现以前，利用烃的催化热裂解（化学气相沉积）制备碳纤维[50]。在这种方法中，惰性载气携带化学前躯体穿过一个具有一定温度的石英管，有机前躯体分解，产物沉积在铝盘中。这种方法可用于纳米管的大批量制备。此外，也可使用两个加热炉，一个用于热解反应，另一个用于前躯体的气化[51,52]。

水平炉是制备碳纤维和碳纳米管最常用的装置。最简单的方式是用于加热盛有基质材料和催化剂的石英管。反应物气体从载有催化剂的基质材料上流过，载有催化剂的基质材料放置在石英管中心可移动的瓷舟内，如图 6.19 所示[53]。水平炉的优势在于加热带没有（或很小的）温度梯度。在大多数情况下，纳米管/纳米纤维的长度可由沉积时间进行控制。

Sharon 和他的研究小组[54]尝试采用天然前躯体的热解合成碳纳米管。他们小组已经从天然的前躯体如松节油和樟脑中合成出各种形式的碳纳米管，如图 6.20 所示[54]。

（3）激光消融

激光技术已成功用于碳纳米管的生长。单壁碳纳米管的产量取决于消融碳靶材的激光束的功率、辐射时间、靶材的组成、惰性气体的压力及流速以及靶材的温度。Wolfgang 等对碳纳米管的制备做了详尽的综述[55]。表 6.1 给出了使用激光消融法合成纳米管（单壁纳米管和多壁纳米管）的一些标准条件。激光蒸发技术的示意图如图 6.21[55]和图 6.22[55]所示。

6.3.2 生长机理

一般而言，通过催化化学气相沉积法生长碳纳米管和纳米纤维，需要纳米微粒催化剂（如 Fe，Co，Ni）、碳原料（如烃或 CO）及加热。生成的细丝的直径通常与金属催化剂的物理维度密切相关。这些过渡金属特殊的性能与它们分解易挥发性碳化合物的催化活性有关，形成亚稳态的碳化物，生成的碳通过金属粒子进行扩散[56]。现在提出的一些有关纳米管和纳米纤维的生长模型已得到各研究组的普遍认可[57~61]。

Baker 等人[57]在 20 世纪 70 年代早期，提出了最被普遍认同的机理。通过碳源的催化分解，他解释了碳纳米丝的生长过程。根据这个机理（如图 6.23），气态烃在金属催化剂与其接触的表面上分解为氢气和碳，并溶入催化剂粒子，溶入的碳通过催化剂颗粒扩散，并且在催化剂颗粒的尾端形成碳纳米丝沉淀。由于烃分解放热，所以催化剂粒子存在着温度梯度。

碳在金属中的溶解性取决于温度，过量的碳沉淀将会出现在粒子后面温度较低的区域，这样，就使得固体丝状物以与催化剂粒子相同的直径方式生长。直到催化剂的顶部中毒或失活，纳米丝的生长才会停止。

基于催化剂和载体之间的相互作用，Baker [58] 和 Rodriguez[62]提出了两种不同的生长模型[如图 6.24（a）和（b）]。载体和催化剂纳米颗粒之间的相互作用可通过在生长温度中的接触角来表征，类似于“疏水的”（弱相互作用）表面和“亲水的”（强相互作用）表面。例如，在 700℃时，Ni 与氧化硅（SiO_2）基质表面有很大的接触角（弱相互作用），因而，这种体系有利于顶部生长[63]，如图 6.24（a）所示。另一方面，据报道 Co 和 Fe 在硅表面有利于底部生长[64,65][如图 6.24（b）所示]，表明在 Co 和 Si 或 Fe 和 Si 之间存在很强的相互作用。因此，催化剂与其载体之间的相互作用是生长模型中的一个重要因素。

6.3.3 碳纳米管的几何构型

碳纳米管的电学性能尤其依赖于纳米管的几何构型。通过沿着对称轴方向或其它方向卷曲一叶石墨片，就可分别获得锯齿形、扶手椅形和手性结构的纳米管，如图 6.25 所示[66]。

圆柱形的管状结构可通过管径 d 和手性角 θ 来描述，如图 6.26 所示[67]。手性矢量 $\boldsymbol{C}=m\boldsymbol{a}_1+n\boldsymbol{a}_2$ 代表纳米管，单位矢量 $\boldsymbol{a}_1$ 和 $\boldsymbol{a}_2$ 代表石墨片。在一个石墨片的平面（一个石墨单片）上，碳原子以六边形结构排列，每个碳原子与相邻的三个碳原子相连。

碳纳米管有许多卷曲方式。当手性角 $\theta=0°$时，纳米管称为“锯齿管”，当 $\theta=30°$时，所卷曲的纳米管称为“扶手椅形纳米管”，这两类纳米管具有高的对称性，且对称面垂直于管轴。当 $0°<\theta<30°$，沿任何角度卷曲的纳米管都属于一个手性管，它们的手性可以是左手型，也可以是右手型。这些手性纳米管具有偏振光光学活性。

碳纳米管是一层卷曲的石墨片，碳纳米管的电子结构类似于石墨烯的结构。你可以想象一种展开的、铺平的纳米管是受周期性边界条件制约的石墨烯的手性矢量，这就是所谓“带折叠”。碳纳米管的电子结构可以视为是石墨烯电子能带结构的一种“带折叠”形式。当这些碳纳米管波矢量的平行线通过某一角度，形成的碳纳米管呈金属性。否则，碳纳米管成为具有带隙大约为 1eV 的半导体，并随管径的增加带隙减小。用这种简单的方法判定（在 $\boldsymbol{C}=m\boldsymbol{a}_1+n\boldsymbol{a}_2$ 的矢量中），如果 $n-m=3\times$整数，可以得出所有的扶手椅型碳纳米管呈显金属性。这样，碳纳米管的电子结构由它们的手性和管径所决定，即就是由它们的手性矢量 $\boldsymbol{C}$ 所决定。

碳纳米管由于质量轻、有弹性、坚硬以及理想的长径比，而具有更大的诱惑力。在具有很强共价键的碳纳米管中，管的形状以及碳原子网络使得碳纳米管成为很特殊的材料。一些计算及测量结果显示了碳纳米管的抗拉强度是钢的 300 倍，其硬度可与金刚石相媲美。

在微电子学中，目前的倾向是要制造体积更小，运算速度更快的器件，由于碳纳米管新颖的、奇特的机械和电子性质，成为满足纳米技术要求的潜在的候选者。碳纳米管的一个最有前景的应用是在场发射设备中作为电子发射器。目前的实验显示，在低电场，碳纳米管具有高电流密度的场发射性能。它们的发射特点及耐久性比其它的电子发射器更优良。碳纳米管最具有前景的应用是作为平板显示器以及微波功率放大器。

6.4 石墨烯

目前对石墨烯研究的热潮可以追溯到 2004 年诺贝尔奖得主 Andre Geim 和 Konstantin Novoselov 发表的论文[68]。作为碳的一种同素异形体的石墨烯 (GR)[69~71]是由碳原子组成的二维蜂窝状网络结构的平面。石墨烯在三维方向上堆叠时形成的材料被称为石墨，人们可以对石墨进行一层一层的剥离。当石墨烯沿某个方向卷曲成圆柱形状或卷曲成一个球（即：由六边形和五边形组成的英式足球）时，这两种碳的同素异形体分别称为碳纳米管及巴克球，如图 6.27 所示[69,72~75]。

6.4.1 石墨烯的性质

石墨烯是一种高度各向异性的材料，一个碳原子在二维平面上通过 sp^2 杂化轨道与三个相邻的碳原子键合形成蜂窝状的晶格网络，原子间距为 0.142nm。沿着石墨晶体的 z-轴，石墨烯中层与层之间是由碳原子的 p 轨道的弱的范德华力所维系的，其层间距为 0.335nm。π(pi)电子位于石墨烯层的上方和下方，并且相互重叠，增强了 C—C 键。

具有商业价值的石墨烯称为石墨烯纳米片，是来自于多层石墨烯堆积形成的晶态或片状的石墨（图 6.28）[76]。然而，人们也已经制备出了单层和双层的石墨烯[77]。

石墨烯的电子迁移率大于 $15000cm^2·V^{-1}·s^{-1}$，而理论值为 $200000cm^2·V^{-1}·s^{-1}$（由于石墨烯中声光子的散射限制）[71]。石墨烯是以空穴和电子作为载流子的一种零重叠的半金属材料。这些载流子能穿过亚微米的距离而不发生散射，这一现象称为弹道传输。然而，石墨烯质量及其衬底会成为限制因素，若以 SiO_2 作为支撑石墨烯的衬底，电子迁移率可能被限制在 $40000cm^2·V^{-1}·s^{-1}$。由于石墨烯 0.142nm 长的碳碳键的强度，石墨烯是最强的材料，它的抗拉强度为 130GPa，其强度是同等厚度钢材的 100 倍以上。石墨烯同样也非常轻（$0.77mg·m^{-2}$），并且具有弹性，在受力之后还能恢复到初始的尺寸。

石墨烯片（厚度 2～8nm）的弹簧常数为 $1～5N·m^{-1}$，杨氏模量（与三维的石墨不同）为 0.5 TPa。单层石墨烯的比表面积约为 $2630m^2·g^{-1}$[78]。

由于上述提及的石墨烯的电性，石墨烯可以吸收 2.3%的可见光。石墨烯具有独特的光学性质，它的带隙值约为 0～0.25eV [79]。它的导热系数大于 $5000W·m^{-1}·K^{-1}$，远远高于其它结构的碳材料，如碳纳米管，金刚石，甚至石墨 ($1000W·m^{-1}·K^{-1}$)。少于 6000 个原子的石墨烯是不稳定的 [80]。尽管石墨也表现出优良的电导率和热导率，但它们是高度各向异性的，即沿着晶体的不同轴，其电导率和热导率不同。

石墨烯具有高的疏水性，然而，一些溶剂，如：二甲基甲酰胺，*N*-甲基-2-吡咯烷酮，二甲基亚砜和γ-丁内酯对于防止石墨烯片之间重新堆叠是有益的[81]。由于氧化石墨烯（GO）存在含氧基团，故其在水中是相对稳定的。其结果，氧化石墨烯（GO）的层间距从石墨烯的 0.335nm（石墨烯的层间距）增大到 0.6nm。与石墨烯相比，氧化石墨烯层间的结合强度要弱得多，超声波用于分离氧化石墨烯片层很有效 [82]。

6.4.2 石墨烯的合成

由于石墨烯是石墨的子单元，所以，最早和最简单的方法就是从块体石墨直接获取石墨烯。应该提及的是，并非所有的石墨性质都相同。石墨分为两种：天然石墨和合成石墨。质量最好的天然石墨具有面内尺寸超过 1mm 的单晶畴。所以，由天然石墨获得的单层石墨烯片具有特殊的晶体品质[83~84]。然而，许多石墨烯的研究工作采用的是人造石墨，即高度有序的热解石墨(HOPG)和 Kish 石墨（注：Kish 石墨是一种特定结构的原生石墨，在被碳饱和的铁液中，石墨以层状生长方式结晶析出。石墨集结于铁液表面，或与渣混合存在于渣层，或离开表面漂浮到四周）。

通过一定压力下烃的热分解制备高度有序的热解石墨 (HOPG)。通过该工艺获得的石墨晶体远厚于天然石墨。较大的尺寸使材料更容易处理和剥离，然而它的面内晶畴远小于天然石墨，尺寸为 1μm 左右。Kish 石墨来自于熔融钢中碳的分步结晶，其晶体的性质介于高度有序的热解石墨（HOPG）和天然石墨之间[84]。这两种石墨烯中的面内晶界影响电子和声子的传输。因此，由这种材料组成的器件更容易制造。权衡之下，由这些石墨烯制造的器件的品质比天然石墨烯（来自天然石墨）制造的器件的品质低得多。晶畴中的缺陷直接降低了理论上预测的石墨烯的最优性能，这就成为合成石墨烯和石墨烯器件的一个主要的工程挑战[85]。其重要性不仅仅在于利用石墨制备石墨烯，而且也体现了利用非石墨制备石墨烯。目前，大部分关于合成石墨烯的研究都集中在获得大面积单晶石墨烯，或者，至少是增加晶畴尺寸和减少晶界数量，其目的是制备出性能接近理论预测值的石墨烯。的确，最近的研究结果表明，采用化学气相沉积法制备的石墨烯结晶度已经超过了天然石墨，约为 1cm 宽的晶体。

（1）微机械剥离法

微机械方法本身简单，不需要专门的设备。将一片胶带置于要剥离的石墨的表面，然后剥掉石墨样品的表面。粘在胶带上的石墨薄片沿着晶体平面的方向优先断裂，留下裸露的原子量别的平坦表面[84]。为了得到少层和单层的石墨烯，将另一个清洁的胶带紧紧压在粘附有石墨薄片的胶带上。撕开这二片胶带，石墨就会进一步被劈开成更薄的石墨片。这个过程不断重复，每次操作都会得到更薄的石墨片。

被称为剥离的自上而下的方法是一种机械的、化学的或者电化学的方法，以破坏石墨中石墨烯层间的范德华力（图 6.29）。然后，可继续将石墨片置于 Si 衬底的 SiO_2 基质上，对石墨烯片加工或制造器件可继续进行（图 6.30）。微机械剥离的耗时步骤可获得单层或少层的石墨烯片。将剥离片置于 300nm 厚的 SiO_2 薄膜上以提高光学对比度[68]，可通过使用光学显微镜来实现。

自从第一篇关于石墨烯的论文发表后，许多其它的研究团队开始利用微机械剥离的方法分离得到石墨烯，并且利用它来制造器件。由于对石墨烯的兴趣和对石墨烯需求的增长，以及它作为一种新材料的潜力和可能对电子产业带来的变革，皆已凸显，研究人员已经意识到了这种简单的方法不能满足未来的需求。因此，其它制备石墨烯的方法也应运而生。然而，为了获得表面更加洁净的高质量石墨烯，微机械剥离仍然是制备石墨烯的主要方法。

由于微机械剥离这一十分简单的方法已经吸引了很多人的注意力，尽管有数十亿美元的巨额研究投入和全球科学政策措施的倡导，但一个仅仅用了些胶带和铅笔的实验竟然产生了诺贝尔奖[88]，但是，很明显，采用胶带的方法制备大量实际应用的石墨烯效率太低了。

（2）液相机械剥离法

在通常的实验室条件下，除了富勒烯，大部分碳元素的存在形式都是完全不溶的。石墨烯的研究者们采用一些液相方法从块体石墨剥离单一的石墨烯片。在一项研究中，科学家将石墨粉在 *N*-甲基吡咯烷酮中进行超声处理，并且离心生成物以去除大量未剥落的石墨片。他们制备了含量为 1%的少层石墨烯的分散液，所得分散液显示了非凡的稳定性，数周至数月后才出现聚集[89]。表面活性剂辅助剥离石墨烯是另一种比较完备的技术，已经应用了许多表面活性剂进行这一研究[90,91]。如通过表面活性剂胆酸钠增强石墨烯在水中的稳定性，其浓度为 0.3mg/mL。实验中利用了长时间的超声处理（长达 400h），随后离心以获得稳定的分散液。分散液的浓度随超声时间而增加，最高质量的分散液是在离心率为 500～2000rpm 之间获得的，采用该方法规模化液相生产石墨烯，具有广泛的应用前景。

溶剂辅助和表面活性剂辅助超声方法已经用于制造器件的超薄碳膜高效生产。但是，即便使用了液相剥离，能够实现的少层石墨烯的最大浓度低于 0.1mg/mL。

（3）化学气相沉积法（CVD）

CVD 生长石墨烯通常是在加热的铜和镍表面上进行的，在这两种金属上的热生长过程颇为相似。使用多种技术通过 CVD 法在镍箔上合成了高质量的少层石墨烯片。在氢气与氩气体积比为 1:10 的混合气体连续流动下，将镍箔（或铜箔）暴露在 900～1000℃氩气环境中，其中的少量氢气用以保持还原气氛，而氩气作为载气为了微调生长石墨烯实验的压力[92]。将甲烷通入氢-氩混合气中，作为碳源的甲烷被基质（镍或铜箔）所吸收。镍-碳固溶体在氩气气氛中冷却，在冷却过程中镍箔上扩散出来的碳会形成石墨烯膜[93]。在非常低的压力下以铜箔作为基质时，石墨烯在形成单层后会自动停止生长[94,95]，如图 6.31 所示。单层生长也是由于甲烷中低的碳浓度，而烃类如乙烷和丙烷则会生长出双层石墨烯片[96]。在常压下，与镍箔类似，CVD 法在铜箔上生长出的是多层石墨烯片[97]。

（4）石墨烯纳米带的化学合成

合成石墨烯最有趣的新方法之一是使用化学技术进行聚合反应，并且，对简单的芳环前驱体进行环化脱氢，从而获得原子水平的石墨烯纳米结构（图 6.32）。这种方法最早由 Mullen 和 Fasel 团队倡导，他们把蒽和苯并菲的衍生物暴露在一个热的金表面，使其在合理设计的几何结构下，组装石墨烯纳米带[99]。最新进展是利用纳米构建功能性器件，以及合成可溶性的和杂原子取代的纳米带。这种在合成化学家和材料科学家之间的跨学科的努力具有很大的前景。自下向上的技术是在原子水平上控制合成纳米材料。现在，这些合成技术的主要缺点是它们仍处于起步阶段，但是，这个研究方向上令人兴奋的进展将会很快出现。

（5）氧化石墨的还原

1859 年，在尝试测量碳原子量的实验中，Benjamin Brodie 首次进行了一些与石墨相关的化学反应[100]。他注意到把石墨反复地浸泡在硝酸和氯酸钾的混合酸中数天后，大量的石墨被氧化，在实验结束时，他得到了“由微透明且明亮的片组成的一种浅黄色的物质”。他继续评价了这种他称之为“彩色酸”的物质的很多性质，也就是我们今天说的氧化石墨（GO）。近一个世纪之后，Hummers 和 Offeman 率先利用石墨，硝酸钠，硫酸和高锰酸钾的混合物高效安全的制得了氧化石墨[101]。Hummers 法仍是实验室制备氧化石墨常用的方法。

氧化石墨（GO）可以通过使用强氧化剂化学剥离石墨而制得，例如：使用发烟硝酸[102]。然而，高度氧化的氧化石墨（GO）应该使用浓硫酸和发烟硝酸的混酸制备[103]。一种替代方法是在浓硫酸中加入高锰酸钾和硝酸钠[104]。一般情况下，在氧化过程中，连接的含氧官能团使氧化石墨（GO）受到结构破坏和出现缺陷，如图 6.33 所示。

氧化石墨(GO)的还原（rGO）通过使用还原剂很容易进行。例如肼还原剂。石墨烯可溶性碎片可通过常用的湿化学方法在硫酸和硝酸的混酸中对微晶石墨进行处理制备，一系列的氧化和剥落步骤生成了在其边缘连有羧基的小石墨烯片[105]，如图 6.33 所示。

后来，通过生物耦合方法将羧基、羟基、磺酸基及氨基等官能团引入了石墨烯[106~108]。考虑到其优良特性，石墨烯已经应用在不同的领域，如纳米电子学、光电学、化学和生化传感、聚合物复合材料、制备 H_2 和储 H_2、插层材料、药物释放、超级电容器、催化和太阳能光伏等[109~114]。

当今，纳米碳材料，包括富勒烯、碳纳米管和纳米孔性碳材料作为 21 世纪的重要材料，由于它们独特的形貌和纳米尺寸，已经广泛应用于工业、环境、能源、IT 和其它领域，如图 6.34 所示[115]。碳材料固有的优势在于碳原子具有与其自身或与其它元素的原子以不同方式结合的非凡能力，在 21 世纪将得到蓬勃发展。

碳材料面临的挑战是如何利用它们固有的力学、热学和电学性质。碳结构的裁剪就是利用 sp^3、 sp^2 和 sp 杂化或六方碳结构，在纳米尺寸上控制碳材料的物理化学性质，其将成为获得具有新颖的和特异功能的新型碳的核心技术。正如一些科学家们所说的那样“碳现在是热点”，目前，科学家们正在掀起碳材料研究的热潮。

可以这样说：“21世纪将是碳的世纪”。

复 习 题

1. 请指出碳的同素异形体以及它们的空间维度。
2. 什么是富勒烯和足球分子？
3. 请说出 C_{60} 的笼结构。
4. 如何使用激光消融法制备 C_{60}？
5. 请简略描述利用电弧放电法制备 C_{60} 的工作原理。
6. 请举例说明什么是富勒烯包合物？
7. 哪种材料可用于制备纳米车的车轮？
8. 什么是单壁碳纳米管？什么是多壁碳纳米管？
9. 如何使用催化热裂解法在水平炉中制备碳纳米管？
10. 简单描述碳纳米管的生长机理。
11. 请叙述使用石墨片卷曲成三种碳纳米管的几何形状。
12. 为什么有人说“21世纪是碳的世纪”？

7. Lithography for Nanofabrication

The nanofabrication has attracted great interest recently due to the potential applications of miniaturization of devices. In general, nanofabrication can be divided into two classes, i.e., bulk nanofabrication and nanoscale patterning. The bulk nanostructure synthesis is important for many applications such as bulk synthesis of nanoparticles, nanotubes, and nanowires, whereas nanopatterning is critical for the production of functional devices. Recently, self-assembly has been used to aid the rational superassembly of two-dimensional (2D) and 3D structures, but traditionally, surface patterning methods such as photolithography have been used to create functional devices. Due to the diffraction limit of light, photolithography is unable to produce the structures in the nanoscale (<100nm) regime. Several new fabrication techniques, including electron-beam lithography, soft lithography techniques, and scanning probe lithography (SPL) techniques have been invented to facilitate the fabrication and patterning of nanostructures.

As the scale of integrated circuits continues to diminish, the scale of a few nanometers will approach in future. It is of great importance to design next generation nanoelectronic devices based on a "bottom-up" approach, namely precise control of the atomic processes on surfaces. Thus, fabrication of low-dimensional structures such as nanoclusters [1], nanowires [2], and nanopatterns [3] has been intensively studied for decades. Technique of these nanostructures through self-organizations is of particular interest to produce uniform structures on the surface of macroscopic material [4].

7.1 Microfabrication by photolithography of ultraviolet light

The silicon integrated circuit is surely one of the wonders of our age and has changed virtually every aspect of our lives. Due to the good mechanical characteristics of silicon, the integrated circuits technology (ICT) appears to be useful for the fabrication of a variety of structures. Silicon micromachining can be subdivided into two main categories: bulk and surface micromachining.

Bulk micromachining refers to the realization of structures out of the single-crystal silicon wafer. Surface micromachining is the techniques fabricated on the surface of single (or poly) crystal silicon wafer and consists of a stack of thin films between 100nm and 10μm thick.

The hallmark of the ICT is assembly of tens of millions individual components on a silicon chip with an area of a few square cm using lithographic techniques as shown in Fig.7.1 [5]. The rapid advances in the number of transistors per chip and the functionality of microel- ectronic circuits are truly phenomenal, and have led to integrated circuits with continuously increasing capability and performance.

Fig.7.1 Integrated circuits on a wafer [5]

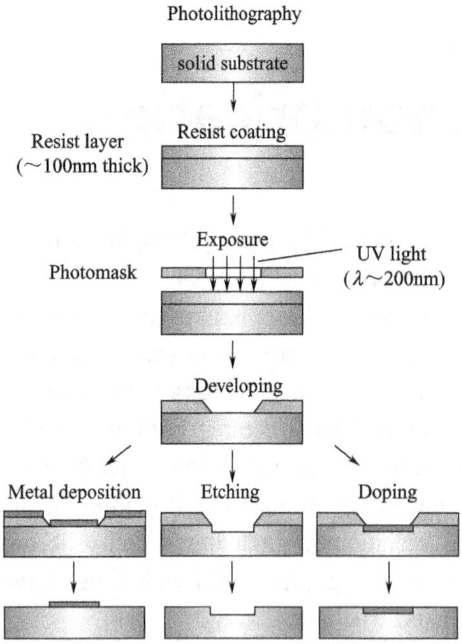

Fig.7.2 Schematic diagram depicting the basic steps of photolithography [6]

Photolithography is a process of using light to create a pattern. Fig.7.2 shows the basic steps of the photolithography technique [6]. In this process, solid surfaces are first coated with a polymer resist layer, and then a specific region on the resist layer is exposed to the ultraviolet light (wavelength about 200nm) through a photographic mask. The molecular bonding was broken (or strengthened) when polymer resist layer was exposed to ultraviolet light. The exposed (or unexposed) resist layer is specifically removed by immersing the substrate into certain solvents, and the polymer resist patterns are left on the specific area of the substrate. The polymer resist pattern can be used as a resist of subsequent etching or other processing steps. By repeating these processes, many complicated microelectronic components can be created on a single substrate [7].

During the last four decades, the semiconductor industry has been extensively working on improving the resolution of photolithography technique so that more components can be integrated in a single chip. This size reduction and higher degree of integration are crucial to improve the performance and power efficiency of the integrated circuits, and effectively reduce the manufacturing cost. Currently, microelectronic chips with feature sizes as small as 90nm are commercially available.

Fig.7.3 shows photolithography of positive and negative photoresist [8]. A polysilicon substrate (wafer) which is used as the ground plane is chemically cleaned to remove particulate matter as well as any traces of organic, ionic, and metallic impurities on the surface. Fig.7.3(a) shows that a thin isolation layer, such as silicon dioxide, is deposited on the surface of polysilicon substrate. Then, a photoresist which should be sensitive to UV light is coated over the SiO_2 layer in Fig.7.3(b). Organic polymer is usually chosen as a photoresist.

It is required that some parts of the silicon dioxide are to be selectively removed and other is to be remained in particular areas on the substrate. For this we need to utilize a mask. The photomask used in the photolithographic phase is a key component. A typical mask is a glass plate that is transparent to ultraviolet (UV) light. The pattern of interest is generated on the glass by depositing a very thin layer of metal, usually chromium or gold. These masks are capable of producing very high quality images of micron and even sub-micron features in Fig.7.3(c). The photomasks are normally prepared by the help of a computer-assisted software platform. Following step is that the mask is placed over the layer of photoresist (PR) in Fig.7.3(d).

Then, UV light is allowed to fall on the mask. The source of UV light is a mercury arc lamp, called a radiator, which has an energy output with spectral peaks at particular wavelengths. Depending upon the complexity of feature, thickness of photoresist layer and property of photoresist materials, the exposing process may be performed once or several times. Therefore, accurate dose of UV light is paramount. When UV light falls on the photoresist in Fig.7.3(e) through

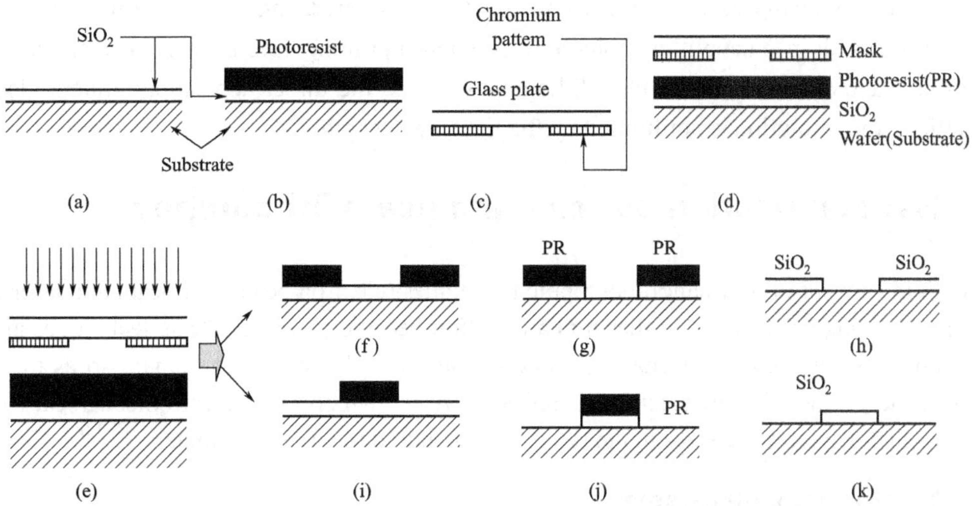

Fig.7.3 Photolithography. (a) A substrate with SiO_2 layer; (b) Additional photoresist layer; (c) The mask—glass plate with chromium pattern; (d) The mask is on the top of the photoresist-SiO_2-wafer layers; (e) UV light falling on the photoresist (PR); (f) After exposure to UV light (positive PR); (g) Third phase of photolithography in which the opening portion of the SiO_2 layer is removed (positive PR case); (h) Final phase of photolithography in which the PR is removed (positive photoresist); (i) After exposure to UV light (negative PR); (j) Third phase of photolithography in which the opening portion of the SiO_2 layer is removed (negative PR case); (k) Final phase of photolithography in which the PR is removed (negative PR) [8]

the mask, a pattern that is similar to the mask is developed on the photoresist layer. This phenomenon is called mask transformation or pattern transformation. The transformation is thus achieved by the combined effect of UV light and the composition of the photoresist that can be either soluble or insoluble after being exposed to UV light in Fig.7.3 (f) and (i). Accordingly, there are two types of photoresist, positive and negative photoresists, respectively. On the one hand, the positive photoresist becomes soluble when exposed to UV light. On the other hand, the opposite occurs to insoluble photoresist. In practice, the positive photoresist is washed away in the region where it was struck by UV light; conversely, the negative photoresist is not. This assures that the resist, which was not exposed to UV light, is washed away forming a negative image of the mask. The process of washing away material is actually called etching.

There are two major parameters to ensure the ultimate resolution of the photolithography: ① wavelength of light and ② thickness of the resist layer. It is usually believed that resolution of the photolithography is approximately $\lambda/2$. Extensive efforts have been paid to use shorter wavelength light so as to improve the resolution. Ultraviolet light with a wavelength of about 160nm has been employed to fabricate~90nm commercial chips. However, it is increasingly difficult to find a proper optics component for shorter wavelength light. On the other hand, the minimum size is also affected by the thickness of the resist layer. High-resolution lithography resist layers are usually thinner than 100nm.

Considering two parameters, the resolution limit of the photolithography technique is believed to be around 100nm. Extensive efforts have been given to develop a next generation lithographic technique that can overcome the resolution limit of photolithography and continue to decrease the size of electronic chips. Possible candidates include X-ray lithography, electron beam lithography, scanning probe-based lithography and soft lithography [9].

These methods demonstrated lithography resolution down to 10nm. However, it is not yet clear which method will eventually replace the photolithography technique. Modern nanodeposition methods such as dip-pen nanolithography or microcontact printing have been extensively developed as possible candidates instead of photolithography. Since the direct deposition strategy does not rely on light, the resolution is not limited by the light wavelength.

7.2 Nanofabrication by scanning beam lithography

With the advent of new nanolithographic techniques (electron beam, ion beam and laser beam, nanoimprint lithography, etc.), further miniaturization became possible. These techniques allowed scaling down of microelectromechanical systems into the submicrometer domain so as to open a door for a new generation of electromechanical devices, namely, nanoelectromechanical systems (NEMS) which have feature sizes in the range from hundred to a few nanometers [10~12].

7.2.1 Electron beam lithography

Electron beam lithography is a standard process to create nanometer-scale structures. The resolution is generally much higher than optical lithography, well below 100 nanometers. The pattern is achieved by a high energy electron beam focused on a thin layer of resist. The resist layer is then developed in a suitable solution, in which either the exposed areas or the unexposed areas are selectively dissolved [13].

Electron beams can be used to write directly onto a substrate using high energy electrons (100keV to 200keV) and a small electron probe size (1nm to 10nm). With electron beams the creation of lines typically 30nm wide can be satisfactorily produced (widths up to 7nm have been reported). The resolution limiting in electron beam lithography strongly depend on the resist properties. Due to its flexibility in creating patterns, Electron-beam lithography is widely used in research or small volume production specialities, and it is the primary technique for mask fabrication. The disadvantages of this technique are the large writing times, causing a low throughput.

Electron resists are polymers. Its behavior is similar to that of a photoresist; that is, a chemical or physical change is induced in the resist by radiation. This change allows the resist to be patterned. Common positive electron resists are polymethyl methacrylate, called PMMA, and poly(1-butene sulfone). Positive electron resists have resolutions of 0.1 micrometer or better. Poly(glycidyl methacrylate-co-ethyl acrylate), called COP, is a common negative electron resist. COP, like a negative photoresist, also swells during development so the resolutions are limited to about 1 micrometer.

7.2.2 Focused ion beam lithography

Focused ion beams (FIB) have become a popular tool for surface modification of materials and functional structure prototyping at the micro- and nanoscale. Focused ion beams with a diameter less than 1μm can be able to create a pattern by using electrostatic lenses to focus a liquid metal ion source onto the substrate, and to deflect it on a surface in a desired pattern, causing implantation and sputtering of the target [14].

When a feature dimension is reduced to about 200nm and below, ion beam writing (IBW) seems particularly important. IBW can be operated in a shallow writing mode using PMMA as an ion beam resist. It can achieve higher resolution than optical, X-ray, or electron beam lithographic techniques because ions have a higher mass and therefore scatter less than electrons. In addition, resists are more sensitive to ions than to electrons. Focused ion beam etching (FIBE) does not employ resist for pattern, but to directly sputter material or directly deposit material on the substrate. FIBE is mainly used for mask repair and offers patterned-doping.

Most of focused ion beam equipment on the market today uses gallium liquid metal ion sources (LMISs) with the service life of up to 2000 μA/h and accelerating voltages in the range of 10～50kV. Other LMISs have typically much shorter service lives, and they include AuPdAs, Si, Co, $Au_{77}Ge_{14}Si_9$, $Co_{36}Nd_{64}$, etc. Commercially available FIB are used mostly for defect repair of optical and X-ray masks, circuit modification, device modification, failure analysis of circuits and devices, sample preparation of transmission electron microscopy (TEM) and so on. Array bridges with lengths of ~30μm produced by 50kV Ga^+ FIB and subsequently etched in KOH solution were fabricated. The narrowest bridge has a width of 80nm as shown in Fig.7.4 [15].

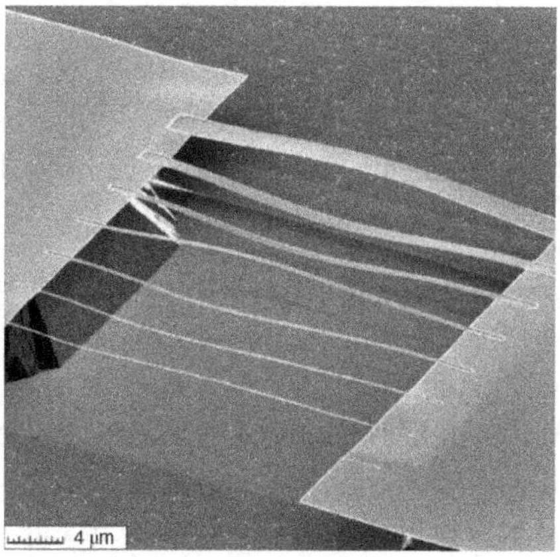

Fig.7.4 Si nanobridges fabricated by KOH etching of a FIB patterned Si (100) surface. The minimum width of 80nm for 30μm long wire was achieved [15]

7.3 Nanoimprint lithography

7.3.1 Nanoimprint lithography

Nanoimprint lithography is also called as "hot embossing lithography". It must meet two conditions, ① a rigid master (stamp) with a pattern on its surface; ② a substrate with a layer of polymer is heated up to the printing temperature which is above the glass temperature of the polymer. Then the stamp is pressed against the substrate via a physical contact. After the master (stamp) was removed from the substrate, the resulting polymer pattern can be fabricated. Fig.7.5 shows a scheme of this manufacture technique [16,17]. For example, to transfer a pattern from a mold into a thin polymer film of PMMA by nanoimprint lithography, it requires heating the polymer film above 110℃.

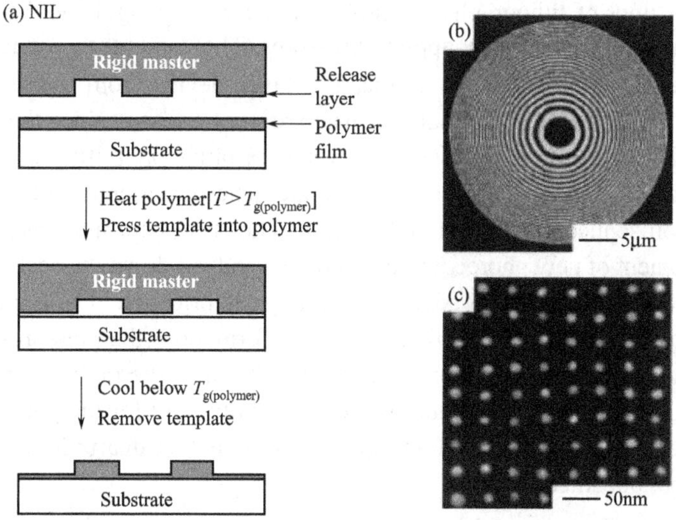

Fig.7.5 (a) Schematic illustration of nanoimprint lithography. The SEM images (b、c) show typical experimental results

Nanoimprint lithography can mold a variety of polymeric materials [Fig.7.5(b),(c)] and pattern feature is about 5nm. This process has been used to pattern components of microelectronic, optical, and optoelectronic devices [18,19]. For example, gate lengths in a metal oxide semiconductor field effect transistor (MOSFET) have been defined by nanoimprint lithography with a minimum feature size as small as 60nm.

7.3.2 Step-and-flash imprint lithography

Step-and-flash imprint lithography (SFIL) is a technique which replicates the topography of a rigid mold using a photocurable prepolymer solution as the molded material. During SFIL [Fig.7.6(a)], a low viscosity, photocurable liquid or solution fills the void spaces of the quartz mold [20]. The solution consists of a low-molecular-weight monomer and a photoinitiator. As UV light radiates to hard mold, the photocurable prepolymer begins to polymerize and harden. Removing the mold leaves a topographically patterned replica on the substrate.

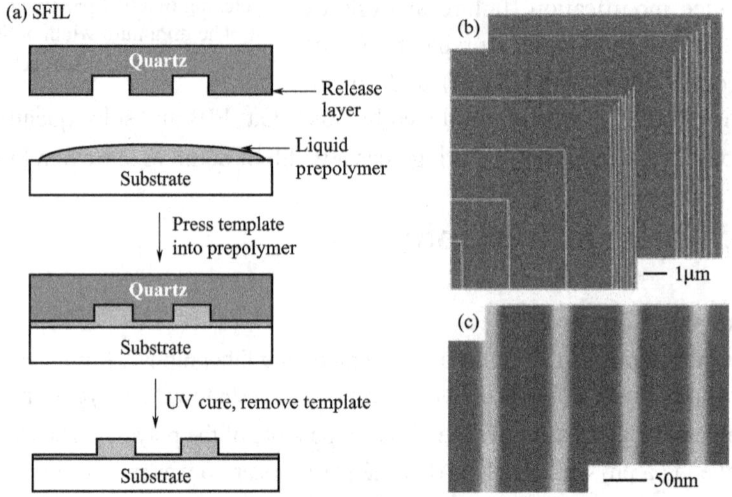

Fig.7.6 (a) Schematic illustration of the procedure for step-and-flash imprint lithography (SFIL). Scanning electron microscopy (SEM) images of (b) 40nm wide lines and (c) 20nm wide lines patterned by step-and-flash lithography [20]

Step-and-flash imprint lithography employs a transparently rigid mold to print pattern at a constant temperature (22℃) with low applied pressures. The comprehensive factors give SFIL an inherent advantage in its potential for fabrication layer device from monolayer to multilayer. Distortions caused by differential thermal expansion are not an issue since the mold and substrate are not heated. The low printing pressure allows imprinting on brittle substrates and reduces distortion caused by flexing of the mold or substrate. The development of SFIL techniques has focused primarily on semiconductor nanofabrication. The successful implementation of SFIL will require the development of new photocurable monomers and cut down the cost of manufacture.

A hard mold offers a number of advantages for nanofabrication. The rigid mold of silicon or quartz retains nanoscale features with minimal local deformation (the pressures for embossing can cause long-range distortions in the substrate). A hard mold is thermally stable for cross-link of polymer precursors. Silicon and quartz molds are chemically inert for the polymer precursors. One of the main differences between silicon and quartz molds is that quartz let ultraviolet and visible light to pass through, but silicon is not.

7.3.3 Microcontact printing

The microcontact printing method was first invented by Kumar and Whitesides at Harvard

University in 1993[21]. The preparation of an elastomer (soft) stamp is one of the most important factors. Pour the mixture of polydimethylsiloxane (PDMS) and curing agent into a template and then heat up to 60℃ to make them solidification. It is easy to separate elastomer stamp of PDMS from the template after the curing. For reliable stamping, good adhesion of ink molecules on the surface is very important. In the air, the stamp surface usually remains hydrophobic making it suitable for molecular ink based on nonpolar solvent. Additional treatment is required for stamping of water-based molecular solution. In this method, soft micrometer scale stamps made of PDMS are utilized to deposit organic molecules onto solid substrates via direct contact like macroscale printing technique as shown in Fig.7.7.

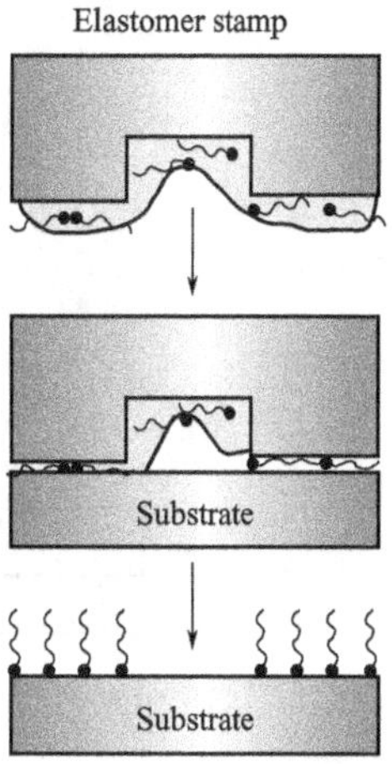

Fig.7.7 Stamping procedure

However, since the stamps are made of soft polymer materials, the bending of stamps caused by swelling of solvent molecules or external pressure is one of the major limiting factors of resolution. In typical applications, since microcontact stamping is a parallel printing method, it can pattern a large surface area very quickly with a micro- or submicrometer scale resolution.

Fig.7.8 illustrates a typical microcontact printing procedure [22,23]. The surface of a PDMS stamp was coated with a continuous layer of gold (about 20nm thick) without an adhesion layer between the gold and the PDMS. This stamp was brought into contact with a substrate coated with a dithiol (e.g., 1,8-octanedithiol). The dithiol forms self-assembled monolayer (SAMs) on the substrate (GaAs in this case) and the exposed thiol groups bond to the gold layer in the regions of contact. Removing the elastomeric stamp left the gold layer onto a GaAs substrate.

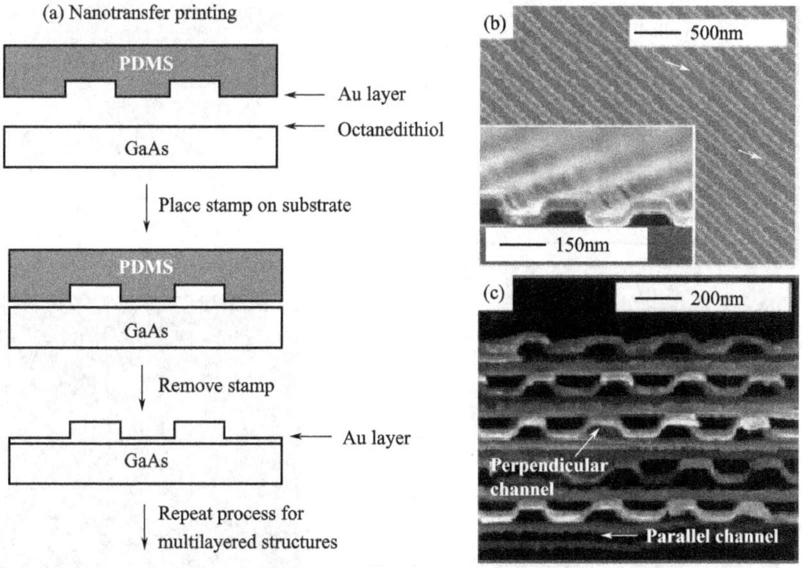

Fig.7.8 (a) Schematic illustration of microcontact printing; (b) SEM images of a 20nm gold layer transferred onto a GaAs substrate; and (c) multilayered stack of 20nm thick layers of gold containing parallel grooves [22,23]

7.4 Scanning probe lithography

Scanning probe lithography such as scanning tunneling microscopy (STM) and atom force microscopy (AFM) has the highest spatial resolution. In terms of structural characterizations, AFM

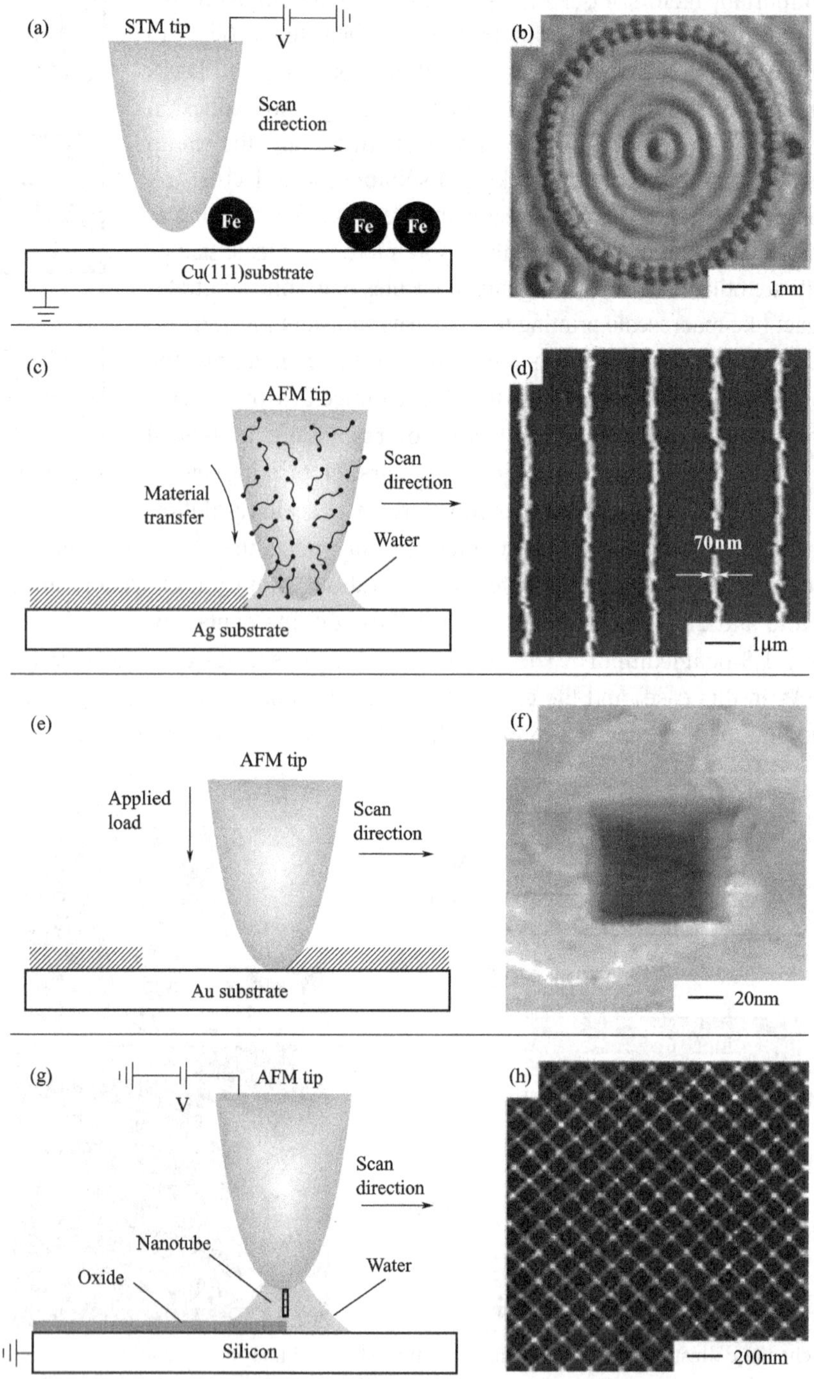

Fig.7.9 Schematic representations of scanning probe lithography. (a) Scanning tunneling microscopy (STM) can position atoms on a surface with high precision to generate patterns, such as (b) a quantum corral of a 48-atom Fe ring formed on Cu enclosing a defect-free region [28]. (c) Dip-pen nanolithography can direct the deposition of SAMs (e.g., 16-mercaptohexadecanoic acid) on Ag as an etch resist to pattern (d) 70nm wide features [29]. (e) Nanoshaving can remove regions of SAMs to pattern features such as (f) a square hole within octadecanethiolate SAMs on Au [35]. Scanning electrochemical oxidation with a carbon nanotube-modified AFM tip can selectively oxidize a surface to pattern (h) 10nm wide (2nm tall) silicon oxide lines spaced by 100nm [38]

can attain a lateral resolution of 1nm and vertical resolution of 0.5nm [24~27]. STM resolution is intrinsically five times higher than AFM [24~27].

An example of these techniques for nanoscale fabrication is described in Fig.7.9. Scanning tunneling microscopy (STM) can position atoms on a surface to generate a high precision pattern, such as a quantum corral of a 48-atom Fe ring formed on Cu enclosing a defect-free region in Fig.7.9 (a) and (b) [28]. This atomic-scale manipulation is interesting scientifically but is not yet a practical technology [29]. Scanning probe lithography is a versatile method for depositing clusters of atoms or molecules onto a surface. One approach to deposit nanoparticles or molecules selectively onto a surface is dip-pen nanolithography (DPN) [30~32]. It utilizes the tip of AFM as a nanoscale pen, molecular substances as ink, and solid substrates as paper. The ink adsorbed onto the AFM tip transfers to the surface of substrate in an arbitrary pattern "written" with the scanning probe [Fig.7.9(c)]. The self-assembled monolayer (SAMs) can also mask the substrate to pattern nanostructures of metals during wet etching, such as Au, Pd, and Ag [Fig.7.9(d)] [33]. This process can fabricate trenches with lateral dimensions from 12nm to 100nm [29].

The spreading of ink on the substrate depends on the humidity, the reactivity of the ink with the substrate, the radius of curvature of the probe, and the linear velocity of the probe [34]. The mechanism of material transfer by DPN (dip-pen nanolithography) is not yet clear. An AFM tip in contact with the surface can remove self-assembled monolayer (SAMs), and this process is referred to as nanoshaving [Fig.7.9(e)] [35]. The SAMs are removed again and again from a substrate to form nanoscale holes [Fig.7.9(f)] or trenches [36]. Scanning probe lithography is also used to modify surface properties of material [37]. Typically, this method can generate about 50nm wide features. A carbon nanotube-modified AFM probe can pattern about 10nm wide lines of silicon oxide on a silicon hydride surface [Fig.7.9(g) and Fig.7.9(h)] [38].

References

[1] G. Renaud, R. Lazzari, C. Revenant, A. Barbier, M. Noblet, O. Ulrich, F. Leroy, J. Jupille, Y. Borensztein, C.R. Henry, J.P. Deville, F. Scheurer, J. Mane-Mane, O. Fruchart, Science ,300 (2003) 1416.

[2] J.H.G. Owen, K. Miki, D.R. Bowler, J. Mater. Sci. ,41 (2006) 4568.

[3] T. Ogino, H. Hibino, Y. Homma, Y. Kobayashi, K. Prabhakaran, K. Sumitomo, H. Omi, Acc. Chem. Res. ,32 (1999) 447.

[4] H. Brune, Surf. Sci. Rep.,31 (1998).

[5] H.S. Nalwa, Encyclopedia of Nanoscience and Nanotechnology, 5(1) (2004) American Scientific Publishers, P163.

[6] H.S. Nalwa, Encyclopedia of Nanoscience and Nanotechnology, 6(1) (2004) American Scientific Publishers, P606.

[7] S. Hong, L. Huang, Encyclopedia of Nanoscience and Nanotechnology, 6 (2004) ,ISBN: 1-58883-062-4 , P605–622.

[8] N.P. Mahalik, Micromanufacturing and Nanotechnology, (2006) Springer-Verlag Berlin Heidelberg, ISBN-10: 3-540-25377-7 (Hardback), ISBN-13: 978-3-540-25377-8 (Electronic).

[9] Y. Xia, G. M. Whitesides, Angew. Chem. Int. Ed. ,37 (1998) 550.

[10] M.L. Roukes, Technical Digest of the 2000 Solid-State Sensor and Actuator Workshop, (2000)Hilton Head Island, SC, Transducer Research Foundation, Cleveland, P4–8.

[11] H.G. Craighead, Science, 290 (2000) 1532.

[12] M.L. Roukes, Phys. World, 14 (2001) 25.

[13] H.S. Nalwa, Encyclopedia of Nanoscience and Nanotechnology, 9 (1) (2004) American Scientific Publishers, ISBN: 1-58883-057-8, P387.
[14] M. Ultaut, Focused ion beams, Handbook of Charged Particle Optics, (1997) CRC Press, Boca Raton, FL, P429–465.
[15] H.S. Nalwa, Encyclopedia of Nanoscience and Nanotechnology, 3 (1) (2004) American Scientific Publishers, P477.
[16] S.Y. Chou, P.R. Krauss, W. Zhang, L. Guo, L.J. Zhuang, Vac. Sci. Technol. B, 15 (1997) 2897.
[17] M.D. Austin, H. Ge, W. Wu, M. Li, Z. Yu, D. Wasserman, S.A. Lyon, S.Y. Chou, Appl. Phys. Lett. ,84 (2004) 5299.
[18] C.M.S. Torres, S. Zankovych, J. Seekamp, A.P. Kam, C.C. Cedeno, T. Hoffmann, J. Ahopelto, F. Reuther, K. Pfeiffer, G. Bleidiessel, G. Gruetzner, M.V. Maximov, B. Heidari, Mater. Sci. Eng. C, 23 (2003) 23.
[19] M. Li, J. Wang, L. Zhuang, S.Y. Chou, Appl. Phys. Lett., 76 (2000) 673.
[20] M. Colburn, S. Johnson, M. Stewart, S. Damle, T.C. Bailey, B. Choi, M. Wedlake, T. Michaelson, S.V. Sreenivasan, J. Ekerdt, C.G. Willson, Proc. SPIE-Int. Soc. Opt. Eng., 3676 (1999) 379.
[21] A. Kumar, G.M. Whitesides, Appl. Phys. Lett. ,63 (1993) 2002.
[22] J.L. Wilbur, A. Kumar, E. Kim, G.M. Whitesides, Adv. Mater., 6 (1994) 600.
[23] A. Kumar, N.L. Abbott, E. Kim, H.A. Biebuyck, G.M. Whitesides, Acc. Chem. Res., 28 (1995) 219.
[24] G. Binnig, Force microscopy, Ultramicroscopy ,42 (1992) 7–15.
[25] G. Binnig, C. Gerber, E. Stoll, T.R. Albrecht, C.F. Quate, Atomic resolution with atomic force microscope, Surf. Sci. ,189 (1987) 1–6.
[26] G. Binnig, H. Rohrer, Scanning tunneling microscopy from birth to adolescence, Angew.Chem Int. Ed. Engl.,26 (1987) 606–614.
[27] F. Ohnesorge, G. Binnig, True atomic-resolution by atomic force microscopy through repulsive and attractive forces, Science ,260 (1993) 1451–1456.
[28] M.F. Crommie, C.P. Lutz, D.M. Eigler, Science, 262 (1993) 218.
[29] H. Zhang, S.W. Chung, C.A. Mirkin, Nano Lett. ,3 (2003) 43.
[30] S. Hong, J. Zhu, C.A. Mirkin, Science, 288 (2000) 1808.
[31] S. Hong, J. Zhu, C.A. Mirkin, Science ,286(1999) 523.
[32] K.S. Ryu, X. Wang, K. Shaikh, D. Bullen, E. Goluch, J. Zou, C. Liu, C.A. Mirkin, Appl. Phys. Lett. ,85(2004) 136.
[33] H. Zhang, C.A. Mirkin, Chem. Mater.,16 (2004) 1480.
[34] S. Rozhok, R. Piner, C.A.J .Mirkin, Phys. Chem. B, 107 (2003) 751.
[35] G.Y. Liu, S. Xu, Y. Qian, Acc. Chem. Res. ,33 (2000) 457.
[36] R. Carpick, W. Salmeron, M. Chem. Rev., 97 (1997) 1163.
[37] J.A. Dagata, J. Schneir, H.H. Harary, C.J. Evans, M.T. Postek, J. Bennett, Appl. Phys. Lett. ,56(1990) 2001.
[38] H. Dai, N. Franklin, J. Han, Appl. Phys. Lett.,73(1998) 1508.

Review questions

1. What is photolithography?
2. What is positive and negative photoresists?
3. Please describe the main procedures of the photolithography.
4. What is the purpose of mask?
5. What are the advantages of electron-beam lithography?

6. What are the commercial purposes of focused ion beams (FIB) lithography?

7. In the manufacture process of focused ion beams (FIB) lithography, do you know usually it needs to employ resists or not? Why?

8. What are rigid mold (stamp) and elastomer (soft) stamp? How to use them, respectively?

9. Can you tell the reasons about why the nanoimprint lithography is also called as "hot embossing lithography"? What are the aims of heating?

10. Pleas expound the working principle of step-and-flash imprint lithography.

11. What kinds of molds (stamps) want to be utilized in microcontact printing technique? Why?

12. What kinds of precision instruments can be used in scanning probe lithography?

13. In the fabraction process of dip-pen nanolithography (DPN), what are real meanings of "pen", "inked", "paper" and "written"?

Vocabulary

atom force microscopy (AFM) 原子力显微镜

chromium ['krəʊmɪəm] n. [化] 铬

corral [kə'rɑːl] n. 环形车阵；畜栏；vt. 把…关进畜栏；捕捉；把…布成车阵

curvature ['kɜːvətʃə] n. 弯曲，曲率

diffraction [dɪ'frækʃn] n. 衍射

dip-pen nanolithography 浸蘸笔纳米蚀刻法，浸蘸笔纳米刻蚀法，

dithiol n. 二硫醇

dose [dəʊs] n. 剂量；一剂，一服；vi. 服药；vt. 给药，给…服药

electron beam lithographic 电子束蚀刻法，电子束刻蚀法

focused ion beam lithography 聚焦离子束蚀刻法，聚焦离子束刻蚀法，

functionality [fʌŋkʃə'nælətɪ] n. 功能；函数性

hallmark ['hɔːlmɑːk] n. 特点，品质证明；vt.使具有…标志

hot embossing lithography 热压蚀刻法，热压刻蚀法，

immerse [ɪ'mɜːs] vt. 沉浸；使陷入

integrated circuits technology (ICT) 集成电路技术

master (stamp) n.硕士；主人；模具(印章)

mercury ['mɝkjəri] n. 水银；水银柱；汞

mercury arc lamp 水银弧光灯；汞弧灯；汞气灯

metal oxide semiconductor field effect transistor 金属氧化物半导体场效应晶体管

microcontact printing 微接触印刷

microelectronic [ˌmaɪkrəʊɪˌlek'trɒnɪk] adj. 微电子的

miniaturization [ˌmɪnɪətʃəraɪ'zeɪʃən] n. 小型化，微型化

mold [məʊld] n.模子；霉菌；vt. 塑造；使发霉；用模子制作；vi. 发霉

nanoimprint lithography 纳米压印蚀刻法，纳米压印刻蚀法

1,8-octanedithiol 1,8-辛烷二硫醇

optoelectronic [ˌɒptəʊɪˌlek'trɒnɪk] adj. 光电子的

paramount ['pærəmaʊnt] adj. 最重要的，主要的；至高无上的；n. 最高统治者

photocurable 光固化的

photographic mask 光掩膜；照相掩模

photoinitiator n.光引发剂

photolithography [ˌfəʊtə(ʊ)lɪ'θɒgrəfɪ] n. 光刻法；影印石版术；照相平印术；

photomask ['fəʊtəumɑːsk] n. 光掩模

photoresist [ˌfəʊtəʊrɪ'zɪst] n. 光刻胶；光致抗蚀剂；光阻材料

poly(butene-1 sulfone) 聚 1-丁烯砜

poly(glycidyl methacrylate-co-ethyl acrylate) 聚(甲基丙烯酸-共-乙基丙烯酸缩水甘油酯)

polydimethylsiloxane(PDMS) n. 聚二甲基硅氧烷

polymethyl methacrylate (PMMA) n. 聚甲基丙烯酸甲酯

resist 光刻胶，抗蚀胶

resolution [rezə'luːʃ(ə)n] n. 分辨率；决议；解决；决心

scanning beam lithography 扫描束光刻，扫描束刻蚀法

scanning probe lithography 扫描探针刻蚀法

scanning tunneling microscopy(STM) 扫描隧道显微镜

slope [sləʊp] n. 斜坡；倾斜；斜率；vi. 倾斜；逃走；vt. 倾斜；使倾斜；扛

spectral ['spektr(ə)l] adj. [物] 光谱的

step-and-flash imprint lithography (SFIL) 步进式闪烁压印蚀刻法，步进式闪烁压印刻蚀法

throughput ['θruːpʊt] n. 生产量，生产能力

wafer ['weɪfə] n. 圆片，晶片；薄片，干胶片；薄饼；vt. 用干胶片封

7. 光刻技术用于纳米制造

近年来，由于器件小型化的应用潜力，人们对纳米制造产生了浓厚的兴趣。一般来说，纳米制造可分为两类：块体材料的纳米制造和纳米图形化。块体纳米结构的合成有许多重要的用途，如纳米粒子、纳米管以及纳米线的块体制造，而纳米图形化对于功能化元件的生产至关重要。自组装近来已用于辅助二维（2D）和三维（3D）结构的超组装。而传统意义上的表面图形化方法，如光刻法一直用于制造功能器件，但由于光的衍射限制，光刻技术难以制造出小于 100nm 的纳米结构材料。而诞生的数种新的制造技术，包括电子束刻蚀技术、软刻蚀技术和扫描探针刻蚀（SPL）技术已经用于纳米制造和纳米结构图形化。

随着集成电路尺寸的不断减小，不久的将来器件的尺寸将会接近数纳米量级。基于“自下而上”的方法，即准确地控制操纵表面原子，设计新一代纳电子器件就显得尤为重要。因而，在数十年中，对低维度结构的制备，如纳米簇[1]、纳米线[2]、纳米图案[3]的制造技术进行了深入的研究。采用这些纳米技术，通过自组装，在宏观材料表面上制造出均匀的纳米结构具有特殊的意义[4]。

7.1 紫外线光刻微制造

硅集成电路无疑是我们这个时代的奇迹，它改变了我们生活的方方面面。由于硅具有良好的力学性能，集成电路技术（ICT）可用于制造各种结构材料。硅的微加工技术可以分成两大类:即块体微加工和表面微加工。

块体微加工是指单晶硅片以外的结构制造技术。而表面微加工技术是在单（多）晶硅片表面上的制造技术，并由 100nm～10μm 厚的薄膜所组成。

集成电路技术（ICT）的标志是将数以千万计的元件组装在几平方厘米的硅芯片上的光刻技术，如图 7.1 所示[5]。每一个芯片上的晶体管的数目快速增加，微电子电路的功能更加显著，导致集成电路的容量和性能更加强大。

光刻技术是利用光制造图案的过程，图 7.2 显示的是光刻技术的基本步骤[6]。在这个过程中，首先在固体表面涂上一层聚合物光刻胶，然后将光刻胶层的一个特定区域通过光掩膜暴露在紫外线下（波长大约 200nm）。当聚合物光刻胶层暴露于紫外线后，分子键被打破（或增强）。通过将基质浸在某种溶剂中，专门去除基质上暴露的（或未暴露的）光刻胶，而聚合物光刻胶的图案就被保留在基质的特定区域上。聚合物光刻胶图案可以作为随后的蚀刻过程或其它处理过程中的光刻胶。重复这些过程，在一个基质上就能够制造出许多复杂的微电子元件[7]。

在过去的四十年里，半导体工业一如既往地致力于提高光刻技术的分辨率，以便在每个芯片上集成更多的元件。这种尺寸的减小和高度集成化，有利于提高集成电路的性能和功率，并能有效地降低制造成本。目前，具有 90nm 特征尺寸的微电子芯片已经进入商业领域。

图 7.3 显示的是涂有正的或负的光刻胶的光刻技术 [8]。对作为底板的一个多晶硅基质（硅片）进行化学清洗，去除表面的粒状物质，以及痕量的有机、离子、金属杂质等。图 7.3（a）给出的是沉积在多晶硅基质上的一个薄的绝缘层，如 SiO_2 绝缘层。接着将一种对紫外线敏感的光刻胶（通常是有机聚合物）涂在 SiO_2 上，如图 7.3（b）所示。

下一步是要将部分 SiO_2 选择性除去，其余部分仍保留在基质的特定区域，为此，我们需要利用一个光掩膜完成此项任务。在光刻阶段使用的光掩膜是一个关键的组件。一个典型的光掩膜是能够透过紫外线的玻璃板；通过在该玻璃板（光掩膜）上涂上一种金属薄层，通常是 Cr 或 Au，制造出一个有价值金属图案。这些光掩膜能够制造出微观和亚微观特征的高质量的图案，如图 7.3（c）所示。光掩膜通常是在一个计算机辅助软件平台上制作的，然后将光掩膜放置在光刻胶的顶部，如图 7.3（d）所示。

紧接着以紫外线照射光掩膜，以汞弧灯作为紫外光源，又称为辐射体，该光源在一定的波长范围，伴随有光谱峰的能量输出。根据图案的复杂性、光刻胶的厚度及光刻胶材料的性质，可能需要进行一次或数次曝光过程。因此，精确剂量的紫外线是极为重要的。当紫外线透过光掩膜而落到光刻胶上时，如图 7.3（e）所示，类似于光掩膜的图案在光刻胶上得到显影，这一现象称为光掩膜转移或图案转移。不管光刻胶成分暴露在紫外线后是可溶还是不可溶，通过紫外线和光刻胶成分的共同作用使这一转移得以实现，如图 7.3（f）和图 7.3（i）所示。因此，有两种光刻胶：正的光刻胶和负的光刻胶。一方面，当曝露于紫外线后，正的光刻胶成为可溶性的，另一方面，负的光刻胶成为不溶性的。在实际中，被紫外线照射过的正的光刻胶区域，就可以被洗掉（刻蚀掉）；相反，负的光刻胶则不能，这就确保了没有暴露在紫外线下的光刻胶被洗掉（刻蚀掉），形成一个反的光掩膜图案。洗掉光刻胶材料的过程实际上称为刻蚀过程。

确保光刻技术最终分辨率的两个主要参数是:①光的波长；②光刻胶层的厚度。通常认为，光刻技术的分辨率大约为$\lambda/2$。人们在利用较短波长的光以改善其分辨率方面，已付出了很大的努力。波长约 160nm 的紫外线用于商业生产，可制造出大约 90nm 线宽的芯片。然而，要找到一个适用于更短波长光的合适的光学器件是相当困难的。另一方面，尺寸最小化也受到光刻胶层的厚度影响，高分辨率的光刻胶层厚度通常小于 100nm。

由于这两个参数的制约，光刻技术的分辨率极限为 100nm 左右。为此，发展新一代的刻蚀技术以突破光刻技术的极限，并能持续减少电子芯片的尺寸还需要进一步的努力。这些刻蚀技术包括 X-射线刻蚀法、电子束刻蚀法、扫描探针刻蚀法、软刻蚀法[9]。

这些方法表明，蚀刻分辨率可小于 10nm。然而，目前还不清楚哪种方法会最终取代光刻技术。先进的纳米沉积方法，如“浸蘸笔”纳米刻蚀或微接触印制方法已经得到广泛的发展，可能会替代光刻技术。因为直接沉积方法并不依赖于光，其分辨率不受光的波长限制。

7.2 扫描束刻蚀纳制造

随着新的纳米刻蚀技术的出现（电子束、离子束和激光束以及纳米压印刻蚀等），器件的进一步小型化成为可能。这些技术将把微电子机械系统缩小到次微米量级，这就为新一代电子机械器件打开了一扇门，即可制造 100nm 至数个纳米特征尺寸范围的纳电子机械系统（NEMS）[10~12]。

7.2.1 电子束刻蚀

电子束刻蚀技术是制造纳米结构的一个标准化过程。其分辨率通常小于 100nm，远高于光刻技术。将高能电子束聚焦在一层薄的抗蚀胶上而获得图案，然后，将抗蚀胶层放进一个合适的溶液中显影，该溶液可选择性地溶解被暴露于电子束区域中的抗蚀胶，或选择性地溶解没有被暴露于电子束区域中的抗蚀胶[13]。

采用 100～200keV 的高能电子和 1～10nm 的小型电子探针的电子束，可以直接在基质上进行“刻写”。用电子束可以制造出令人满意的 30nm 的典型线宽（7nm 的线宽也有所报道）。电子束刻蚀法的分辨极限主要取决于刻蚀胶的性质。在创作图案过程中，由于它的灵活性，电子束刻蚀法广泛用于研究或小批量生产小产品，它的关键技术是制造光掩膜。这种技术的不足之处是制造耗时长和生产能力低。

电子抗蚀胶是聚合物。它的性质类似于光刻胶，也就是，当抗蚀胶被电子束辐照时，会产生化学或物理变化，在这些变化中，抗蚀胶形成图案。通常正电子抗蚀胶是聚甲基丙烯酸甲酯（PMMA）和聚（1-丁烯砜）。正电子抗蚀剂的分辨率是 0.1μm 或更高。称为 COP 的聚（甲基丙烯酸共乙基丙烯酸缩水甘油酯）是一种常见的负电子抗蚀剂，COP 像负光刻胶一样，在显影的过程中也会发生膨胀，所以，它的分辨率极限为 1μm。

7.2.2 聚焦离子束刻蚀

聚焦离子束（FIB）已经成为在微、纳米尺寸上，对材料表面改性和功能化结构原型设计的一种普遍使用的工具。直径小于 1μm 的聚焦离子束，利用静电透镜将液态金属离子源聚焦在基质上，然后偏转光束在一个理想的图案上，通过离子注入或靶溅射[14]而创造图案。

当特征尺寸减小到 200nm 或更小时，离子束刻写（IBW）就显得非常重要。以 PMMA（聚甲基丙烯酸甲酯）作为离子束的抗蚀胶，离子束刻写以浅的刻写模式进行。它能达到比光学的、X 射线、电子束刻蚀技术更高的分辨率，由于离子具有较高的质量，因此散损失的离子比电子少。此外，抗蚀胶对离子比对电子更加敏感。目前，聚集离子束刻蚀（FIBE）不是用抗蚀胶制造图案，而是直接向基质上溅射或沉积材料。聚焦离子束刻蚀主要用于光掩膜的修补和提供图形化掺杂。

目前，市场上大多数聚焦离子束设备是使用寿命长达 2000μA/h、加速电压范围为 10～50kV 的镓液态金属离子源（LMISs）。其它液态金属离子源的服役寿命更短。它们包括 AuPdAs、Si、Co、$Au_{77}Ge_{14}Si_9$、$Co_{36}Nd_{64}$ 等。商业使用的聚焦离子束主要用于光学的和 X-射线的掩膜缺损修复、电路改造、设备改造、电路和设备的故障分析、透射电子显微镜（TEM）的样品制备等。通过 50 kV 的 Ga^+聚焦离子束，然后在 KOH 溶液中进行刻蚀，可以制造出长度约 30μm、宽度为 80nm 的最窄“桥”，如图 7.4 所示[15]。

7.3 纳米压印刻蚀技术

7.3.1 纳米压印刻蚀

纳米压印刻蚀又称为“热压刻蚀”。它必须满足两个条件：①在其表面上雕有图案的刚性模具（图章）；②涂有一层聚合物的基质被加热到高于聚合物玻璃化温度的印刷温度。然后通过一个物理接触，将印章压在基质上。从基质上移走印章后，可以得到聚合物图案。图 7.5 是这种制造工艺的示意图[16, 17]。例如，通过纳米压印刻蚀技术，将 PMMA 聚合物薄膜加热到 110℃以上，就可以把一个模具（印章）图案转移到 PMMA（聚甲基丙烯酸甲酯高分子）聚合物薄膜上。

纳米压印刻蚀技术能够制备各种各样的高分子材料[图 7.5（b）、（c）]，图案的特征尺寸约为 5nm。该技术已经用于微电子、光学、光电子器件的图案化结构中 [18,19]。例如，在一个金属氧化物半导体场效应晶体管（MOSFET）中的门长度已经由纳米压印刻蚀法定义 60nm 为最小的特征尺寸。

7.3.2 步进式闪烁压印光刻

步进式闪烁压印光刻（SFIL）是利用光固化预聚物溶液作为成型材料，复制刚性模具图案的一种技术。在步进式闪烁压印光刻过程中[图 7.6（a）]，石英模具的空腔由低黏度、光固化液体或溶液所填充 [20]，该溶液是由一种低分子量的单体和一种光引发剂所组成。当紫外线辐射刚性模具时，光固化预聚合物开始聚合并硬化，移去模具，在基质上留下了模具图案的复制品。

步进式闪烁压印光刻是采用一个透明的刚性模具，在恒定的温度（22℃）和施加的低压下，进行图案印制。综合因素赋予了步进式闪烁压印光刻法一个得天独厚的潜在优势，那就是制备单层及多层膜器件。由微分热膨胀引起的变形不再是问题，因为不用加热模具和基质。低的印制压力允许在脆的基质上压印，并且减少模具或基质的弯曲变形。步进式闪烁压印光刻技术的发展主要集中在半导体纳米制造。步进式闪烁压印光刻技术要成功地推广，还需要开发新的光固化单体以及需要降低制造成本。

硬模具用于纳米制造有许多优势。硅或石英的刚性模具在保持纳米尺度时具有最小的变形（用于压印的压力能够引起基质的长期变形）。硬模具对于聚合物前驱体的交联是热稳定的。硅和石英模具对聚合物前驱体是化学惰性的。硅模具与石英模具的最大区别在于石英模具能使紫外线和可见光透过，而硅模具则不能。

7.3.3 微接触印制

微接触印制技术是在 1993 年，由哈佛大学的 Kumar 和 Whitesides 首次发明的[21]。弹性体（软）印章的制备是最重要的因素之一。将 PDMS（聚二甲基硅氧烷）和固化剂的混合物倒入模板，然后，加热至 60℃使其固化，等凝固后，可很容易地将 PDMS（聚二甲基硅氧烷）弹性体印章与模板分开。要进行可靠的印制，印章表面上存在附着力强的“墨水”分子是非常重要的。在空气中，印章表面一直保持疏水性，使其适合于在非极性溶剂中进行分子印制。对于在水基分子溶液中的印制需要额外的处理。在该方法中，由 PDMS 制作的柔性微米量级的印章，像宏观印刷技术一样，通过直接接触，可在固体基质上沉积有机分子，如图 7.7 所示。

然而，因为印章是由软的聚合物材料制成的，溶剂分子的膨胀或外部压力造成印章的弯曲，是该方法分辨率受到限制的主要因素之一。在典型的应用中，因为微接触印制是一个平行的印刷技术，它可以在一个大的表面上很快地形成微米或亚微米分辨率的图案。

图 7.8 给出了一个典型的微接触印制过程 [22, 23]。在一个 PDMS（聚二甲硅氧烷）印章的表面涂上一层 Au（约 20nm 厚），在金层与 PDMS 之间没有黏结层。将该印章与涂有二硫醇（如 1,8-辛烷二硫醇）的一个基质接触，二硫醇在基质（GaAs）上自组装形成单层膜，暴露的硫醇官能团在接触的区域与金层发生键合。移去弹性体印章，在 GaAs 基质上会留下金层。

7.4 扫描探针刻蚀

扫描探针刻蚀具有最高的空间分辨率，如扫描隧道显微镜（STM）、原子力显微镜（AFM）。就结构特点而言，AFM 可以获得 1nm 的横向分辨率和 0.5nm 的纵向分辨率 [24~27]。STM 的分辨率是 AFM 的 5 倍多 [24~27]。

这些纳米制造技术的实例如图 7. 9 所示。扫描隧道显微镜能够操纵原子在一个基质的表面上以产生高清晰的图案，例如：在 Cu 的基质上，48 个 Fe 原子围成一个无缺陷区域的量子围栏，如图 7.9（a）和（b）所示 [28]。这种原子尺度的操纵在科学上是十分有意义的，但目前尚未成为一项实用技术 [29]。扫描探针刻蚀是将原子簇或分子簇沉积在某种表面的一种通用技术。将纳米颗粒或分子选择性地沉积到表面上的一种途径是“浸蘸笔”纳米刻蚀（DPN）技术 [30~32]。该方法是利用 AFM 针尖作为一支“纳米笔”、分子物质作为“墨水”、固体基质作为“纸张”。吸附于 AFM 针尖上的“墨水”用扫描探针在基质表面“书写”成所要的图案，再将该图案转移到表面[图 7.9（c）]。在湿法蚀刻过程中，自组装的单层（SAMs）也可以覆盖于基质上以绘制金属纳米结构图案，如 Au、Pd 和 Ag [图 7.9（d）][33]，这种工艺可以制备横向尺寸为 12～100nm 的沟槽[29]。

“墨水”在基质上的扩散取决于湿度、“墨水”与基质的反应活性、探针的曲率半径以及探针的线速度[34]。通过 DPN（“浸蘸笔”纳米刻蚀）材料的转移机理目前尚不清楚。与表面接触的一个 AFM 针尖能够去除自组装单层，该过程称为“纳米剃削”[图 7.9（e）][35]。SAMs (自组装单层膜)反复地从基质上去除而形成纳米孔穴[图 7.9（f）]或沟槽[36]。扫描探针刻蚀法也用于改善材料的表面性质[37]。一般来说，这种方法能够制造大约 50nm 宽的形貌特征。碳纳米管改性的 AFM 探针能够在氢化硅表面刻蚀出 10nm 线宽的氧化硅图案 [图 7.9（g）和图 7.9（h）][38]。

复 习 题

1. 什么是光刻法？
2. 什么是正的光刻蚀剂？ 什么是负的光刻蚀剂？
3. 请描述光刻法的主要步骤。
4. 掩膜的目的是什么？
5. 电子束刻蚀的优点是什么？
6. 聚焦离子束（FIB）刻蚀法的商业目的是什么？
7. 在聚焦离子束（FIB）的制备工艺中，你知道通常需要使用刻蚀胶还是不需要？为什么？
8. 什么是刚性模具（印章）？什么是弹性（软）印章？怎样分别使用它们？
9. 为什么纳米压印光刻法又称为“热压印刻蚀法”？加热的目的是什么？
10. 请阐述步进式闪烁压印刻蚀法的工作原理。
11. 在微接触印制法中，需要使用什么种类的模具（印章）？为什么？
12. 在扫描探针刻蚀法中，需要使用什么种类的精密仪器？
13. 在“浸蘸笔”纳米刻蚀（DPN）法中，“笔”、“墨水”、“纸张”、“书写”的真正含义是什么？

8. Nanotechnology for Production of Hydrogen by Solar Energy

Renewable energy can be defined as one type of energy sources which can be provide light, electricity and heat without polluting the environment. Energy generation from fossil fuels has been identified as the main reason of environmental pollution. The obvious advantage of renewable energy is that no fuel is required, which eliminates the emission of carbon dioxide. The current global energy problem can be returned to insufficient fossil fuel supplies and excessive gas emissions resulting from increasing fossil fuel consumption. It was reported that the present petroleum consumption was 10^5 times faster than the nature can create and at this huge rate of consumption, the world's fossil fuel reserves will be diminished by 2050 [1~3]. Also, it is interesting to mention that the global demand for energy is predicted to be approximately 30 and 46TW of power by 2050 and 2100, respectively [3]. Fossil fuels are crude oil, coal and natural gas. They are not renewable, once burnt they are gone for ever. These sources supply more then 90% of our energy demand, but carry a steep environmental cost. For example, the concentration of CO_2 in the environment has increased from (280nm to 370ppm) over the past 150 years. It's expected to pass (550 ppm) in this century [4~6]. However, due to the huge demand of energy and lesser availability of fossil fuels there is a shift toward renewable energy sources. Examples of these sources of energy include solar, wind, biomass, hydrogen and geothermal energies. These clean sources can be used instead of conventional fossil fuels and nuclear energy. Nowadays, there is a huge interesting this subject since it is expected to provide 50% of the world's primary energy by 2040. Moreover, renewable energy can be play a crucial role to reduce gas emissions to the environment by about 70% during 2050. From the other side, a new science field which is called nanotechnology, has been used to improve renewable energy sources.

8.1 Hydrogen economy

The interest in the use of hydrogen is based on its clean burning qualities, its potential for domestic production, and high efficiency of FCV (fuel cell vehicles) about 2～3 times that of conventional gasoline ICE (internal combustion engines). Hydrogen is one of the most abundant elements on earth and when used in a fuel cell, the sole by-product is water.

Fig.8.1 illustrates the 'hydrogen economy' as a net-work of primary energy sources linked to multiple end-uses through hydrogen as an energy carrier. Hydrogen can be produced using any of the primary energy sources such as fossil, nuclear, and renewable. However, in order to implement the hydrogen economy and completely replace fossil fuels, there are significant technical challenges that need to be overcome in each of the following areas: (1) Production, (2) Storage and (3) Conversion to electrical energy and heat at the point of use. All the technical barriers hinge on understanding and controlling atomic and molecular level processes at the interface between hydrogen and various materials. Nanoscale materials would appear ideally suited for overcoming these barriers because of their enormous surface areas. The performance of catalysts, hydrogen storage materials, and electrode assemblies for fuel cells can all be enhanced by the use of nanoscale processes and architectures.

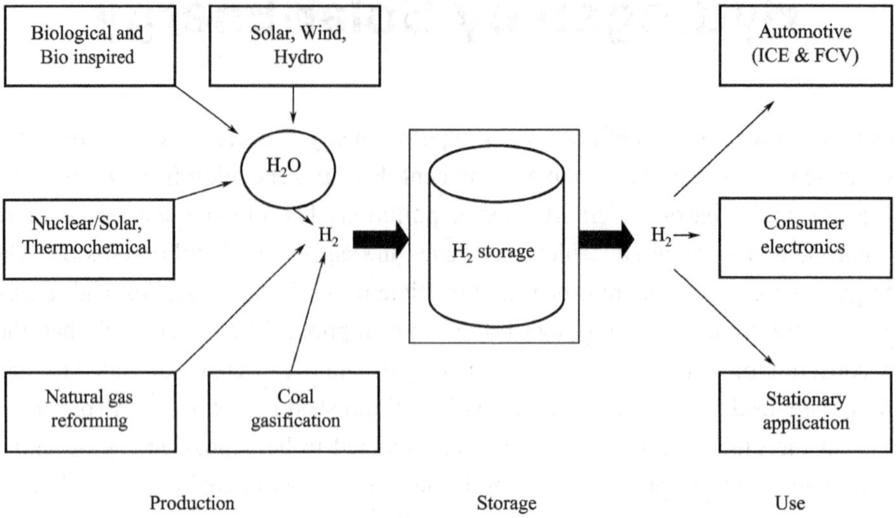

Fig.8.1 Hydrogen economy as a network of primary energy sources linked to various end uses through hydrogen as an energy carrier [3]

8.2 Hydrogen production

Although hydrogen is an attractive replacement for fossil fuels, it does not occur in nature as the molecule H_2. Rather, it occurs to bind in chemical compounds such as water or hydrocarbons that must be chemically transformed to yield H_2. Hydrogen can be produced using diverse domestic resources including fossil fuels, such as natural gas and coal, nuclear, hydroelectric, biomass, and renewable sources such as wind, solar, and geothermal as illustrated in Fig.8.2.

Steam reforming of methane is a process in which hydrogen can be produced from methane in natural gas using high temperature steam. In a similar liquid reforming process liquid fuels such as ethanol are reacted with steam at high temperature to produce hydrogen. Electrolysis uses electricity generated by wind, solar, geothermal, nuclear, or hydroelectric power to split water into hydrogen and oxygen. In gasification, coal or biomass is converted into gaseous components and then into synthesis gas, which is reacted with steam to produce hydrogen. Thermochemical water splitting uses high temperatures generated by solar concentrators or nuclear reactors to drive chemical reactions that split water to produce hydrogen. In photobiological process, microbes, such as green algae, consume water in the presence of sunlight producing hydrogen as the by-product. Similarly, photocatalytic and photoelectrochemical systems produce hydrogen from water using special semiconductors and energy from sunlight.

Therefore, Hydrogen as an energy carrier for its renewable and environmental-friendly characteristics may be produced by renewable energy sources via a variety of pathways. It is one of the fabrication methods to produce hydrogen by photocatalysis of solar energy.

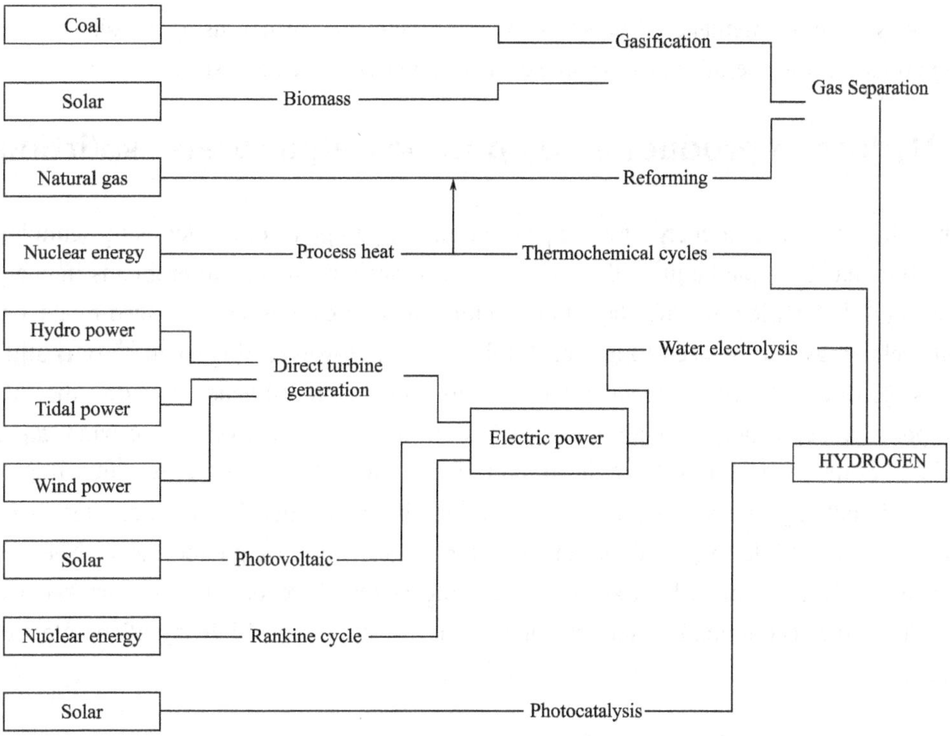

Fig.8.2 Primary energy sources and their use in hydrogen production[3]

8.3 Conversion of solar energy

Solar energy is free and rather available in many parts around the world. In just 1 year, the sun can provide the earth with 15000 times more energy than those of fossil fuels actually needed during the year. The sun's energy arrives on Earth as radiation distributed across the color spectrum from infrared to ultraviolet as shown in Fig.8.3.

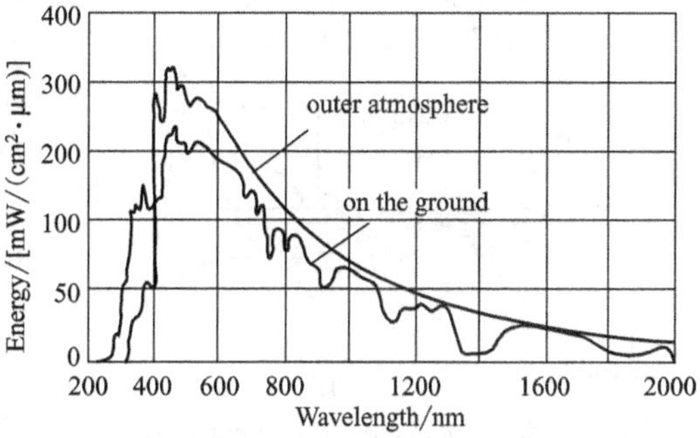

Fig.8.3 Spectrum of the solar energy observed on earth

The conversion of solar energy into useful energy forms can generally be divided into thermal and photonic processes. In solar thermal processes, solar energy is first converted to heat, which can either be used directly, stored in a thermal medium (e.g., water or dry rocks) or converted to mechanical and/or electrical energy by an appropriate machine (e.g., a steam turbine for the generation of electricity). In solar photonic processes, the solar photons are absorbed directly into an absorber that may convert the photon energy to electricity in a photovoltaic cell or convert the photon energy to chemical energy in an endergonic chemical reaction, such as solar water-splitting to produce hydrogen and oxygen.

The importance of hydrogen energy has recently been recognized again due to its new renewable characteristics. However, hydrogen is industrially produced by steam reforming of hydrocarbons such as methane. Production of hydrogen from water using a solar energy is a challenge because it is one of routes to solve energy and environmental issues [7].

8.4 Hydrogen production by photocatalytic water splitting

The utilization of solar energy for the production of hydrogen from water using semiconductor photocatalysts has been paid a great deal of attention because of the global problems in energy and environment [8,9]. A variety of TiO_2-based nanosemiconductor catalysts synthesized by nanoscience and nanotechnology can be used to approach this aim [10]. However, this technology is still in the research stage due to the cost associated with its low conversion efficiency. Besides, the TiO_2 as a photoresponse material only strongly absorb UV light due to its relatively large band gap energy of ~3.2eV. Some approaches have been paid to extend its photoresponse and improve its photocatalytic activity. These approaches can be classified into two broad categories, one is the addition of chemical additives and the other is the modification of photocatalyst. The chemical additives of electron donors which are holes scavengers react irreversibly with the valence band holes avoiding the recombination and thus increasing the quantum efficiency of the electron/hole separation.

8.5 Loading metal over TiO_2

One method is design of photocatalytic systems using semiconductor particles or powders, as shown in Fig.8.4. The loading of small amounts of Pt on the TiO_2 dramatically increased the reaction rate through an enhancement of the charge separation of the photo-generated electrons and holes [11]. It has been generally considered that Pt loading is effective for the photocatalytic evolution of H_2 from water by UV light irradiation since the presence of Pt significantly reduces the over-potential for H_2 production [12].

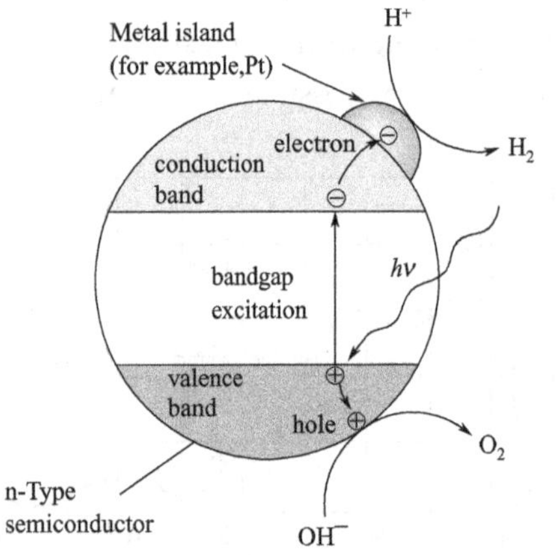

Fig. 8.4 Semiconductor particulate systems for heterogeneous photocatalysis [11]

When the TiO_2 was irradiated, the electron from the valence band of TiO_2 leaps into the conduction band, it leaves behind a hole in the valence band. The hydroxyl ion from the ionization of water is oxidated by photogenerated hole to release O_2, while hydrion is reduced by photogenerated electron on the Pt nanoparticle to liberate H_2.as shown in equations of (1)~(3).

$$TiO_2 + 2h\nu = 2e^- + 2h^+ \text{ (excitation of } TiO_2 \text{ by light)} \quad (1)$$

$$2h^+ + H_2O = 1/2O_2 + 2H^+ \text{ (on the surface of } TiO_2\text{)} \quad (2)$$

$$2e^- + 2H^+ = H_2 \text{ (on the surface of Pt)} \quad (3)$$

Ni doped mesoporous TiO_2 with crystalline framework and high specific surface area has been prepared by sol-gel associated with hydrothermal methods. Pt as cocatalyst from precursors of

$H_2PtCl_6 \cdot 6H_2O$ is photodeposited in situ on a photocatalyst. Photocatalytic hydrogen evolution is performed under the conditions of a high pressure Hg lamp as the light source. The reaction cell is connected to a closed gas circulation system and the products of gaseous phase is analyzed with an on-line thermal conductivity detector (TCD) gas chromatograph (NaX zeolite column, nitrogen as carrier gas) [13].

8.6 Development of visible-light-driven photocatalysts

Development of visible-light-driven photocatalysis to produce hydrogen from water using solar light has been urged the improvement of conversion efficiency of light-energy. The strategy for the design of photocatalytic materials with visible light response is classified into three categories as shown in Fig.8.5 [14]. Strategy (a) is the doping of foreign elements into active photocatalysts with a wide band gap to form donor levels in the forbidden band. Strategy (b) is to find materials that have a stable valence band in a more negative position than the valence band consisting of O_{2p} orbitals as seen in many oxide photocatalysts. Strategy (c) is the control of the energy structure by making solid solutions between semiconductor photocatalysts with wide and narrow band gaps.

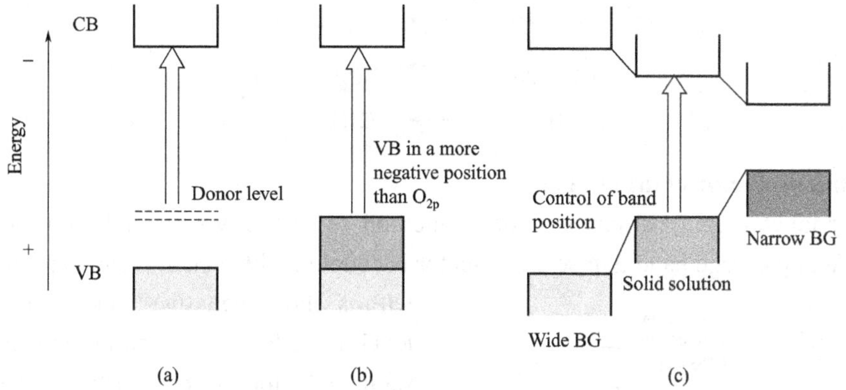

Fig. 8.5 Band engineering for design of photocatalyst materials with visible-light response. (a) Donor level formation by doping of foreign elements, (b) new stable valence band formation in a more negative position, and (c) a controllable band formation by making a solid solution [14]

8.6.1 Loading Cr^{3+} over titanate nanotubes

The nanocomposites by loading Cr^{3+} over titanate nanotubes (TNTs) are prepared via hydrothermal and impregnating processes. All of chromium ion-doped TNTs nanocomposites display much higher activities of H_2 evolution than that of pure TNTs sample.

The reaction mechanism of photocatalytic hydrogen production by water-splitting under visible light irradiation is proposed as shown in Fig.8.6 [15]. After Cr^{3+} doped into TNTs with a wide band gap of 3.3eV, the donor level (2.3eV) is formed in the forbidden band of TNTs. When a photon energy is absorbed by 3d electron of Cr^{3+} under visible light irradiation, the photogenerated electron (e^-) from donor level (Cr^{3+}) migrates to the conduction band of TNTs and simultaneously leaves behind a positive hole (h^+) in the donor level (Cr^{3+}) as shown in Eq. (8.1). Subsequently, The OH^- from the ionization of H_2O is oxidized by h^+ to produce oxygen and H^+ in Eq. (8.2).

At the same time, the sacrificial reagent (hole scavengers), SO_3^{2-}, reacts with photogenerated hole (h^+) to enhance separation efficiency of the electron–hole pairs as described in Eq.(8.3) and (8.4) [16~18]. Since sacrificial reagent (electron donor) is consumed in photoca- talytic reaction, continual addition of sacrificial reagent is necessary to sustain hydrogen production. The photogenerated electron (e^-) combines with H^+ deriving from the ionization of H_2O to form hydrogen in Eq.(8.5). Such, the photogenerated electron- hole pairs will be effectively separated and the photocatalytic efficiency can be improved due to donor levels provided by doping of Cr^{3+} into the TNTs.

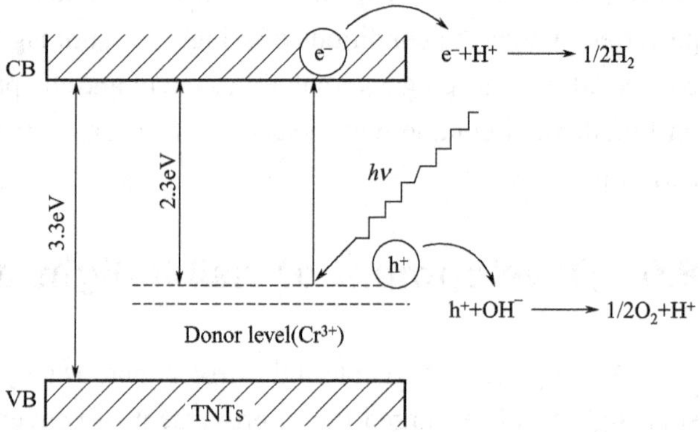

Fig.8.6 Mechanism of photocatalytic hydrogen production [15]

$$Cr^{3+} \xrightarrow{h\nu} h^+ + e^- \quad (8.1)$$

$$h^+ + OH^- \longrightarrow 1/2O_2 + H^+ \quad (8.2)$$

$$SO_3^{2-} + H_2O + 2h^+ \longrightarrow SO_4^{2-} + 2H^+ \quad (8.3)$$

$$2SO_3^{2-} + 2h^+ \longrightarrow S_2O_6^{2-} \quad (8.4)$$

$$H^+ + e^- \longrightarrow 1/2H_2 \quad (8.5)$$

8.6.2 Semiconductor composition

Semiconductor composition is another method to utilize visible light for production of hydrogen. When a large band gap semiconductor is coupled with a small band gap semiconductor with a more negative conduction band (CB) level, the CB electrons from the small band gap semiconductor can be injected to the large band gap semiconductor. Thus, a separation of electron and hole is achieved as shown in Fig.8.7 [19].

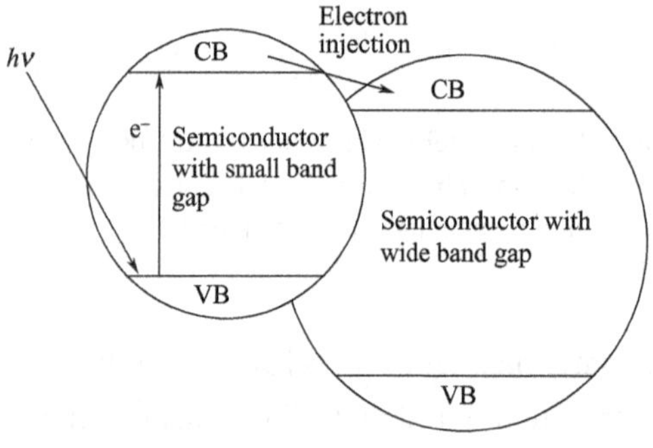

Fig.8.7 Electron injection in composite semiconductors [19]

The process is similar to dye sensitization. The difference is that electrons are injected from one semiconductor to another semiconductor, rather than from excited dye to semiconductor. Successful coupling of the two semiconductors for photocatalytic water-splitting hydrogen production under visible light irradiation can be achieved when the following conditions are met:① semiconductors should be photocorrosion free;② the small band gap semiconductor should be able to be excited by visible light;③ the CB of the small band gap semiconductor should be more negative than that of the large band gap semiconductor;④ the CB of the large band gap semiconductor should be more negative than electrode potential of H_2 and ⑤ electron injection should be fast and efficient.

(1) $CdS-TiO_2$ nanotubes nanocomposites

For enhancing the efficiency of light conversion, it is most important to find semiconductor with a suitable band gap as a photocatalyst to produce hydrogen by water splitting under visible light irradiation. One of the most well-known semiconductor photocatalyst, CdS, has been used for hydrogen evolution from aqueous solutions containing sacrificial reagents of Na_2S/Na_2SO_3 due to its narrow band gap of approximately 2.4eV [20,21]. However, CdS is prone to produce photocorrosion in the reaction. One effort has been made to improve the stability of metal sulfide, by coupled CdS nanoparticles with TiO_2 nanotubes. The CdS/TiO_2 NTs composite was prepared via ion-exchanging and precipitation reaction. The results revealed that CdS/TiO_2 NTs composite consists of uniform anatase nanotubes in average tubular diameter of about 15nm with homogenously loaded hexagonal phase CdS nanoparticles in average particle size of about 8nm as TEM shown in Fig.8.8 [22].

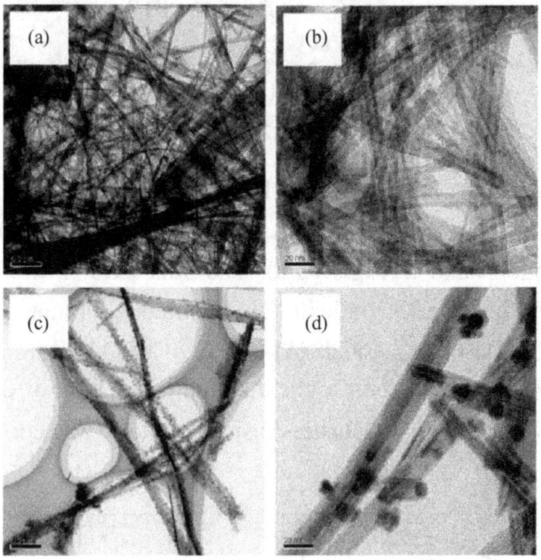

Fig.8.8 TEM images of samples. (a) $Na_2Ti_3O_7$ nanotubes; (b) $Na_2Ti_3O_7$ nanotubes; (c) CdS/TiO_2 NTs and (d) CdS/TiO_2 NTs [22]

(2) CdS and Al-HMS nanocomposites

The CdS/Al-HMS photocatalyst was firstly synthesized by use of cheap sodium aluminate ($NaAlO_2$) as aluminum source to get aluminosilicate mesoporous molecular sieve (Al-HMS) with high specific surface area, and subsequently assemble the CdS clusters inside the channels of Al-HMS via the sequential reactions of ion exchange and sulfurization. Meanwhile, a series of Pt- and Ru-loaded CdS/Al-HMS photocatalysts were synthesized by incipient wetness impregnation of the CdS/Al-HMS with solutions of $H_2PtCl_6 \cdot 6H_2O$ and $RuCl_3 \cdot xH_2O$ in a mixture of benzene and ethanol (4:1 *V*/*V*), respectively [23] and the activities of hydrogen production were evaluated in an aqueous formic acid solution under visible light irradiation. Formic acid, as a usual pollutant in water resulting from some industrial processes, was chosen as a model pollutant to achieve the dual goals of eliminated pollutant (form acid) with simultaneous production of hydrogen in the photocatalytic process [24].

Some structural information of CdS incorporation into Al-HMS can be obtained from high resolution TEM images in Fig.8.7 [25~28]. The Al-HMS samples in Fig.8.9(a) and (b) show wormhole like mesoporous framework with average pore diameter of 3～4nm and the run-through of wormhole-like channels are similar to HMS [25]. After incorporating CdS clusters inside the channels of Al-HMS, the distinguishing feature of the framework for Al-HMS is still remained in Fig.8.9 [25~28]. This suggests that the average size of the CdS nanoparticles is close to the pore diameter of wormhole-like mesoporous framework. Some authors reported that the extremely small particles of CdS at low loading levels can be formed within the zeolite cages [26,27].

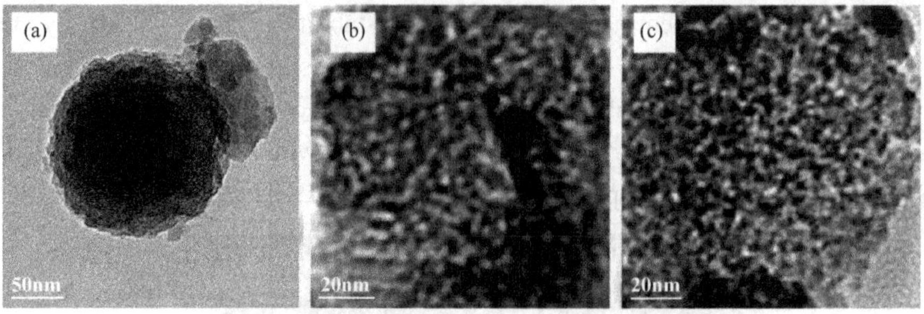

Fig.8.9 Images of transmission electron microscopy
(a) Al-HMS; (b) Al-HMS and (c)CdS/Al-HMS [28]

The growth of CdS nanoparticles inside Al-HMS is further characterized by diffuse reflectance UV–visible spectra as shown in Fig.8.10 [29]. It can be observed that the pure Al-HMS sample doesn't show any absorption band in Fig.8.10(a). The bulk CdS shows the maximum absorption edge of 585nm. The absorption edge is gradually blue-shifted to 540nm, 528nm, 518nm and 509nm for the samples of 21CdS/Al-HMS, 16CdS/Al-HMS, 4CdS/Al-HMS and 2CdS/Al-HMS from Fig.8.10(e)～(b), respectively, resulting from the quantum size effect of CdS nanoparticles. Peng et al. reported that the location of CdS nanoclusters inside zeolite Y was confirmed by the blue-shifted reflection UV–visible spectra [30].

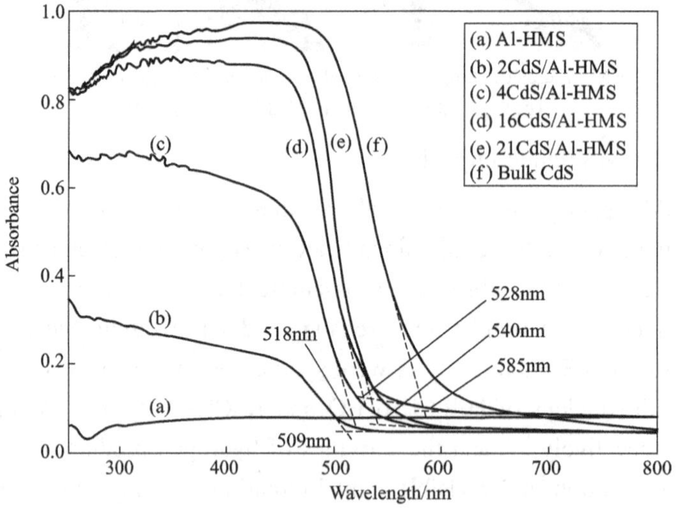

Fig.8.10 Diffuse reflectance UV–visible spectra of samples [29]

Meanwhile, all of Ru-supported CdS/Al-HMS nanocomposites show much higher photocatalytic activities of H_2 evolution than that of CdS/Al-HMS sample. It can be observed the 0.99Ru/21CdS/Al-HMS nanocomposite displays the highest quantity of H_2 evolution by photocatalytic degradation of formic acid under visible light irradiation for 6 h as exhibited in Fig.8.11 [29].

(3) CdS and ZnS composites

Some authors reported that photocatalytic properties of CdS could be modified by mixing with wide band gap semiconductors of chalcogenides such as ZnS [31~35].

Recently, we find that the rate of hydrogen production intensively depends on the size of the band gap of the solid solution of $Cd_{1-x}Zn_xS$ that can be adjusted by mixing different amount of CdS and ZnS. The band structure-controlled solid solution of $Cd_{1-x}Zn_xS$ was designed in our present work. A set of $Cd_{1-x}Zn_xS$ (x = 0～0.92) photocatalysts was prepared by coprecipitation method and was calcined at 723K under N_2 atmosphere, and the catalysts with different band gaps for hydrogen production in the sacrificial reagents of 0.1mol/L Na_2S/0.04mol/L Na_2SO_3 were also examined [36].

Fig.8.12 shows the diffuse reflectance UV-Vis absorption spectra of the various solid solutions of $Cd_{1-x}Zn_xS$, as well as ZnS and CdS. It is seen that ZnS only has an absorption edge located about 397nm, whereas the absorption edge of each solid solution was gradually red shifted as the amount of cadmium increased and distributed between those of CdS and ZnS in Fig.8.10. The phenomena should be attributed to a band transition of $Cd_{1-x}Zn_xS$ solid solution. The band gaps of these photocatalysts were estimated from the onset of the absorption edges to be 2.20～3.12eV (x = 0～0.92). The positions of conduction band and valence band for each photocatalyst were calculated using the following formula [37].

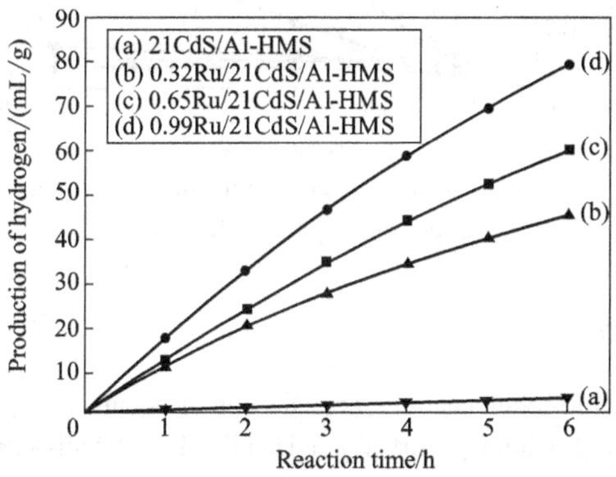

Fig.8.11 Activities of hydrogen production for Ru-loaded CdS/Al-HMS samples under visible light irradiation for 6 h [29]

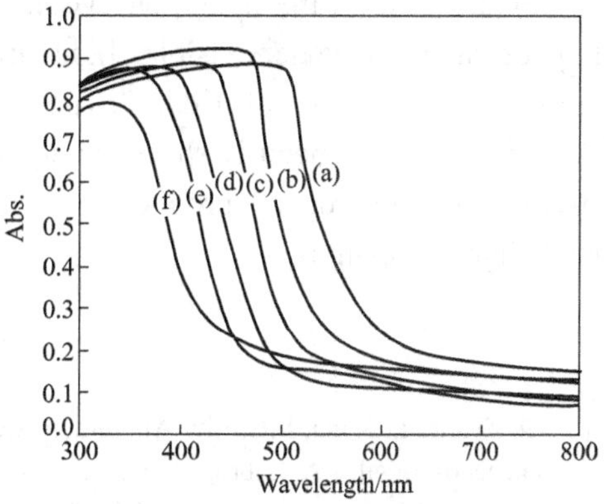

Fig.8.12 UV–Vis spectra of photocatalysts: (a) $Cd_{0.73}S$; (b) $Cd_{0.62}Zn_{0.16}S$; (c) $Cd_{0.54}Zn_{0.37}S$; (d) $Cd_{0.36}Zn_{0.56}S$; (e) $Cd_{0.18}Zn_{0.77}S$ and (f) $Zn_{0.92}S$ [36]

We took $Cd_xZn_yS_z$ as an example:

$$E_{cs}^{0}=E^{e}-X+1/2E_{g} \tag{8.6}$$

$$X=[x_{Cd}^{x}\,x_{Zn}^{x}\,x_{S}^{x}]^{1/(x+y+z)} \tag{8.7}$$

$$x_{Cd}=1/2(A_{Cd}+I_{Cd}) \tag{8.8}$$

$$x_{Zn}=1/2(A_{Zn}+I_{Zn}) \tag{8.9}$$

$$x_{S}=1/2(A_{S}+I_{S}) \tag{8.10}$$

Where A is the electron affinity energy, I the ionization energy, E_{cs}^{0} the conduction band potential, E^{e} is constant as 4.5eV, E_g represents the band gap of catalyst.

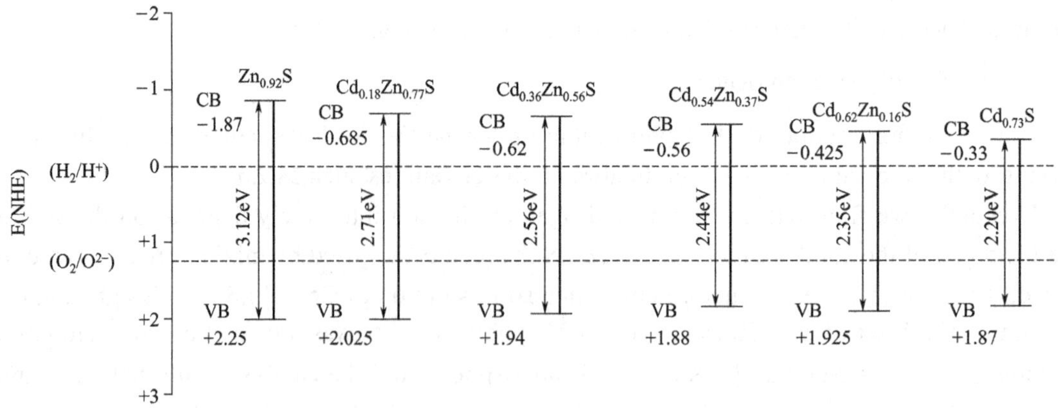

Fig.8.13 The conduction and valence band potentials of the photocatalysts [36]

After calculating, it can be found that the positions of conduction band and valence band for the solid solution of $Cd_{1-x}Zn_xS$ were shifted toward more negative potential and more positive potential with the increase of zinc as shown in Fig.8.13, respectively. The band gap position of the $Cd_{1-x}Zn_xS$ solid solutions can be adjusted by changing the ratio of the compositions of CdS to ZnS.

Fig.8.14 shows the hydrogen evolutions of photocatalysts under visible light irradiation. The catalysts of $Cd_{0.62}Zn_{0.16}S$ and $Cd_{0.54}Zn_{0.37}S$ simultaneously show the highest activities of hydrogen production under visible light irradiation.

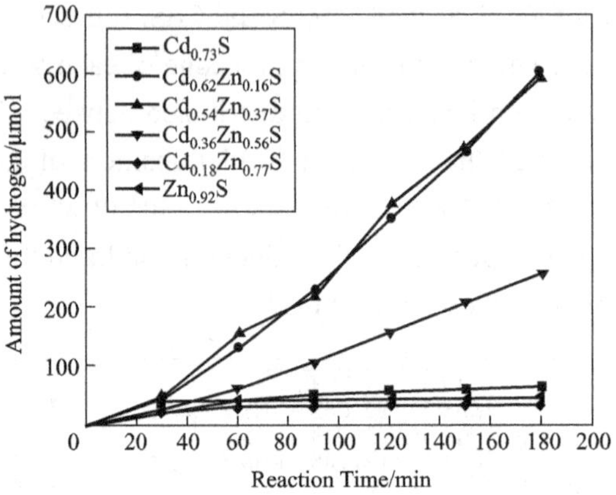

Fig.8.14 Performance of photocatalytic hydrogen production under visible light [36]

References

[1] Y.J. Zhang, L. Kang, L.C. Liu, Alkali-activated cements for photocatalytic degrada-tion of organic dyes, in: F. Pacheco-Torgal, J.A. Labrincha, C. Leonelli, A. Palomo,P. Chindaprasirt (Eds.), Handbook of Alkali-activated Cements, Mortars andConcretes, Woodhead Publishing, UK, 2015, pp. 729-777.

[2] A.K. Hussein, Applications of nanotechnology in renewable energies-A comprehensive overview and understanding Renewable Sustainable Energy Rev. 42 (2015) 460-476.

[3] U. Sahaym, M. Norton, Advances in the application of nanotechnology in enabling a 'hydrogen economy', J. Mater. Sci. 43 (2008) 5395.

[4] K. Satyanarayana, A. Mariano, J. Vargas,A review on microalgae, a versatile source for sustainable energy and materials, Int. J. Energy Res. 35 (2011) 291-311.

[5] A. Demirbas, Global renewable energy projections, Energy Sources Part B 4 (2009) 212-224.

[6] H. Lund, Renewable energy strategies for sustainable development, Energy 32 (2007) 912-919.

[7] A. Kudo, Int. J. Hydrogen Energy, 31 (2006) 197-202.

[8] T. Takata, A. Tanaka, K. Domen, Catal Today, 44(1998) 17-26.

[9] F. Barbir, Energy, 34 (2009) 308-312.

[10] A. Patsoura, D.I. Kondarides, X.E.Verykios, Catal . Today, 124 (2007) 94-102.

[11] A.L. linsebigler, G.Lu, J.T. Yates, Chem. Rev., 95 (1995) 735.

[12] M. Matsuoka, M. Kitano, M. Takeuchi, K. Tsujimaru, M. Anpo, J.M. Thomas, Catal .Today, 122 (2007) 51-61.

[13] D.Jing, Y.J. Zhang, Chem. Phys. Lett. ,415 (2005) 74-78.
[14] I.Tsuji, H. Kato, H. Kobayashi, A. Kudo, J. Am. Chem. Soc. ,126 (2004) 13406-13413.
[15] Y.J. Zhang, Y.C. Wang, W. Yan, T. Li, S. Li, Y.R. Hua, Appl. Surf. Sci. ,255 (2009) 9508-9511.
[16] N. Buhler, K. Meier, J. Reber, J. Phys. Chem. ,88 (1984) 3261-3268.
[17] M.A. Fox, T.L. Pettit, Langmuir, 5 (1989) 1056-1061.
[18] M. Sathish, R.P. Viswanath, Catal. Today, 129 (2007) 421-427.
[19] M. Ni, M.K.H. Leung, D.Y.C. Leung, K.Sumathy, Renew. Sustain Energy Rev., 11 (2007) 401-425.
[20] M. Matsumura, S. Furukawa, Y. Saho, H.Tsubomura, J. Phys .Chem., 89 (1985) 1327-1329.
[21] J.F. Reber, K.Meier, J. Phys. Chem., 90 (1986) 824-834.
[22] Y.J. Zhang, W. Yan, Y.P. Wu, Z.H. Wang, Mater. Lett. ,62 (2008) 3846-3848.
[23] Y.J. Zhang, A. Maroto-Valiente, I. Rodriguez-Ramos, Q. Xin, A. Guerrero-Ruiz, Catal.Today, 619 (2004) 93-95.
[24] Y.J. Zhang , L. Zhang, Appl. Surf. Sci. ,255 (2009) 4863-4866.
[25] W. Zhang, T.R. Pauly, T.J. Pinnavaia, Chem. Mater. ,9 (1997) 2491.
[26] Y. Wang, N. Herron, J. Phys. Chem. ,91 (1987) 257.
[27] N. Herron, Y. Wang, M.M. Eddy, G.D. Stucky, D.E. Cox, K. Moller, T. Bein, J. Am.Chem. Soc. ,111 (1989) 530.
[28] Y.J. Zhang, L. Zhang, Desalination, 249 (2009) 1017-1021.
[29] Y.J. Zhang, L. Zhang, S. Li, Int .J .Hydrogen Energy, 35 (2010) 438-444.
[30] H. Peng, S.M. Liu, L. Ma, Z.J. Lin, S.J. Wang, J .Cryst .Growth, 224(2001) 274-279.
[31] S. Kohtani, A. Kudo, T. Sakata, Chems .Phys .Lett., 206 (1993) 166-170.
[32] G. Chandra, A.M. Roy, Int. J. Hydrogen Energy, 21 (1996) 19-23.
[33] A.M. Roy, G.C. De, J.P. Photobiol A: Chem. ,157 (2003) 87-92.
[34] M. Subrahmanyam, V.T. Supriya, P.R. Reddy, Int. J. Hydrogen Energy, 21 (1996) 99-106.
[35] A. Koca, M.Sahin, Int. J. Hydrogen Energy, 27 (2002) 363-367.
[36] C.J. Xing, Y.J. Zhang, W.Yan, Int. J. Hydrogen Energy, 31 (2006) 2018-2024.
[37] S.R. Morrison, Elctrochemistry at semiconductor and oxidized metal electrodes, (1980)Plenum New York.

Review questions

1. Please describe the main conversion modes of solar energy.

2. Why can loading small amounts of Pt on the TiO_2 increase photocatalytic production of H_2 activity?

3. How many kinds of strategy for the design of photocatalysts with visible light response are proposed in the chapter?

4. Please illustrate the working principle of electron injection in composite semiconductors.

5. Hwo to distinguish nanotubes and nanorods?

6. What is the band gap of semiconductor?

7. What is red-shifted?

8. Which direction (toward short wavelength or long wavelength) is shifted for the UV–Vis spectra of photocatalysts when the semiconductor particles decrease? Why?

9. Why can photocatalytic activities be improved by doping CdS into ZnS?

Vocabulary

affinity [ə'fɪnɪtɪ] n. 密切关系；吸引力；姻亲关系；类同

aluminate [ə'lju:mɪneɪt] n. 铝酸盐

aluminosilicate [əˌljʊmənoˈsɪləkɪt] n. [化] 铝硅酸盐；铝矽酸盐

aluminum [ə'lu:mɪnəm] n. 铝

anatase ['ænəteɪs] n. [矿] 锐钛矿
aqueous ['eɪkwɪəs] adj. 水的，水般的
assemble [ə'semb(ə)l] vt.收集;装配；vi. 集合，聚集
benzene ['benzi:n] n. [化] 苯
cadmium ['kædmɪəm] n. [化] 镉
chalcogenide ['kælkədʒənaɪd] n. 氧属化物；硫族化物
chromatograph [krəʊ'mætəgrɑ:f] vt. 用色谱法分析；n. 套色版；色层分析计
chromium ['krəʊmɪəm] n. [化] 铬
circulation [sɜ:kjʊ'leɪʃ(ə)n] n. 流通，传播；循环；发行量
convert [kən'vɜ:t] vt. 使转变；转换…；使…改变信仰；vi. 转变，变换；改变信仰；
degradation [ˌdegrə'deɪʃ(ə)n] n. 退化；降格，降级；堕落
donor ['dəʊnə; -nɔ:] n. 捐赠者；供者；赠送人；adj. 捐献的；经人工授精出生的；给予体
dual ['dju:əl] adj. 双的；双重的；n. 双数；双数词
electrolysis [ˌɪlek'trɒlɪsɪs; ˌel-] n. [化学] 电解，[化学] 电解作用；以电针除痣
enhance [ɪn'hɑ:ns; -hæns; en-] vt. 提高；加强；增加
ethanol ['eθənɒl] n. 乙醇
formic ['fɔ:mɪk] adj. 蚁酸的；蚁的；甲酸的
framework ['freɪmwɜ:k] n. 框架，骨架；结构，构架
hexagonal [hek'sægənəl] adj. 六边的；六角形的
homogenous [hə'mɒdʒɪnəs] adj. 同质的；同类的
hydrocarbon [ˌhaɪdrə(ʊ)'kɑ:b(ə)n] n. 碳氢化合物
hydrogen ['haɪdrədʒ(ə)n] n. 氢
hydrothermal [haɪdrə(ʊ)'θɜ:m(ə)l] adj. 热液的；热水的
impregnate ['ɪmpregneɪt] vt. 使怀孕；灌输；浸透；adj. 充满的；怀孕的
incipient [ɪn'sɪpɪənt] adj. 初期的；初始的；起初的；发端的
infrared [ɪnfrə'red] n. 红外线；adj. 红外线的
ionization [ˌaɪənaɪ'zeɪʃən] n. 离子化
irreversible [ɪrɪ'vɜ:sɪb(ə)l] adj. 不可逆的；不能取消的；不能翻转的
lamp [læmp] n. 灯，照射器；vt. 照亮；vi. 发亮
methane ['mi:θeɪn; 'meθeɪn] n. [化] 甲烷，沼气
migrate [maɪ'greɪt; 'maɪgreɪt] vi. 移动；随季节而移居；移往；迁移；vt. 使移居；使移植
nitrogen ['naɪtrədʒ(ə)n] n. [化] 氮
orbital ['ɔ:bɪt(ə)l] adj. 轨道的；眼窝的
oxygen ['ɒksɪdʒ(ə)n] n. 氧气，氧
photoactivity [ˌfəutəuæk'tivəti] n. 光敏；光活；感光
photonic [ˌfoʊ'tɑnɪk] adj. 光激性的
photonics [fo'tɑnɪks] n. 光电；光子学
photovoltaic [ˌfəʊtəʊvɒl'teɪɪk] adj. 光电伏打的，光电的
precipitation [prɪˌsɪpɪ'teɪʃ(ə)n] n. 沉淀，沉淀物；冰雹；坠落；鲁莽
reagent [rɪ'eɪdʒ(ə)nt] n. 试剂；反应物
reflectance [rɪ'flekt(ə)ns] n. [物] 反射比
sacrificial [sækrə'fɪʃ(ə)l] adj. 牺牲的；献祭的
scavenger ['skævɪn(d)ʒə] n. 食腐动物；清道夫；[化] 清除剂；拾荒者
semiconductor [ˌsemɪkən'dʌktə] n. 半导体
sensitization [ˌsɛnsɪtɪ'zeʃən] n. [化] 敏化作用；促进感受性；感光度之增强
sieve [sɪv] n. 筛子；滤网；不能保守秘密的人；vt. 筛；滤
sodium ['səʊdɪəm] n. [化] 钠
split [splɪt] vt. 分离；使分离；劈开；离开；vi. 离开；被劈开；断绝关系；n. 劈开；裂缝；adj. 劈开的
steam [sti:m] vt. 蒸，散发；用蒸汽处理 n. 蒸汽；精力 vi. 蒸，冒水汽 adj. 蒸汽的
sulfurization [ˌsʌlfəraɪ'zeʃən] n. 硫化；硫化作用；磺化作用
titanate ['taɪtəneɪt] n. 钛酸盐
ultraviolet [ʌltrə'vaɪələt] adj. 紫外的;紫外线的；n. 紫外线辐射，紫外光
wormhole ['wɜ:mhəʊl] n. 虫孔；蛀洞
zeolite ['zi:əlaɪt] n. 沸石

8. 纳米技术用于太阳能制氢

可再生能源是一种可以提供光、电、热等能量而不污染环境的能源。化石能源产生的能量是目前环境污染的主要原因。可再生能源的明显优势在于不需要燃料，这就消除了二氧化碳的排放。目前全球的能源问题归因于化石燃料供应不足以及消耗化石燃料所导致的过量的气体排放，从而导致化石能源消耗量的增加。据报道，目前的石油消耗速度比自然界生成的石油量的 10^5 倍还快，以目前如此巨大的消耗速度，到 2050 年，全球化石燃料的储量将进一

步减少[1~3]。另外，预测全球对化石燃料的需求在2050年及2100年分别约为30TW及46TW[3]。化石燃料包括原油，煤和天然气，它们是不可再生的，一旦燃尽就不复存在。这些能源提供了全球90%以上的能量需求，但随之而来的是严峻的环境代价，例如，环境中CO_2浓度在过去的150年间已经从280ppm增长到了370ppm。本世纪可能会超过550ppm[4~6]。然而，由于对能源的巨大需求以及化石能源较低的可用性，对能源需求的趋势可以转向可再生能源。这些能源包括太阳能、风能、生物质能、氢能和地热能。这些清洁能源可以用来代替传统的化石能源和核能。当今，对可再生能源的巨大兴趣在于预测到2040年将提供全球50%的一次性能源。此外，可再生能源在减少气体排放方面能够起到关键作用，预计2050年能减少70%的气体排放。另一方面，纳米技术已经用于改善可再生能源。

8.1 氢能经济

使用氢能的益处基于其清洁燃烧的品质，氢能的优势在于就地生产，氢燃料电池汽车的效率大约是传统汽油内燃机汽车的2～3倍。氢是地球上含量最丰富的元素之一，并且用于燃料电池时其唯一的副产物是水。

图8.1显示了作为来自于一次性能源的“氢经济”，是以氢作为能量载体，具有多种最终用途。可以使用任何一次性能源如化石、核能及可再生能源制备氢。然而，为了实现“氢经济”和完全替代化石燃料，下列每个领域都存在需要攻克的关键性技术挑战：①产生；②储存；③转化成电能和热能使用。这些技术壁垒取决于了解和控制氢与各种材料之间界面处的原子及分子过程。因为纳米材料很大的比表面积，完全适合用来克服这些壁垒。可以通过采用纳米工艺和设计以提高催化剂、储氢材料以及燃料电池的电极组装等性能。

8.2 产氢

虽然氢有望代替化石燃料，但是，自然界中的氢不以H_2分子的形式存在。而是以化合物的形式存在于水、烃中，通过化学转化才能获得氢气。可以通过各种当地资源，包括化石燃料如：天然气、煤、核能、水力发电、生物质能，可再生能源如：风能、太阳能和地热能制备氢气，如图8.2所示。

甲烷的蒸汽重整工艺是利用天然气中的甲烷和高温蒸汽生产氢气的方法。在类似的液体重整工艺中，乙醇与高温蒸汽反应产生氢气。利用风能、太阳能、地热能、核能或水力发电产生的电能电解水使之分解成氢气和氧气。在气化方面，煤或生物质首先转化成气态组分，然后转化为合成气，合成气再与蒸汽反应产生氢气。热化学水分解是使用由太阳能聚光器或核反应堆产生的高温以驱动化学反应分解水产生氢气。在光生物过程中，微生物如绿藻，在阳光下消耗水产生副产物氢气。同样地，光催化和光电化学系统则利用特殊的半导体和太阳能分解水产氢。

因此，作为能源载体的氢，由于可再生性及环境友好的特点，可以利用可再生能源经由多种途径进行制备，采用太阳能光催化制氢是其制备途径之一。

8.3 太阳能转换

太阳能是一种免费的且在世界很多地方都可以充分利用的能源。太阳一年提供给地球的能量，比人类一年实际所需的化石能源所提供的能量多15000倍。太阳的能量以辐射的方式到达地球，其光谱分布从红外到紫外，如图8.3所示。

太阳能转换成有用的能源形式，通常可分为热过程和光过程两种。在太阳能的热过程中，太阳能首先转换为热，热可以被直接利用，也可存储在热介质中（如水或干燥的岩石中），也可以利用适当的机器将其转换为机械能或电能（如蒸汽机发电）。在太阳能的光过程中，太阳的光量子直接被一个吸收器所吸收，吸收器可将光子能在一个光伏电池中转化为电能，或将光子能在一个吸收能量的化学反应中转化为化学能，如太阳能分解水制氢和制氧。

近年来，由于氢的可再生性的特点，氢能的重要性再次得到广泛的认可。但是，工业上是通过烃重整（如甲烷）制取氢。利用太阳能以水为原料制氢是一种挑战，因为它是解决能源问题和环保问题的途径之一[7]。

8.4　光催化分解水制氢

因为全球的能源和环境问题，采用半导体光催化剂，利用太阳能从水中制取氢已经得到了足够的重视[8,9]。利用纳米科学及纳米技术合成的一系列 TiO_2 基的纳米半导体催化剂可以达到该目的[10]。然而，由于低的转换效率及成本问题，这种技术仍然处于研发阶段。除此之外，由于约 3.2eV 的宽带隙，作为一种光响应材料，TiO_2 仅能吸收紫外线。人们已经采用一些方法以延伸它的光响应范围，并改进它的光催化活性。这些方法可分为两大类：一类是加入化学添加剂，另一类是对光催化剂进行改性。空穴牺牲剂电子给体的化学添加剂与价带上的空穴发生不可逆的反应，避免电子-空穴的复合，从而增加电子-空穴的分离效率。

8.5　TiO_2 上负载金属

一种方法是利用半导体微粒或粉末设计光催化系统，如图 8.2 所示。在 TiO_2 表面上负载少量的 Pt，可提高光生电子和光生空穴的电荷分离，以显著地提高反应速率[6]。通常认为 Pt 的负载对于紫外线辐照下的光催化分解水制氢是奏效的，因为 Pt 的存在大大地降低了产氢的过电势 [11]。当 TiO_2 接受光照时，TiO_2 价带电子获得光能而跃迁进入导带成为光生电子（e^-），在价带上留下了光生空穴（h^+）；水电离出的氢氧根离子（OH^-）被光生空穴所氧化，并释放出 O_2，而水电离出的氢离子被光生电子在铂（Pt）纳米离子上还原，并释放出 H_2，如方程式(1)～(3)所示。

通过溶胶-凝胶法及水热法制备出 Ni 掺杂的具有晶体结构及大比表面的介孔 TiO_2。来自于 $H_2PtCl_6\cdot 6H_2O$ 前驱体的作为共催化剂的 Pt 被原位沉积在光催化剂上，光催化产氢是在高压汞灯为光源的条件下进行。反应池与一个闭合的气体循环系统相连，气相产物由在线的热导池（TCD）气相色谱（NaX 分子筛柱，N_2 作为载气）进行检测[7]。

8.6　可见光驱动的光催化剂的发展

可见光驱动的光催化剂太阳能分解水制氢，加速了光能转换效率的改善。设计具有可见光响应的光催化材料的战略分为三类，如图 8.5 所示[14]。方法（a）是将外来元素掺杂进入具有宽带隙的活性光催化剂中，在该催化剂的禁带中形成施主能级。方法（b）是寻找具有稳定价带的，且价带比 O_{2p} 价带位置更负，如许多的氧化物光催化剂。方法（c）是采用宽带隙的半导体和窄带隙的半导体制备固溶体，以控制材料的能隙结构。

8.6.1　在钛酸盐纳米管上负载 Cr^{3+}

通过水热法和浸渍法，制备出 TNTs 负载 Cr^{3+} 的纳米复合材料。与纯的纳米管样品相比，

Cr^{3+}掺杂的纳米复合材料显示出较高的产氢活性。

可见光催化分解水制氢的反应机理如图8.6所示[15]。当Cr^{3+}掺杂进入带隙为3.3eV的TNTs（钛酸盐纳米管），在钛酸盐纳米管的禁带中形成了施主能级（2.3eV）。在可见光辐照下，当一个光子的能量被Cr^{3+}的3d轨道上的电子所吸收，施主能级（Cr^{3+}）的光生电子（e^-）可跃迁至TNTs的导带，同时在施主能级（Cr^{3+}）上留下一个带正电的空穴（h^+），如方程（8.1）所示。然后，水中电离的OH^-被h^+氧化为氧气和H^+，如方程（8.2）所示。

同时，牺牲剂（空穴清除剂）的SO_3^{2-}与光生空穴（h^+）反应，提高了电子-空穴对的分离效率，如方程(8.3）和(8.4）所示[16~18]。由于牺牲剂（电子给体）在光催化反应中被消耗掉，不断地添加牺牲剂以维持产氢是必不可少的。光生电子（e^-）与水中电离的H^+结合形成氢气，如方程(8.5）所示。这样，由于在钛酸盐纳米管中掺杂Cr^{3+}提供了施主能级，光生电子-空穴对得到有效的分离，且提高了光催化效率。

8.6.2 半导体复合材料

半导体复合是另一种利用可见光制氢的方法。当一个宽带隙的半导体与一个具有更负导带(CB)的窄带隙的半导体复合后，来自窄带隙半导体的导带电子可以注入到宽带隙的半导体中，这样，就可实现大量电子-空穴的分离，如图8.7所示[19]。

该过程与光敏化相似。不同之处在于电子是从一个半导体注入到另一个半导体，而不是从染料的激发态到半导体。若使可见光催化分解水制氢的两种半导体成功复合，需满足以下条件：①半导体不发生光腐蚀；②窄带隙的半导体能被可见光激发；③窄带隙半导体的导带（CB）应比宽带隙的更负；④宽带隙半导体的导带（CB）应比H_2的电极电势更负；⑤电子的注入要快速高效。

（1）CdS-TiO_2纳米管纳米复合材料

为了提高光转换的效率，最重要的是寻找一种可作为可见光分解水制氢催化剂的适当带隙的半导体。一种最为熟知的CdS半导体光催化剂，因为大约2.4eV的窄带隙，已经用于含有牺牲剂Na_2S/Na_2SO_3的水溶液制氢体系[20,21]。然而，CdS在反应中易于发生光腐蚀。提高金属硫化物稳定性的一种方法是将CdS纳米颗粒与TiO_2纳米管进行复合，通过离子交换和沉淀反应可以制备出CdS/TiO_2 NTs复合材料。结果表明：CdS/TiO_2 NTs复合材料是由管径约15nm且均匀的锐钛矿型纳米管和粒径约8nm的六方相CdS纳米粒子复合而成，如图8.8所示[22]。

（2）CdS和Al-HMS纳米复合材料

CdS/Al-HMS光催化剂首次合成得到，先利用廉价的偏铝酸钠作为铝源制得高比表面积的铝硅酸盐介孔分子筛（Al-HMS），接着，经过离子交换和硫化反应，将CdS分子簇组装进入Al-HMS分子筛的孔道，从而制备出CdS/Al-HMS复合材料。通过初湿浸渍法，分别将$H_2PtCl_6·6H_2O$和$RuCl_3·xH_2O$溶于苯和乙醇（4:1，体积比）中，对CdS/Al-HMS进行浸渍，可以制备出一系列负载Pt或Ru的CdS/Al-HMS光催化剂[23]，产氢活性可通过可见光催化分解甲酸溶液实验进行评价。以来自一些工业废水中的常见污染物甲酸作为模型污染物，以实现消除污染物，并同时产氢的双重目的[24]。

高分辨率透射电镜照片能够获取有关CdS掺入Al-HMS的结构信息，如图8.9所示[25~28]。图8.9（a）和图8.9（b）中，Al-HMS样品展示了平均孔径为3～4nm蠕虫状的介孔结构，并且贯通的蠕虫状的通道与HMS相似[19]。当CdS簇掺入Al-HMS的孔道后，Al-HMS的结构特点仍然保留，如图8.9所示[28]。这表明CdS纳米粒子的平均粒径与蠕虫状的介孔结构的孔径很接近。一些作者也报道了在低负载量时，CdS微粒可进入沸石的笼状结构中[26,27]。

CdS 纳米粒子在 Al-HMS 结构内的生长进一步用漫反射紫外-可见光谱进行表征，如图 8.10 所示[29]。从图中可以看出，纯的 Al-HMS 样品没有显示出任何吸收带，如图 8.10（a）所示。块体 CdS 具有 585nm 的最大吸收边。对于样品 21CdS/Al-HMS、16CdS/Al-HMS、4CdS/Al-HMS 和 2CdS/Al-HMS，由于 CdS 纳米粒子的量子尺寸效应，它们的吸收边均发生蓝移，分别为 540nm、528nm、518nm 和 509nm，如图 8.10(e)～(b) 所示。Peng 等报道了漫反射紫外-可见光谱的蓝移可以证实 CdS 纳米粒子位于 Y 沸石内[30]。

同时，与 CdS/Al-HMS 样品相比，Ru 负载的 CdS/Al-HMS 纳米复合材料具有更高的光催化产氢活性，其中 0.99Ru/21CdS/Al-HMS 纳米复合材料在可见光辐照 6h 后，展示出最高的光催化降解甲酸制氢活性[29]，见图 8.11。

（3）CdS 和 ZnS 复合材料

一些作者报道了 CdS 的光催化性质通过与宽带隙的硫属化合物 ZnS 混合而得到了改善[31~35]。

最近研究发现氢气产率主要由 $Cd_{1-x}Zn_xS$ 固溶体的带隙大小决定，可通过改变 CdS 和 ZnS 的含量调节带隙宽度。我们设计了带隙结构可控的 $Cd_{1-x}Zn_xS$ 固溶体，通过共沉淀法，且在氮气氛围中经 723K 煅烧，合成了一系列 $Cd_{1-x}Zn_xS$（x = 0～0.92）的光催化剂，在 0.1mol/L Na_2S/0.04mol/L Na_2SO_3 牺牲剂体系中，分别考察了这些不同带隙催化剂的产氢活性[36]。

图 8.12 是各种 $Cd_{1-x}Zn_xS$ 固溶体、ZnS 以及 CdS 的漫反射紫外-可见吸收光谱图。从图中可见，ZnS 在约 397nm 处有一个吸收边，随着 Cd 含量的增加，固溶体的吸收边逐渐发生红移，并分布在 CdS 和 ZnS 的吸收边之间，这种现象应归于 $Cd_{1-x}Zn_xS$ 固溶体中带隙转变过程。这些光催化剂的带隙用初始吸收边进行估算，分别为 2.20～3.12eV （x = 0～0.92），而光催化剂的导带与价带的位置可采用以下公式计算[37]。

我们以 $Cd_xZn_yS_z$ 为例说明。

其中，A 是电子亲和能；I 是电离能；E_{cs}^0 是导带电势；E^e 是常数，等于 4.5eV；E_g 是催化剂的带隙。

通过计算，发现随着 Zn 含量的增加，固溶体 $Cd_{1-x}Zn_xS$ 的导带和价带位置分别朝着更负的电势和更正的电势方向移动，如图 8.13 所示。固溶体 $Cd_{1-x}Zn_xS$ 的带隙位置可以通过改变 CdS 与 ZnS 的成分比率进行调节。

图 8.14 分别是不同光催化剂在可见光辐照下的产氢结果。其中 $Cd_{0.62}Zn_{0.16}S$ 和 $Cd_{0.54}Zn_{0.37}S$ 两种催化剂同时显示出最高的产氢活性。

复　习　题

1. 请描述太阳能的主要转换方式。
2. 为什么在 TiO_2 上负载少量的 Pt 可提高 TiO_2 的产氢活性？
3. 本章中提及了多少种关于可见光响应的光催化剂的设计方法？
4. 请举例说明复合半导体中电子注入法的工作原理。
5. 如何区分纳米管和纳米棒？
6. 什么是半导体的带隙？
7. 什么是红移？
8. 当半导体粒径减小时，光催化剂的紫外-可见吸收光谱向哪个方向（向短波还是向长波方向）移动？为什么？
9. 在 ZnS 中掺杂 CdS 为什么能够改善其光催化活性？